DIY Science

ILLUSTRATED GUIDE TO
ASTRONOMICAL
WONDERS

First Edition

Robert Bruce Thompson & Barbara Fritchman Thompson

O'REILLY®

BEIJING · CAMBRIDGE · FARNHAM · KÖLN · PARIS · SEBASTOPOL · TAIPEI · TOKYO

ILLUSTRATED GUIDE TO ASTRONOMICAL WONDERS

by Robert Bruce Thompson & Barbara Fritchman Thompson

Published by Make:Books, an imprint of Maker Media, a division of O'Reilly Media, Inc.
1005 Gravenstein Highway North, Sebastopol, CA 95472.

O'Reilly books may be purchased for educational, business, or sales promotional use. For more information, contact our corporate/institutional sales department: 800-998-9938 or *corporate@oreilly.com*.

Print History
October 2007
First Edition

Publisher: Dale Dougherty
Associate Publisher and Executive Editor: Dan Woods
Editor: Brian Jepson
Creative Director: Daniel Carter
Designer: Alison Kendall
Production Manager: Terry Bronson
Indexer: Patti Schiendelman
Cover Photography: Steve Childers

ISBN-10: 0-596-52685-7
ISBN-13: 978-0-596-52685-6

To John Dobson, whose creativity and engineering skills made large telescopes affordable and ubiquitous.

—Robert Bruce Thompson and Barbara Fritchman Thompson

CONTENTS

Preface

We wish someone else had written this book years ago. We could have used it ourselves when we started observing the night sky. Instead, we had to deal on our own with the same two problems that every beginning amateur astronomer faces: which objects to observe and how to find them.

There are any number of observing lists, of course, some of them much better suited for beginning and intermediate observers than others. The Astronomical League (www.astroleague.org) has several observing lists suitable for beginners (along with many others suited only for advanced observers). RASC, the Royal Astronomical Society of Canada (www.rasc.ca), publishes an excellent list for intermediate observers. As useful as these lists are, when we were getting started what we really wanted was a consolidated list by constellation that included a wide variety of astronomical objects of all types for both telescopic and binocular observing. So for this book we made our own list by merging the best observing lists for beginning and intermediate astronomers.

We started, of course, with the famous Messier List, which includes 110 of the best and brightest objects visible in the night sky, and is universally recommended as the best first list for beginners to pursue. As a follow-on to the Messier List, we added the RASC Finest NGC Objects list, which includes 110 of the best and brightest non-Messier objects. For binocular observing, we added the Astronomical League Binocular Messier List and Deep-Sky Binocular List, which together include all of the best objects visible with binoculars for mid-northerly observers. For those who observe from light-polluted urban locations, we added the Astronomical League Urban Observing List, which includes both deep-sky objects and multiple stars. Finally, we added the Astronomical League Double Star List, which includes many of the finest multiple stars visible in the night sky. Taken together, these six lists total nearly 400 of the best objects visible in the night sky.

With that out of the way, we turned next to addressing the problem of how to find these objects. The first time we observed these objects, we did it the hard way, using general star charts and planetarium software running on our notebook to track down each object, one by one. Wouldn't it be nice, we thought, if we had a book that listed each object, with directions on how to find it and a description of what it should look like in our telescope or binocular? Oh, and how about individual large-scale finder charts for each object, showing where that object lay in relation to nearby stars and how to move our finder scope to center the object in its cross-hairs? And, as long as we were making a wish-list, we thought it would be nice to have photographs of the objects.

There are a lot of astronomy field guides out there. Some are excellent, but none of them offered everything we wanted when we were getting started. Most beginner guides cover too few objects. Once you have observed that limited list—typically just the Messier List with perhaps a handful of "extra" objects—you've outgrown that book. We decided instead to include enough objects to keep even a frequent observer busy for a long time. Even if you observe every dark, clear night, it should take you at least a year, and more probably two or three years, to observe all of the objects in this book.

By the time you finish this book, you'll have long since graduated from beginner status to have become an intermediate to advanced observer. Instead of depending on others for help, you'll find that other members of your astronomy club have started coming to you for help and advice. You'll also have completed the requirements for the RASC Finest NGC Objects certificate as well as for the certificates of five Astronomical League Observing Clubs—the Messier Club, the Binocular Messier Club, the Deep-Sky Binocular Club, the Urban Observing Club, and the Double Star Club, which is half of the ten AL club lists you need to finish to qualify for the coveted Astronomical League Master Observer certificate.

Look Before You Leap

As tempting as it may be to turn immediately to the constellation chapters and start observing, if you intend to pursue the Astronomical League and RASC certificates you should read at least Chapter 1. The various Astronomical League observing clubs have specific rules and requirements for observing and logging objects, which differ from club to club.

How this Book is Organized

This book is primarily a field observing guide. Originally, we intended it to be purely a collection of constellation chapters focused on those objects in the Astronomical League and RASC lists we chose to cover. But several of our advisors pointed out that we needed to at least touch on the fundamentals of observing and equipment, so we added a pair of concise narrative chapters on these topics.

Chapter I, *Introduction to DSO Observing*, tells you what you need to know to get started observing Deep-Space Objects (DSOs), even if you've never used a telescope before.

Chapter II, *Observing Equipment*, provides a quick overview of the equipment you'll need to observe DSOs, from binoculars and telescopes to eyepieces and accessories to charts and planetarium software.

The bulk of this book is made up of the constellation chapters, 50 of them, listed alphabetically. (The remaining 38 constellations contain no objects from the lists this book covers; many of these 38 are too far south to be visible to observers at mid-northern latitudes anyhow.) Each constellation chapter includes a summary table of the featured objects in that constellation and an overview chart that shows the constellation as a whole and the surrounding constellations. A section is devoted to each object, with a detailed description of how to find the object and what it looks like. We provide a detailed finder chart for each object, and for most objects we include a DSS image.

Acknowledgments

We decided to write this book during a conference call with Mark Brokering and Brian Jepson, our publisher and editor at O'Reilly. We'd just finished writing our first astronomy book for O'Reilly, Astronomy Hacks, which is full of tips and tricks about observing but is not a field guide to the night sky. After reading Astronomy Hacks, Mark and Brian had both become interested in pursuing astronomy as a hobby. Brian had just bought an 8" Orion Dobsonian telescope, and mentioned that he needed help to figure out which objects to observe and how to find them. Mark commented, "You know, we should do a book about that." And so we did.

In addition to Mark, Brian, and the O'Reilly production staff, who are listed individually on the copyright page, we want to thank our technical reviewers. Gene Baraff, Steve Childers, Jim Elliott, Sue French, Geoff Gaherty, and Paul Jones have, among them, more than 100 years of observing experience. Despite that, all of them remember clearly the frustrations that are faced by all inexperienced observers. We asked them to read our manuscript, bringing their experience to bear, but at the same time trying to look at it through the eyes of a new observer. They've done an excellent job, making numerous useful suggestions, all of which helped make this a better book. Any errors that remain are ours alone.

We produced all of the charts in this book with MegaStar planetarium software, which is published by Willmann-Bell (www.willbell.com). Before we decided to use MegaStar, we tried literally dozens of planetarium programs, free and commercial. With the exception of MegaStar, none of them offered the level of control we needed to produce the charts for this book. If you want a top-notch planetarium program, check out MegaStar. We think you'll like it as much as we do.

Last, but certainly not least, we want to thank Brian McLean and Lynn Kozloski at the Space Telescope Science Institute (STScI) for granting us permission to use the DSS images that illustrate most of the objects covered in this book.

How to Contact Us

We have verified the information in this book to the best of our ability, but you may find things that have changed (or even that we made mistakes!). As a reader of this book, you can help us to improve future editions by sending us your feedback. Please let us know about any errors, inaccuracies, bugs, misleading or confusing statements, and typos that you find anywhere in this book.

Please also let us know what we can do to make this book more useful to you. We take your comments seriously and will try to incorporate reasonable suggestions into future editions. You can write to us at:

Maker Media
1005 Gravenstein Hwy N.
Sebastopol, CA 95472
(800) 998-9938 (in the U.S. or Canada)
(707) 829-0515 (international/local)
(707) 829-0104 (fax)

Maker Media is a division of O'Reilly Media devoted entirely to the growing community of resourceful people who believe that if you can imagine it, you can make it. Consisting of Make Magazine, Craft Magazine, Maker Faire, and the Hacks series of books, Maker Media encourages the Do-It-Yourself mentality by providing creative inspiration and instruction.

For more information about Maker Media, visit us online:
MAKE: www.makezine.com
CRAFT: www.craftzine.com
Maker Faire: www.makerfaire.com
Hacks: www.hackszine.com

To comment on the book, send email to:
bookquestions@oreilly.com

The web site for The Illustrated Guide to Astronomical Wonders has more information, such as errata and plans for future editions. You can find this page at:
www.makezine.com/go/astrowonders

For more information about this book and others, see the O'Reilly web site:
www.oreilly.com

To contact one of the authors directly, send mail to:
barbara@astro-tourist.net
robert@astro-tourist.net

We read all mail we receive from readers, but we cannot respond individually. If we did, we'd have no time to do anything else. But we do like to hear from readers.

Thank You

Thank you for buying *Illustrated Guide to Astronomical Wonders*.
We hope you enjoy reading it as much as we enjoyed writing it.

Author Bios:

Robert Bruce Thompson is the author or co-author of numerous on-line training courses and books about computers, science, and technology. He got started in amateur astronomy as a teenager. In 1966, he built his first telescope, a 6" Newtonian reflector, grinding his own mirror with materials purchased from Edmund Scientific. Robert is a co-founder and president of the Winston-Salem Astronomical League (www.wsal.org) and is currently pursuing the Astronomical League Master Observer certificate.

Barbara Fritchman Thompson is the co-author of several computer and technology books. She began observing the night sky in early 2001, and has since observed and logged hundreds of astronomical objects. Barbara is a co-founder and treasurer of the Winston-Salem Astronomical League, and is currently pursuing the Astronomical League Master Observer certificate.

i

Introduction to DSO Observing

Until the 1970s, most amateur astronomers spent most of their time observing Solar system objects—the moon, planets, and comets. Nowadays, although Solar system objects remain popular observing targets, many amateurs devote most of their time to observing *DSOs* (*Deep-Sky Objects* or *Deep-Space Objects*). Multiple star observing has also become popular. (Although multiple stars are technically DSOs because they lie far outside the Solar system, many astronomers reserve the term DSO for remote objects other than multiple stars.) In this chapter, you'll learn what you need to know to get started observing multiple stars and DSOs successfully.

Multiple Stars

A *multiple star* (*MS*) is two or more stars that appear to be in close proximity. A multiple star pair is often called a *double star* or *binary star*. A multiple star triplet is sometimes called a *triple star* or *trinary star*. Multiple star systems with more than three stars are simply called multiple star systems.

Many multiple stars are so closely separated and/or so distant from Earth that they appear to be single stars, even in the largest telescopes. There are two types of these multiple stars that are revealed as multiples with telescopes or binoculars (this is called *splitting a multiple star*):

Physical multiple star—A *physical multiple star* is one in which the stars are close together physically, actually orbiting each other. Professional astronomers identify physical multiples

by making exact measurements of the relative motions of the member stars with a series of observations, sometimes over several years (or decades). Depending on the mass and separation of the member stars, their orbital periods may be anything from a few seconds to millions of years. (Physical multiples with very short orbital periods are so close together physically that they cannot be split.)

Optical multiple star—An *optical multiple star* is one in which the component stars are independent of each other and are at greatly differing distances, but due to a chance alignment appear close together from our viewpoint on Earth. Optical multiples are easily discriminated by observation, because the motions of the component stars are independent of each other.

Degrees, Arcminutes, and Arcseconds

Astronomers specify the apparent extent (size) of and separation between celestial objects by using units of angle. The fundamental unit of angle is the degree (°), where 360° make a complete circle. So, for example, the angular distance between a point on the horizon and a point directly overhead (at zenith) is 90°, or one quarter of a complete circle.

Although 1° may seem to be a small subdivision, astronomers observe very small patches of sky, and so need much finer gradations. (The field of view in a typical amateur telescope is often less than 1°, sometimes much less.) For that reason, astronomers divide each degree into 60 arcminutes (abbreviated ' and often called simply *minutes*) and each arcminute into 60 arcseconds (abbreviated " and often called *seconds.*)

So, for example, the full moon, whose extent is about 0.5°, can also be described as having an extent of 30'. It could even be described as having an extent of 1,800", although that would be comparable to telling someone that you run 126,720 inches every morning, rather than the more usual (and equivalent) denomination of the distance as two miles.

Degrees, minutes, and seconds are often mixed. For example, if two objects are separated by 1.25°, that separation may be described as 1.25°, 1°15', or 75', all of which are different ways of describing exactly the same distance.

The reason for the shift to DSO observing is easy to understand. When I began observing in the mid-60s, the typical amateur instruments were 60mm refractors and 6" Newtonian reflectors, neither of which had sufficient *aperture* (size of the primary lens or mirror) to provide satisfying views of any but the brightest DSOs. In the 1970s, John Dobson revolutionized amateur astronomy by inventing the Dobsonian mount. Suddenly, large aperture became affordable. Telescopes of 8", 10", and even 12" aperture soon became commonplace sights at star parties, and many amateurs bought or built 18", 24", 30", and larger scopes. The race was on to observe DSOs, the "faint fuzzies" that were now easily within the reach of these larger telescopes. —*Robert*

Find Polaris, Split It, and Sketch It

The bright star Polaris lies just 0.74° from the north celestial pole, which means that as Earth rotates on its axis once every 24 hours, Polaris describes a circle only 1.48° in diameter. For practical purposes, Polaris appears to be fixed in the same position in the sky at any hour of the night on any night of the year, with all of the other stars rotating around it as Earth turns on its axis. It's important to be able to identify Polaris because it corresponds so closely with the north celestial pole, upon which celestial coordinates are based.

The mean altitude of Polaris is the same as your geographic latitude. For example, if you are standing on the equator (0°), Polaris lies (give or take 0.74°) at altitude 0° on the north horizon. If you are standing on the north pole (90°N latitude), Polaris is straight overhead, at 90° elevation. If you are at 45°N latitude,

Polaris lies at 45° elevation, halfway between the horizon and zenith. Our regular observing sites are located at about latitude 36°N, so for us Polaris is always at about 36° elevation, or just over a third of the way from the horizon to zenith.

If the Big Dipper (the Plough for our British friends) is up, the easiest way to locate Polaris is to use the two "pointer stars" that form the side of the dipper's bowl opposite the handle. Those two stars—Merak, diagonally opposite where the handle connects to the bowl, and Dubhe—form a 5.4° south-north line. Extend the line from Merak to Dubhe by about five times the distance that separates them, and look for a prominent star. That star is Polaris. If you know which direction is geographic north, you can also identify Polaris by looking straight north and looking for a prominent star at the proper elevation.

Measuring Things by Hand

You can estimate distances in the sky by extending your arm toward the object and using your hand to judge angular distances. At arm's length, the tip of your little finger subtends almost exactly 1°. Your thumb at its widest point subtends about 2°, and the distance from the tip of your thumb to the first joint is about 3°. The middle three fingers used for the Boy Scout salute subtend about 5°. One fist width is about 10°, and the distance between the tips of your spread first and little fingers is about 15°. These values are remarkably consistent among men, women, and children, because people with larger hands usually also have correspondingly longer arms.

There's one more way you can use your hand to estimate angular distances, but this one varies a bit from person to person. If you spread your hand completely, the distance from the tip of your thumb to the tip of your little finger may be anything from 18° to 25° or a bit more. You can find out where your own hand "fits" on that continuum by using the Big Dipper.

The distance from Alkaid (the last star in the handle) to either Dubhe or Merak (the two "pointer" stars on the far end of the bowl) is 25.6°. The distance from Alcor/Mizar (the naked-eye double at the bend of the handle) to Dubhe or Merak is just under 20°. The distance from Alkaid to Phad is about 18°. Try it yourself the next time you're out under the night sky and the Big Dipper is up.

FIGURE i-1.

Robert's sketch of Polaris

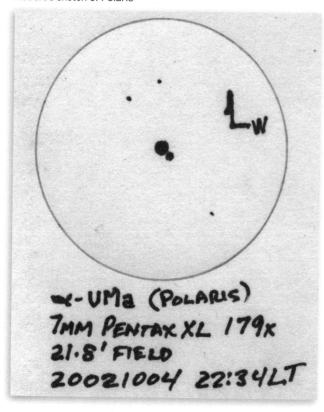

α-UMa (POLARIS)
7MM PENTAX XL 179x
21.8' FIELD
20021004 22:34LT

Once you've located Polaris, you might as well observe it. To the naked eye or with a binocular, Polaris is a bright (second magnitude or m2.0; see the upcoming note on Stellar Magnitudes for an explanation of this number) warm-white star, but otherwise nothing special. That changes with higher magnification. Center Polaris in your finder scope, put your highest magnification (lowest focal length) eyepiece in the focuser, and focus the scope. In the eyepiece, Polaris is now glaringly bright—as bright or brighter than Venus appears to the naked eye. Very close to that bright star, called Polaris A, is another, yellowish and much dimmer companion star, called Polaris B. That companion star is about m8.5, so much dimmer than m2.0 Polaris A that it's easy

to overlook. It lies 18.4" southwest of Polaris A, at position angle (PA) 218°.

Congratulations! You've just logged your first multiple star, and this one happens to be one of those required to complete the Astronomical League Double Star Club list. One of the requirements of the Double Star Club is that you sketch each of the double stars you observe. If you're artistically-challenged, don't despair. Robert isn't much of an artist, either—he flunked finger-painting in kindergarten—but even his limited drawing skills suffice to meet the requirements of the Double Star Club, as shown in Figure i-1.

To begin, draw a small circle in your log book or log sheet. Something around 2" to 2.5" in diameter is about right. (We often use the rim of a soft drink can as a template.) In the center of the circle, draw a dot to represent the primary star of the multiple star system, with the size of the dot representing the brightness of the star—larger equals brighter. Add an appropriately-sized dot in the correct location for the companion star or stars in the multiple star system, and then add appropriately-sized dots for any other reasonably prominent stars that appear in the eyepiece. Somewhere in the field, insert an arrow pointing north, with a perpendicular line pointing either east or west, and label it appropriately. (Depending on the scope you're using, the view may be inverted and/or reversed; that's fine as long as you label the directions correctly on your direction arrow.) Label the drawing to indicate the object it represents, the eyepiece you used (with its magnification and true field of view), and the date and time of the observation. (We always use local date and time, but many observers use UTC instead. Either is fine as long as you're consistent.) That's all there is to it.

With one exception, the stars in a multiple star system are labeled alphabetically by decreasing brightness. The brightest star (called the *primary*) is labeled A, the next brightest star (called the *secondary* or *companion*) is labeled B, and so on. The exception is the stars in the Trapezium in the Sword of Orion, the four brightest stars of which are labeled A to D from west to east. In addition to the magnitude of their components, multiple stars are characterized in two ways: separation and position angle, both of which are described next.

Feet and Minutes, Inches and Seconds

For historical reasons, astronomers use both traditional and metric units, often mixed together willy-nilly. For example, the specifications for a telescope may list its aperture as 8 inches and its focal length as 2,032 mm, or an eyepiece may be listed as having a 2-inch barrel diameter and a focal length of 27 mm. Such mixed units are so common in amateur astronomy that no one thinks twice about them.

But astronomers also frequently use angular measurements, which are denominated in degrees (°), (arc)minutes ('), and (arc)seconds ("), to specify the size or distance between celestial objects. Therein lies a potential for confusion, because (arc)minutes and feet are both abbreviated with the ' symbol, and (arc)seconds and inches with the " symbol. The easiest way to avoid confusion is to remember that if you're referring to an object in the night sky, ' means (arc)minutes and " means (arc)seconds; if you're referring to an object on the ground, ' means feet and " means inches.

Unless, of course, you're timing a transit or other celestial event, in which case " means seconds (that's clock time rather than angular separation or distance). Or unless you're referring to the right ascension coordinate of a celestial object, which is denominated in hours, minutes, and seconds (all of time). Fortunately, right ascension values are abbreviated using h for hours, m for minutes, and s for seconds. Otherwise, we'd all be insane.

STELLAR MAGNITUDES

The apparent brightness of a star is determined by its inherent brightness and its distance from us, and is specified as its *apparent visual magnitude*. The brighter the star, the lower its magnitude. A hundred-fold difference in brightness is defined as exactly five magnitudes. For example, a 1st-magnitude (m1; don't confuse the lowercase m with the uppercase M, which designates one of the Messier objects—you'll learn all about those soon) star appears exactly 100 times brighter than a 6th-magnitude (m6) star. One magnitude translates to a brightness difference of about 2.51 times, so an m2 star appears to be about 2.51 times brighter than an m3 star, and an m9 star appears about 2.51 times brighter than an m10 star. The brightest stars in the night sky are 0th-magnitude; the dimmest visible to the unaided human eye under very dark conditions are 6th- to 7th-magnitude.

In this book, we list the apparent visual magnitude of stars. Other reference works may specify different types of magnitude. For example, a star that is bright visually may be dimmer photographically at various wavelengths. Although *photographic magnitude* is significant scientifically, it is irrelevant for visual observing. Similarly, some works specify the *absolute magnitude* of stars, which describes their inherent brightness. An inherently bright star (low absolute magnitude) that is far from us appears dim (high apparent magnitude), while an inherently dim star that is very near us appears bright, so there is no direct correlation between the absolute magnitude of a star and how bright it appears to us.

On star charts, including those in this book, the size of a star indicates its apparent visual magnitude relative to the other stars shown on the chart, with brighter stars being larger. (In the eyepiece of a telescope, all stars are the same size, tiny points of light of different brightness.) The size of the dot that corresponds to a particular magnitude depends on the scale of the chart. For example, on a large-scale chart (one that shows lots of sky), the largest dots may represent stars of magnitude 0 or 1. On a detailed finder chart that shows only a tiny area of sky, large dots are used for the brightest stars shown on that chart, which may be only magnitude 5 or dimmer.

Separation—*Separation* is the apparent distance between the components, and is stated in *seconds of arc* (*arcseconds*, or "). The smallest separation that is splittable depends on many factors, including aperture, magnification, seeing (atmospheric stability), and the relative brightness of the components. Larger aperture and higher magnification make it easier to split close pairs, and it is easier to split a pair of similar brightness than one in which the components differ greatly in magnitude. Many double stars are separated widely enough that they can be split easily with a binocular. A few (like Alcor-Mizar in Ursa Major) can be split with the naked eye.

Atmospheric stability, called *seeing*, puts an absolute limit on splitting very close doubles, because even the largest telescope using extremely high magnification can resolve detail no finer than the seeing allows. (Observing under poor seeing conditions is like trying to read a newspaper at the bottom of a swimming pool.) The best observing sites on Earth routinely have sub-

Visual Multiples

A multiple star like Porrima, for which actual orbital motion (and therefore changes in separation and position angle) has been observed visually, is formally known as a *visual multiple*. We just call them "fast movers." To account for these changes over time, most multiple star lists specify the dates of the observations for the separations and position angles given.

Clean Versus Dirty Splits

Technically, in order to split a double star, you must be able to see that double as two distinct points of light with darkness separating them, which is called a *clean split*. That's not always possible for a very close double. Poor seeing, mediocre or dirty optics, too little aperture, or insufficient magnification may make it impossible to achieve a clean split. In such cases, it may still be obvious that you are looking at a double star, because the star appears elongated or dual-lobed. We log such an instance as a *dirty split*.

Declination and Right Ascension

Terrestrial coordinates are specified by latitude and longitude, both of which are denominated in degrees, minutes, and seconds. Celestial coordinates use the same concept, but celestial latitude is called *declination* and celestial longitude is called *right ascension*.

Declination is exactly analogous to terrestrial latitude and is also denominated in degrees, minutes, and seconds. An object that lies exactly on the celestial equator has a declination of 0° 0' 0", just as an object that lies on Earth's equator has a latitude of 0° 0' 0". There is one difference between the way latitude and declination are specified. On Earth, latitudes are specified as north or south. Celestial coordinates are instead signed. An object that lies at the north celestial pole has a declination of +90° 0' 0", and an object at the south celestial pole has a declination of -90° 0' 0".

Right ascension is analogous to terrestrial longitude, but is ordinarily denominated differently. Although it is possible (and correct) to specify right ascension values in angular degrees, minutes, and seconds, it's customary to specify those values as times, denominated in hours, minutes, seconds, and tenths of seconds. Valid RA coordinates range from 0h0m0.0s to 23h59m59.9s. (Just as a clock never reaches 24:00 hours, RA never reaches 24h0m0.0s.) RA values are specified in clock times rather than angular units because those clock times correlate directly to the times that a particular object rises and sets.

arcsecond seeing, but most locations seldom exceed one- or even two-arcsecond seeing. This means that multiple stars with smaller separations can be visually split only by space-based telescopes like Hubble.

Position angle—*Position angle* (*PA*) specifies the direction of the secondary from the primary, using *equatorial coordinates*. For example, if the secondary star is located dead East of the primary, the PA is 90°. If the secondary is South of the primary, the PA is 180°; West is 270°, and North is 0°. Although professional astronomers use special instruments to determine exact position angles, amateur astronomers generally log position angles estimated to the nearest 5° or so.

When a multiple star has more than two components, the separations and position angles are specified individually for each named pair. For example, the catalog data for a triple star might specify a 42" separation and 45° position angle for the A-BC pair—which may be written in the form A(BC)—with a 4" separation and 315° position angle for the BC pair. For that system, the center of the BC pair lies 42" NE of the A star, and the C star lies 4" NW of the B star.

The separation and position angle of most multiple star systems changes very slowly relative to human life spans, but there are exceptions. For example, Porrima (γ Virginis), one of the doubles on the Astronomical League's Double Star list, is listed at 3.6" separation and PA 293°. Those values were correct when the list was compiled in about 1994. But by 2005, as the two stars reached the near extreme of their 168.68-year orbit, separation reached a minimum at 0.3" with a PA of about 160°. By early 2006, the separation had increased to just under 1" with the PA about 85°. In 2089, the separation of Porrima will reach its

maximum of more than 6" at a PA of about 325°, as the two components reach the far extreme of their orbit.

Most multiple stars are unremarkable other than for being multiples. Their components are often dim and show no color. There are many exceptions, though, of which Albireo (β Cygni) is the best-known and, to our eyes at least, the most striking. The m3.1 primary shines golden, in stark contrast to the bluish m5.1 companion. Many other doubles also provide striking color contrasts, from deep red, orange, and yellow to bright blue or even violet.

There are numerous multiple star catalogs, each of which uses its own labeling method for multiple stars. Among those you're most likely to encounter are multiple stars from the *Aitken Double Star Catalog* (abbreviated *ADS*), the *Otto Struve Double Star Catalog* (OΣ) and *Otto Struve Double Star Catalog Supplement* (OΣΣ), the *Wilhelm Struve Double Star Catalog* (*STF* or Σ), and the *Winchester Double Star Catalog* (*WDS*). A particular double star often has multiple catalog designations. For example, Albireo in Cygnus, considered by most observers to be the most beautiful double star of all, was cataloged by Wilhelm Struve as STF 43 (or Σ 43) and also by Aitken as ADS 12540.

You don't actually need printed copies of multiple star catalogs. Most published star charts include the most interesting multiple stars, and all but the most basic planetarium software charts hundreds to thousands of multiple stars.

Planetarium Software

If you have a notebook or PDA, you can use it to run planetarium software right at the telescope. In addition to displaying object positions in real time, such software allows you to zoom in and out, place finder circles and eyepiece circles right on the charts, display only selected objects, look up object data instantly, and so on. (Of course, you'll need to cover the display with a sheet of ruby-red plastic film to preserve your night vision.)

There are dozens of planetarium programs available for Windows, Mac OS X, Linux, and even Palm OS. We've tested many of these, and we've come to have strong preferences. Among our favorites is Cartes du Ciel (Windows, Linux, and Mac OS X using X11), written by Swiss astronomer Patrick Chevalley. CdC is full-featured, flexible, powerful, and free for the download. We suggest you try it first. Chances are you'll never need anything more.

Among commercial programs, Starry Night (Windows, Mac OS X) is a popular choice. SN is available in several levels, from an inexpensive basic version to the $250 Pro Plus version. SN is popular among Mac OS X users, where the selection of planetarium programs is limited, and among those who have "outgrown" the capabilities of Cartes du Ciel. TheSky6 (Windows, Mac OS X) is another popular commercial planetarium program that has a strong following among amateur observers. TheSky6 Student Edition ($49) is too limited to be of much use to serious observers. TheSky6 Serious Astronomer Edition ($129) is a full-featured planetarium program, similar in features and functionality to Starry Night Pro ($150). TheSky6 Professional Edition ($279) adds a lot of bells and whistles, few of which we find useful, and is comparable to Starry Night Pro Plus.

We've used all of these programs, but eventually we put all of them aside in favor of the most powerful, flexible planetarium program we know of. The $129 Megastar (Windows only) simply blows away its competition when it comes to features needed by serious amateur astronomers. At first glance, Megastar appears simpler and cruder than many of its more polished competitors. It's not until you've used it for a while that you realize that this really is a program written by astronomers for astronomers. With all of the other programs we've used, we'd eventually discover that the program couldn't do something we wanted to do. With Megastar, that's never happened.

For links to these programs and many others, visit astro.nineplanets.org/astrosoftware.html. Chapter ii discusses planetarium software in detail.

IT'S ALL GREEK TO US

It's important for amateur astronomers to know the Greek alphabet because Greek letters are commonly used on charts and in reference materials. For example, although you may know the brightest star in the constellation Lyra by its common name, Vega, it's also often shown in charts and other reference materials under its Bayer designation of α-Lyrae, or alpha Lyrae. Understanding the Bayer designation is even more important for the many stars that have no common names, and so are most commonly listed primarily or solely by their Bayer designations.

Although the lower-case Greek letters are far more commonly used in astronomy, you might come across upper-case Greek letters in some situations. For example, the list of double stars compiled by F. G. Wilhelm Struve are designated STF or **Σ** (capital sigma, the Greek S, for "Struve"), while those of his son Otto Wilhelm Struve's original list are designated STT or **ΟΣ** (capital omicron-sigma, the Greek OS, for "Otto Struve") and those of his supplemental list are designated STS or **ΟΣΣ** (capital omicron-sigma-sigma, the Greek OSS, for "Otto Struve Supplemental").

LOWER	UPPER	NAME	PRONUNCIATION
α	A	alpha	AL-fuh
β	B	beta	BEET-uh
γ	Γ	gamma	GAMM-uh
δ	Δ	delta	DELT-uh
ε	E	epsilon	EPP-sih-lawn
ζ	Z	zeta	ZEET-uh
η	H	eta	EET-uh
θ	Θ	theta	THEE-uh
ι	I	iota	YOTT-uh
ϰ	K	kappa	CAP-uh
λ	Λ	lambda	LAM-duh
μ	M	mu	MEE
ν	N	nu	NEE
ξ	Ξ	xi	KSEE (KS as in foX)
ο	O	omicron	AW-muh-kron
π	Π	pi	PEE
ϱ	P	rho	ROE
σ, ς	Σ	sigma	SIG-muh
τ	T	tau	TAF (AF as in cALF)
υ	Y	upsilon	EEP-sih-lawn
φ	Φ	phi	FEE
χ	X	chi	KHEE (KH as in Scottish loCH)
ψ	Ψ	psi	puh-SEE
ω	Ω	omega	aw-MEG-uh

FLAMSTEED NUMBERS AND OTHER STELLAR DESIGNATIONS

By preference, most amateur astronomers designate stars by their common names, if a common name exists and is well known. The second-best choice is to use the Bayer Greek-letter designation, if that star has a Bayer letter. Of course, only naked-eye stars (and not all of them) *have* common names or Bayer designations. For stars that have neither well-known common names nor Bayer letters, another alternative is to use the Flamsteed number.

Flamsteed numbers were assigned to most bright stars in each constellation, progressing from west to east (with increasing Right Ascension). That is, the westernmost star in each constellation was assigned the Flamsteed number of 1. Moving eastward, each succeeding bright star was assigned the next higher Flamsteed number. (The highest Flamsteed number assigned is 139-Tauri in the constellation Taurus.)

With a few exceptions (such as 61-Cygni), most amateur astronomers don't know the Flamsteed numbers for many stars. For example, any astronomer recognizes the brightest star in the constellation Orion by its common name, Rigel. Many also recognize that star by its Bayer designation, β-Orionis. But few would recognize Rigel by its Flamsteed number, 19-Orionis. Still, it's necessary to use the Flamsteed number when the star in question has no common name or Bayer designation.

But what if a star has no common name, Bayer designation, *or* Flamsteed number? In that case, the only alternatives are to identify the star by specifying its exact coordinates or to use a catalog number from one of the "deeper" catalogs, which include thousands of stars too dim to be seen by the naked eye. Although there are many such catalogs, the catalog numbers you're most likely to find in printed charts and planetarium programs are those from the *Henry Draper Catalog* (*HD*) and the *Smithsonian Astrophysical Observatory Star Catalog* (*SAO*). HD catalogs more than 300,000 stars, including all stars down to ninth magnitude and some to tenth magnitude. SAO catalogs more than 250,000 stars to ninth magnitude. That means that any star of ninth magnitude or brighter has both HD and SAO numbers, either of which identifies the star unambiguously.

All of this means that any one star may have many catalog numbers and other identifies. For example, Albireo may also be identified as β-Cygni, 6-Cygni, HD 183912, and SAO 87301. For bright stars like Albireo, no one uses anything but the common name or the Bayer designation. But what if you want to refer to the eighth magnitude star that lies about 21.6 arcminutes west of Albireo? That star has no common name or Bayer letter. It doesn't even have a Flamsteed number. But it is included in the HD and SAO catalogs, so you can locate it with your planetarium program or a detailed star chart as HD 183560 or SAO 87268.

DSO Sizes

The apparent size of DSOs is specified in degrees(°), minutes('), and seconds (") of arc, where one degree equals 60 minutes, and one minute equals 60 seconds. For example, the full moon subtends about 0.5°, which can also be specified as 30 arcminutes (30') or 1800 arcseconds (1800"). The largest DSOs, such as M 31, cover several degrees of sky. The smallest, such as many planetary nebulae, may be only a few arcseconds in apparent size.

With very few exceptions, we specify the extent of objects in arcminutes. For example, if the extent of a nebula is 2° x 1.5°, we ordinarily specify that as 120 x 90 arcminutes. Similarly, if the extent of a planetary nebula is 48" x 36", we specify that as 0.8' x 0.6'

The maximum field of view of typical amateur telescopes at low magnification ranges from less than 1° (for a large SCT) to perhaps 5° (for a small, wide-field refractor). Binoculars and optical finders have fields of view ranging from 2.5° or less (high-magnification giant binoculars) to perhaps 6° to 8° for standard models. At maximum useful magnification, a typical amateur telescope may have a true field of view of 0.2° (12') or less. Accordingly, the largest DSOs may not fit within the field of view of a telescope even at low magnification, while the smallest DSOs may appear nearly stellar even at very high magnification.

DSO Catalog Numbers

DSOs are identified by various catalog numbers. About a third of the objects covered in this book are members of the Messier catalog, compiled by Charles Messier in the late 18th century. Messier's catalog includes 110 objects, numbered M 1 through M 110 roughly in order of their discovery dates. Most of the remaining objects covered in this book are members of the *New General Catalog* (NGC) or *Index Catalog* (IC, a supplement to the NGC), which were compiled in the 19th and early 20th centuries and include thousands of objects.

The Messier, NGC, and IC catalogs are all general catalogs, which means they include many types of DSOs—open star clusters, globular star clusters, various types of nebulae, galaxies, and other objects. In addition to these and other general catalogs, various specialized catalogs exist. These catalogs may, for example, include only open star clusters or only planetary nebulae.

One object may belong to several catalogs. For example, Messier cataloged an open star cluster in the constellation Auriga as his object M 38. That open cluster was later cataloged as NGC 1912 and in the Collinder Catalog of Open Clusters as object Collinder 67 (Cr 67). Messier numbers are by far the best known identifiers, and are used almost exclusively when referring to objects that are in the Messier Catalog even though they appear in other catalogs. Objects that are not in the Messier catalog are most often referred to by their NGC or IC numbers, even if they are also members of one or more specialized catalogs. A few objects, such as Markarian 6 (Mrk 6) have neither a Messier number nor an NGC or IC number. These objects are referred to by their specialized catalog numbers for lack of a more familiar alternative.

DSO Types

Astronomers categorize DSOs as one or more of several types, which are described in the following sections. As you read about each object type, we suggest that you fire up Stellarium (www.stellarium.org), Google Sky (earth.google.com/sky) or another planetarium program that provides photo-realistic rendering of astronomical objects and look at an example of each object. The view on your computer will be much brighter and more detailed than what you'll actually see in even a large telescope, but it will at least give you an idea of the differences between types of objects.

Open Clusters (OC)

Open clusters (*OC*), sometimes called *galactic clusters* in older books, are groups of stars that are loosely bound to each other by their mutual gravity. Open clusters range from a dozen or so member stars to many thousands. More than 1,100 open clusters are known in our own Milky Way Galaxy. By some estimates, the total number of open clusters in our galaxy exceeds 100,000, but most of them are invisible to us, obscured by distance, clouds of dust and gas, and the galaxy core itself.

In astronomical terms, most open clusters are relatively young objects and have short expected lifetimes. The youngest open clusters, such as the group of stars around the Trapezium in Orion, are still in the process of being born. Average open clusters range in age from a few tens of millions of years to a few hundred million. Although the oldest open clusters are several billion years old, nearly as old as the youngest globular clusters (see the following section), it's unusual for an open cluster to reach even a billion years old before it loses most or all of its members.

As an open cluster moves through space, it is subject to external gravitational influences strong enough to break the weak gravitational bonds that bind the cluster members. Eventually, an open cluster degenerates into a collection of stars moving in the same general direction through space, but no longer gravitationally bound to each other. Such a collection is known as a *stellar association* or a *moving group*. The best known moving group comprises several of the stars that make up the Big Dipper portion of Ursa Major.

ALL ABOUT DSO MAGNITUDE

Stars are so distant from us that they are merely infinitely small points of light. DSOs—such as star clusters, nebulae, and galaxies—are different. They have finite sizes, also called *extents*. When magnitude is specified for a DSO, it is an *integrated magnitude*, which means that if all the light from that DSO were concentrated into a single stellar point, that star would have the indicated magnitude. Because DSOs have extent, their light is not concentrated into a stellar point, but is instead distributed across the extent.

Accordingly, a DSO may have a misleadingly bright magnitude, when that DSO is in fact quite dim visually. For example, consider three DSOs, all of which are magnitude 9. The first is a tiny planetary nebula, about 1 arcminute in diameter. The second is a galaxy, about 2.5 arcminutes in diameter. The third is an emission nebula about 6.25 arcminutes in diameter.

The tiny planetary nebula is brightly visible even in a small telescope, because that magnitude-9 amount of light is concentrated in a very small area. The galaxy is 2.5 times larger linearly, or or 6.25 times larger areally. Assuming the light is distributed evenly, the galaxy is 6.25 times (two magnitudes) dimmer visually than the planetary nebula, and will appear much dimmer in the telescope eyepiece. The emission nebula is 2.5 times larger linearly (6.25 times areally) than the galaxy, which dilutes the light by a further two magnitudes.

This diluting effect with increasing extent is taken into account by specifying a *surface brightness* for extended objects. In this example the planetary nebula, being very small, may have a surface brightness about equal to its integrated magnitude, or 9 for both. The galaxy may have a surface brightness of only 11.0, two magnitudes dimmer than its integrated magnitude. The emission nebula may have a surface brightness of only 13.0, four magnitudes dimmer than its integrated magnitude. (Actually, surface brightness versus integrated magnitude is extremely complex. There are many other factors involved. But the important thing to remember is that DSO magnitude doesn't necessarily have much to do with how easily visible a DSO is. Some DSOs with relatively dim magnitudes are easily visible in small telescopes, while other DSOs with relatively bright magnitudes are extremely dim even in a large scope.)

For an example of how deceptive magnitude can be, go outside one night when Andromeda is up and look for the Andromeda Galaxy (M 31; remember, the uppercase M designates a Messier object, and the lowercase m designates magnitude). If your site is very dark and you are fully dark-adapted, you may just barely be able to catch a glimpse of M 31 with your naked eye as a small, dim patch of haze, but don't be surprised if you can't see it at all. New observers are often puzzled because most catalogs list the magnitude of M 31 as 4.0 (m4.0), give or take. An m4.0 star is easily visible from even a moderately dark site, so why isn't M 31 just as easy to see?

The reason for the difference is that the surface brightness of that m4.0 star is also 4.0, since the star is an infinitely small point of light. Conversely, the light from M 31 is spread across a significant area of sky, reducing its surface brightness to about 12.9, thousands of times dimmer than the star.

Open clusters differ dramatically in appearance. The sparsest open clusters have only a few widely-scattered member stars, and are often difficult to discriminate from the surrounding star field. The richest, tightest open clusters appear similar to loose globular clusters.

The brightest open clusters are visible to the naked eye, and so have been known since antiquity. Many of them, including the Double Cluster in Perseus (shown in Figure i-2), the Pleiades (M 45; The Seven Sisters) in Taurus, and the Beehive (M 67; Praesepe) in Cancer, are the subjects of myths and legends.

Open clusters are categorized by their *Trumpler Classification*, which takes into account concentration and detachment magnitude range, and richness, as follows:

Concentration

I. Detached; strong concentration toward center
II. Detached; weak concentration toward center
III. Detached; no concentration toward center
IV. Not well detached from surrounding star field

Range in Brightness

1. Small range in brightness
2. Moderate range in brightness
3. Large range in brightness

Richness

p. Poor (less than 50 stars)
m. Moderately rich (50–100 stars)
r. Rich (more than 100 stars)

For example, both members of the Double Cluster, shown in Figure 1-2, are classified I 3 r, because both are well-detached clusters with strong central concentrations, have member stars that vary over a wide range of magnitudes, and have many more than 100 member stars.

Conversely, M 38, shown in Figure i-3, is classified II 2 r, or, by some sources, II 2 m. M 38 is a class II because, although it is well detached, it shows only weak central concentration. The member stars vary over a moderate range of magnitudes, and the cluster contains about 100 stars, which puts M 38 on the borderline between medium and rich.

Like all open clusters, M 45 (shown in Figure i-4) originated as a cloud of dust and gas that provided the raw material for the stars that make up the cluster. As an open cluster ages and moves through space, the remaining dust and gas gradually dissipates. In some young open clusters, vestiges of that dust and gas remain, and shine softly blue by reflected star light. Other open clusters, including M 45, just happen to be passing through an area of space that contains dust clouds that reflect the light of stars that are members of the open cluster. Still other open clusters are embedded in emission nebulae or dark nebulae. The Trumpler Classification for such clusters includes a final "n" to indicate that nebulosity is associated with the cluster. Although this nebulosity may be difficult to detect visually, it is often evident in photographs.

M 45 provides an excellent illustration of why open cluster classifications are a matter of opinion. Based on this image and the preceding explanation of concentration, you might assign M 45 to class II (weak concentration) or even class III (no concentration). In fact, Trumpler himself classified M 45 as II 3 r (missing the nebulosity). But many current sources, including Sky Catalog 2000, classify M 45 as I 3 r n.

Most of the open clusters in this book are members of the *NGC* (*New General Catalog*) or *IC* (*Index Catalog*). Others, including a few prominent ones, were not included in these general DSO catalogs, but were later added to one or more specialty open cluster catalogs, including the *Collinder* (*Cr*), *Kemble*, *Markarian* (*Mrk*), *Melotte* (*Mel*), *Stock* (*St*), and *Trumpler* (*Tr*) catalogs. It's not important to know much about these supplemental catalogs, other than to understand that some open clusters have no NGC or IC designations and so are identified only as members of one or more of these supplemental catalogs.

Open Cluster Nebulosity

An "n" following the Trumpler Classification means there is *nebulosity* associated with the cluster. Nebulosity appears visually as a cloud or haze of undifferentiated light, which may range from quite bright to so dim that it lies below the threshold of visibility. Nebulosity may represent literal clouds of gas or dust interspersed with the cluster stars and illuminated by their light, or it may be a visual artifact caused by the presence of many stars that are too dim to be resolved individually, but together are bright enough to give the visual impression of faint nebulosity.

Green Stars?

Although the color contrast of doubles sometimes makes one of the components appear to be green, there actually aren't any green stars. Cool stars emit almost no light in the green, blue, and violet parts of the spectrum, and so appear red, orange, or yellow to our eyes. (Oddly, those colors are ordinarily described as "warm" colors.) Hot stars emit primarily at the short wavelengths of blue and violet light, and so appear to have a blue or violet tinge. Stars that burn in the middle temperature range associated with green light also emit a large percentage of their light in the upper and lower portions of the visible spectrum, and therefore appear to be white.

ABOUT THE IMAGES

Unlike many astrophotographs, which show color and detail that are much too dim to actually be seen in the eyepiece, the images we've selected for this chapter show objects pretty much as they actually look in the eyepiece of a moderately large telescope from a dark site on a night with excellent transparency. Our thanks to our observing buddy Steve Childers for providing these images.

FIGURE i-2.

NGC 869 (left) and NGC 884, the Double Cluster (image courtesy of Steve Childers)

FIGURE i-3.

The open cluster M 38 (image courtesy of Steve Childers)

FIGURE i-4.

M 45, the Pleiades, an open cluster (image courtesy of Steve Childers)

Globular Clusters (GC)

Globular clusters (*GC*) are ancient objects. The best recent estimates put the age of the oldest globular clusters at more than 13 billion years old, nearly as old as the universe itself. In fact, confusion reigned for a time when the best estimate of the age of globular clusters was older than the best estimate for the age of the universe itself that had been arrived at by other means. Nowadays, the oldest globular clusters are thought to be only a few hundred million years younger than the universe itself.

The population of a globular cluster may range from about 10,000 stars to several million. (In terms of star count and total mass, the largest globs approach the size of the smallest "dwarf" galaxies.) Because globulars contain so much mass in such a limited volume of space—typically a sphere of between 10 and 30 light years diameter—the members of a globular cluster are very tightly bound to each other by their mutual gravity.

With very few exceptions, the stars that make up globular clusters are very old stars with masses no more than twice that of our own Sun, Sol. Stars of higher mass have long since blown apart as supernovae or passed through the nova stage and become white dwarfs. Globulars may contain a few young stars,

known as *blue stragglers*, that are thought to result from stellar collisions within the extremely dense core of the cluster.

Globular clusters are relatively rare, at least within our own Milky Way galaxy. About 200 Milky Way globular clusters are known, most of which are concentrated near the galactic center. Most of the globular clusters that are easily visible in amateur instruments are in the summer constellations—particularly Sagittarius, Scorpius, and Ophiuchus—because when we look toward those constellations we are looking toward the galactic center. Most of the few exceptions, such as the globular cluster M 79 in the winter constellation Lepus, are thought to be what amounts to intergalactic immigrants—clusters that originally belonged to other galaxies that have been subsumed by the Milky Way.

Because they are all tight, more-or-less spherical groups of stars, globular clusters show less visual variation than open clusters. But globs do differ, albeit more subtly than open clusters. Figure i-5 shows M 13 (NGC 6205), the most impressive globular cluster visible from mid-northern latitudes.

Globular clusters are classified on the 12-point *Shapley-Sawyer Concentration Class* scale, with Class I globular clusters most concentrated and Class XII clusters least concentrated. Class I and II globulars, such as M 75 (NGC 6864) in Sagittarius, are very tight indeed. Class XI and XII globulars, such as NGC 6749 in Aquila, look more like a tight open cluster than a typical globular cluster.

Move South

Those of us in the northern hemisphere are unfortunate when it comes to viewing globs. M 13 is a spectacular glob, certainly, but it pales in comparison to the best the southern hemisphere has to offer, notably Omega Centauri (NGC 5139) and 47 Tucanae (NGC 104). Actually, we have been able to view Omega Centauri from our home, although at declination -47° 29', it never rises higher than about 6.5° above our southern horizon—just enough to give us a taste of what we're missing. At declination -72° 05', 47 Tucanae is simply impossible for us, unless we take a trip far south.

FIGURE i-5.

M 13, a magnificent Class V globular cluster in Hercules (image courtesy of Steve Childers)

Bright Nebulae (BN)

A *bright nebula* (*BN*) is a cloud of gas and dust that emits light of its own or reflects starlight. There are several classes of bright nebulae, each of which is described in the following sections.

Reflection Nebulae (RN)

A *reflection nebula* (*RN*) emits no light of its own, but shines by reflected starlight. Reflection nebulae consist of very cool, relatively dense clouds of dust intermixed with molecular (non-ionized) hydrogen gas. With very few exceptions, such as the embedded nebulosity in M 45 (see Figure i-4), pure reflection nebula are small, of very low surface brightness, and best viewed at low to medium power in large telescopes.

Emission Nebulae (EN)

An *emission nebula* (*EN*) emits light of its own. All emission nebulae comprise clouds of atomic (ionized) gases that surround or are near very hot stars, which emit large amounts of high-energy ultraviolet radiation. This UV radiation is absorbed by the gas atoms, temporarily bumping their electrons to higher energy levels. As the electrons fall back to lower energy levels, that lost energy is emitted as light at specific wavelengths. Most emission nebulae actually combine emission and reflection components, because that portion of the gas cloud close enough to the energizing star is stimulated into producing its own light, while more distant and cooler parts of the gas cloud are not ionized and therefore shine only by reflected starlight.

Most emission nebulae appear red photographically, because most of their light is emitted by excited hydrogen on the Hydrogen alpha (H-α) line at 656 nm, which is in the deep red part of the visible spectrum, to which the human eye is relatively insensitive. Fortunately, most emission nebula contain gases other than hydrogen. For visual observing, the most important of these is doubly-ionized atomic oxygen, which emits at the Oxygen-III (O-III) wavelengths of 496 nm and 501 nm. By coincidence, these two wavelengths happen to be very close to the peak sensitivity of our night-adapted eyes.

Why We See Blue-Green as Gray

In most cases, the blue-green O-III light is too dim to trigger our color-sensitive cones, and so we see this light only in the shades of gray revealed by our rods. There are exceptions, though. The O-III light from some emission nebulae, notably M 42 (shown in Figure i-6), is just bright enough in medium and larger telescopes to trigger our cones, revealing the nebulosity as a greenish-gray haze. Young observers, whose eyes are more sensitive, often report seeing green, blue, and even red tinges in M 42.

Planetary Nebulae (PN)

A *planetary nebula* (*PN*) is a special type of emission nebula. A planetary nebula is the ruins of an old, red giant star that has exploded into short-lived prominence as a nova and then subsided into a small, hot white dwarf star surrounded by a shell of expanding gas. Planetary nebulae were given that name in the 18th century by astronomer William Herschel, who thought many of the small planetary nebulae he observed resembled the planets Jupiter, Saturn, and Uranus.

On a cosmic scale, planetary nebulae have very short lifetimes, as little as a few thousand years. This is true for two reasons. First, the white dwarf star that illuminates the planetary nebula has

burned out and no longer has any continuing source of energy. It shines only from residual heat, which dissipates relatively quickly. (The central star of many planetaries is so dim that it's invisible visually even in large instruments.) Second, the expanding shell of gas that forms the planetary nebula dissipates as it expands, and soon reaches a distance at which the relatively small amounts of energy emitted by the central star are no longer capable of stimulating emission.

Planetary nebulae vary greatly in apparent size, depending on their actual sizes and their distance from us. The smallest planetaries are measured in arcseconds and may appear almost stellar, even at high magnification in large telescopes. The largest, such as the Helix Nebula (NGC 7293) in Aquarius, have extents comparable to mid-sized open clusters.

Most large planetaries have relatively low surface brightnesses, and are best viewed at low magnification. There are exceptions, such as the Dumbbell Nebula (M 27, NGC 6853) in Vulpecula, shown in Figure i-7. At six or seven arcminutes extent, M 27 is quite large, but its surface brightness is high enough to benefit from high magnification. Conversely, many small planetaries have relatively high surface brightnesses, sometimes remarkably high, and so are best viewed at high magnifications. The visual magnitude and photographic magnitude of a planetary nebula are often considerably different, with the nebula appearing much brighter visually than it does in photographs.

Planetary nebulae also vary widely in shape, both because their actual shapes may differ significantly and because the angle at which we see them affects the apparent shape. For example, although we might expect the gases ejected from a nova to assume a generally spherical shape, some planetaries, including M 27 and the Little Dumbbell Nebula (M 76) in Perseus, are believed actually to have the bi-lobed structure evident in Figure i-7. Similarly, the annular Ring Nebula (M 57) in Lyra, shown in Figure i-8, appears as a soft celestial smoke ring, but its gaseous shell is thought to be a cylinder which we happen to be looking at end-on.

Most planetary nebulae emit most of their visible light at the blue-green O-III wavelengths. In fact, the surface brightness of some planetaries is high enough to stimulate our cones, so those planetaries have a bluish or blue-green appearance visually, particularly in telescopes of moderate to large aperture.

FIGURE i-6.

M 42, the Great Orion Nebula, a magnificent emission nebula, with M 43 above it (image courtesy of Steve Childers)

FIGURE i-7.

*M 27, the Dumbbell Nebula,
a planetary nebula (image
courtesy of Steve Childers)*

FIGURE i-8.

*M 57, the Ring Nebula, a
planetary nebula (image
courtesy of Steve Childers)*

Supernova Remnants (SN, SR, or SNR)

A supernova is to a nova as a hydrogen bomb is to a firecracker. A supernova is an unimaginably stupendous explosion. When a star goes supernova, it may for short time shine more brightly than all of the other billions of stars in its galaxy combined. A *supernova remnant* (*SN*, *SR*, or *SNR*) is what remains of that cataclysm. The best-known supernova remnant is the Crab Nebula, M 1 (NGC 1952) in Taurus, shown in Figure i-9.

The titanic energies released by a supernova mean that supernova remnants shine by a different process than planetary nebulae. Rather than the line emissions of planetaries, supernova remnants emit broadband radiation, not just across the visible spectrum, but in the Radio Frequency (RF) and X-ray parts of the spectrum as well.

FIGURE i-9.

M 1, a supernova remnant (image courtesy of Steve Childers)

Galaxies (Gx)

Galaxies are island universes, unimaginably large and unimaginably distant from us. The smallest galaxies, called *dwarf galaxies*, shine with the light of a few million suns, and are little more massive than the largest globular clusters. The largest galaxies span hundreds of thousands of light years in extent and have the mass of trillions of suns.

Our own galaxy, the *Milky Way*, is above average in both size and mass, as is our sister galaxy and near-twin, the *Great Andromeda Galaxy*. Andromeda, which is the closest major galaxy to our own, is about 2.9 million light years (ly) distant. That means that the light by which we see it actually left Andromeda 2.9 million years ago, so we are literally looking into the far-distant past when we view the Andromeda Galaxy.

Figure i-10 shows Andromeda (M 31), with two of its companion galaxies, M 32 and M 110, which actually orbit M 31. M 32 (NGC 221) is visible as the bright "fuzzy star" below and right of the core of M 31. M 110 (NGC 205) is the hazy streak visible to the right of center near the top of the image. All three of these galaxies, along with the Milky Way and others, are members of our *Local Galaxy Group*.

FIGURE i-10.

M 31, the Great Andromeda Galaxy (image courtesy of Steve Childers)

Organizing Your Observing Activities

In order to make the most of your observing time, it's important to plan and organize your observing sessions. Otherwise, you'll find yourself out under the night sky just spinning your wheels— observing the same familiar objects repeatedly or waving your scope around hoping to find something interesting. The best way to organize your observing sessions is to "work" the lists of objects systematically, marking off each object as you observe it.

But which lists and which objects should you be observing? Comprehensive general catalogs of astronomical objects like the *New General Catalog* (*NGC*) and the *Index Catalog* (*IC*) include literally thousands of objects, the vast majority of which are not even visible in typical amateur telescopes. Fortunately, you don't need to sort the wheat from the chaff yourself. The *Astronomical League* (www.astroleague.org) and the *Royal Astronomical Society of Canada* (www.rasc.ca) have compiled several lists of objects suitable for beginning and intermediate amateur astronomers.

Astronomical League and RASC Observing Lists

The *Astronomical League* (www.astroleague.org) sponsors numerous observing clubs, each of which is devoted to a particular aspect of astronomical observing. Each observing club publishes a list of objects, some or all of which must be observed and logged to meet the requirements of that club. Some observing clubs have special requirements, such as sketching the objects; using only a binocular; or observing the objects from a light-polluted urban location. Completing the list for a particular observing club entitles anyone who is a member of the AL or an AL affiliate club to an award, which includes a certificate and in some cases a lapel pin.

There is a great deal of overlap in the object lists of the various AL clubs, so observing one object may gain you credit for two or more clubs. For example, the open cluster M 38 in the constellation Auriga appears on the observing lists of the Messier Club, the Binocular Messier Club, and the Urban Observing Club.

- The Messier Club allows you to count M 38 if you observe it anywhere, anyhow.

- The Binocular Messier Club requires that you observe M 38 with a binocular.

- The Urban Observing Club requires that you observe M 38 from a light-polluted urban location.

So, if you observe M 38 with a binocular from an urban location, you've met the requirements to count M 38 towards completion of all three of these clubs. That's not to say that this catch-all approach is the best way to appreciate M 38. From a light-polluted urban location, M 38 in a binocular is just a dim smudge. Using your telescope reveals at least some of the splendor of M 38, even from a bright backyard. But when you get out to a dark site, take a moment to locate M 38 again. With your binocular from a dark site, M 38 is a beautiful open cluster, much more impressive than the poor view from the urban site. With your telescope, M 38 is spectacular.

This book covers the following observing lists:

Messier Club

We think the *Messier Club* is the best starting point for nearly any novice DSO observer. Most of the Messier Objects are relatively bright and easy to find (although many will seem impossibly dim and difficult to locate when you're first starting out). The Messier list includes 110 objects, which you may observe with a telescope, binocular, or the naked eye from any location for credit toward club requirements. After you observe and log any 70 of the Messier objects, you qualify for the standard Messier Club certificate. When you have observed and logged all 110 Messier objects, you qualify for the Messier Club Honorary certificate and lapel pin.

Binocular Messier Club

Pursuing the *Binocular Messier Club* list is an excellent way for beginning observers to develop their binocular observing skills. Many beginning observers work this list with their binoculars at the same time they work the standard Messier list with their telescopes.

There are actually two lists for this club, one for those using standard 35mm or 50mm binoculars, and a second for those using 70mm or larger giant binoculars. Each list is divided into groups by difficulty. The list for standard binoculars includes 76 objects, 42 of which are rated Easy, 18 rated Tougher, and 16 rated Challenge. The list for giant binoculars is a superset of the standard list, and includes 102 objects, 58 rated Easy, 23 rated Tougher, and 21 rated Challenge. To qualify for the Binocular Messier Club certificate and lapel pin, you need only observe and log any 50 of these objects.

Urban Observing Club

Although AL formerly categorized the *Urban Observing Club* as an introductory club, they have since moved it to their Telescopic group. We believe that was a good decision, because locating objects under light-polluted urban conditions can be extremely challenging, particularly for beginning observers.

The UO Club actually has two lists, one that includes 87 DSOs and a second that includes 12 multiple stars and one variable

---

Find it Yourself

With very few exceptions, the AL observing clubs require that you locate objects manually, without using any computerized aid such as a go-to telescope or digital setting circles. Some even prohibit the use of manual setting circles. (All clubs permit using planetarium software on a notebook computer or PDA, as long as the telescope is not controlled by the computer.) That is not to say that you cannot use a telescope *equipped* with go-to or DSCs to meet the requirements of these observing clubs. Simply turn off the computerized functions while you are working an observing club list, and find the objects manually by star-hopping.

Think Ahead

Consider logging not just the information required by the particular clubs whose lists you are currently working, but any additional information required by other clubs you may work in the future. Some of the advanced AL club lists include many of the same objects covered by this book, but require you to log additional information. For example, the AL Globular Cluster Club requires you to estimate the Shapley-Sawyer Concentration Class of each globular cluster you observe. The AL Open Cluster Club requires you to log the Trumpler classification of each open cluster, and to sketch any 25 of the open clusters on that list. Capturing this additional information now gives you a head start if you decide to pursue those advanced lists. Visit the AL web site for details about the requirements of these advanced clubs.

Easy, Tougher, Challenge

In this book, we flag all 102 of the objects on the larger Binocular Messier Club as candidates for the club award, without discriminating by aperture size or difficulty. Take care while you are pursuing this list to choose objects that are bright enough to be visible with your binocular under the conditions at your observing site.

As a point of reference, Barbara and Robert, using their 50mm binoculars from moderately dark sites with excellent transparency, found all of the objects on the 50mm Easy list were indeed relatively easy to locate and view. Those labeled as Tougher we found little or no more difficult than the dimmer objects on the Easy list. With only one or two exceptions, we found the Challenge objects to be impossible from a moderately dark site, and extremely difficult from a very dark site, even when the transparency was excellent. For the more difficult objects in particular, you will find that mounting your binocular on a tripod rather than hand-holding it makes the objects much easier to see.

star, which is the only variable star covered in this book. To qualify for the Urban Observing Club certificate, you must locate, observe, and log all 100 objects on the two lists from an urban site, which the club defines as a site that is sufficiently light-polluted that the Milky Way is invisible to the unaided eye. This AL club is one of the few that does not explicitly prohibit using a go-to telescope or digital setting circles to locate objects.

Deep Sky Binocular Club

The *Deep Sky Binocular Club* is intended as a follow-on for those who have completed the Binocular Messier Club list. Some of the 60 objects on the Deep Sky Binocular list—such as the Double Cluster, Hyades (Melotte 25), and α-Perseii Association (Melotte 20)—are very bright and easy to locate. Others are considerably more challenging, and are clearly visible only under reasonably dark skies with good transparency. To qualify for the Deep Sky Binocular Club certificate and lapel pin, observe and log all 60 of the objects on the club list.

Double Star Club

The *Double Star Club* introduces new observers to the best 100 double and multiple stars visible in the night sky. For some, double stars become a primary observing activity and a life-long passion. Although we're definitely DSO folks, we enjoy observing double stars on occasion, particularly on nights that aren't quite clear enough for hunting down the faint fuzzies. Some people are put off by the requirement to sketch (gasp!) each double star on the list.

RASC Finest NGC List

The *RASC Finest NGC List* includes 110 non-Messier DSOs, and is considered by most experienced observers to be the best follow-on list for beginners who have completed the Messier list. Overall, objects on the RASC list are somewhat fainter and sometimes a bit harder to find than Messier objects, but there are exceptions. (The best/brightest objects on the RASC list are better/brighter than the worst/faintest objects on the Messier list.) The Royal Astronomical Society of Canada offers a certificate to anyone who completes this list.

Ordering Your Observing

There are three common ways to order your observing activities: by list, by constellation, or by object type. Each has advantages and disadvantages.

Most novice observers tend to focus on working a particular list or lists. For example, a novice may decide to get started by pursuing the Messier Club list and the Binocular Messier club list. One advantage of this strategy is that it's possible to get results relatively quickly. (In fact, most years on a new moon weekend in late March or early April, it's possible to run a Messier Marathon and observe all 110 Messier objects in one night.) Another advantage is that most of the Messier objects are relatively bright and easy to find. That allows a novice to enjoy some early success, which is important, and to develop the skills needed to locate and observe more difficult objects.

One disadvantage of pursuing only one or two lists at a time is that you pass up a great many objects that are in the same vicinity as your list objects. For example, if you're working the Messier list in Cassiopeia and have logged M 52 and M 103, you've finished with Cassiopeia and move on to the next constellation with Messier objects. But Cassiopeia contains many other interesting objects that are members of other lists, all of which you've skipped over. Which brings up the other disadvantage of pursuing only one or two lists: it's easy to run out of objects to look for.

In contrast, most experienced observers tend to "work a constellation." By that, we mean pursue all of the interesting objects in a particular constellation, regardless of which list or lists those objects are members of. This "constellation sweeping" method is the one we recommend for beginning observers, and is the basis for how we structured this book. The main advantage to this method is that you are working a small area of sky, with which you soon become intimately familiar, making it easier to locate the fainter objects in the constellation. The other nice thing about working constellations is that you can do so under less than perfect observing conditions. If there are clouds covering Cassiopeia, for example, you can just go work Draco. The only real disadvantage to constellation sweeping is that it takes longer to complete each individual list, so it's not suitable for those who are into the immediate gratification of quickly earning certificates for completing those lists.

The final method, working by object type, is sometimes used by some advanced amateurs. For example, our observing buddy Paul Jones, an extremely experienced observer, often goes on "kicks." For several observing sessions in a row, Paul may observe only planetary nebulae or only globular clusters. The advantage of this method is that it allows you to compare and contrast objects of the same type. The disadvantage is that you soon find yourself looking for some very faint, obscure objects.

CONSTELLATIONS BY SEASON

CONSTELLATION	MIDNIGHT CULMINATION	CONSTELLATION	MIDNIGHT CULMINATION
Canis Major	1 January	Lyra	2 July
Gemini	4 January	Sagittarius	5 July
Monoceros	5 January	Aquila	12 July
Puppis	9 January	Sagitta	17 July
Lynx	20 January	Vulpecula	26 July
Cancer	30 January	Delphinus	31 July
Hydra	9 February	Capricornus	5 August
Sextans	21 February	Aquarius	26 August
Leo Minor	24 February	Lacerta	28 August
Leo	1 March	Pegasus	1 September
Ursa Major	11 March	Pisces	27 September
Corvus	28 March	Sculptor	27 September
Coma Berenices	2 April	Cepheus	29 September
Canes Venatici	7 April	Andromeda	30 September
Virgo	12 April	Cassiopeia	9 October
Boötes	30 April	Eridanus	14 October
Libra	9 May	Cetus	15 October
Corona Borealis	19 May	Aries	20 October
Draco	24 May	Triangulum	23 October
Serpens	3 June	Perseus	7 November
Scorpius	3 June	Taurus	30 November
Ophiuchus	11 June	Auriga	9 December
Hercules	13 June	Orion	13 December
Cygnus	29 June	Lepus	13 December
Scutum	1 July	Camelopardalis	23 December

Seeing in the Dark

Dark adaption refers to how well your eyes have adjusted to the darkness. The quality of your night vision gradually goes up the longer you are in darkness, and instantly goes down when you are exposed to anything other than red light.

Each of the objects covered in the constellation chapters of this book is a member of one of 50 constellations. Each of these constellations is covered in its own chapter, all of which use the same format. A chapter begins with a summary table that covers the important details of the constellation, including the date of midnight culmination (the date when the object reaches its highest elevation at midnight and is therefore best placed for observing during the evening and early morning hours), prominent objects that belong to it, and the constellations that border it. Here is the summary table for the constellation Andromeda.

NAME: Andromeda (an-DROM-eh-duh)

SEASON: Autumn

CULMINATION: 9:00 p.m., late November

ABBREVIATION: And

GENITIVE: Andromedae (an-DROM-eh-dye)

NEIGHBORS: Ari, Cas, Lac, Peg, Per, Psc

BINOCULAR OBJECTS: NGC 205 (M 110), NGC 221 (M 32)...

URBAN OBJECTS: NGC 221 (M 32), NGC 224, (M 31), NGC...

The chapter continues with a brief introduction and a full-page, large-scale chart of the constellation that shows the general position of each featured object. All of the featured objects in the constellation are summarized in two tables, one for DSOs and another for multiple stars. Table i-1 is the DSO summary chart for Andromeda.

The columns in the DSO table include the following information:

Object

The catalog number of the object in question. For most objects, this is the *NGC* (*New General Catalog*) number or the *IC* (*Index Catalog*) number. Some objects are not cataloged in the NGC or IC, but are members of other specialty catalogs, which are noted in the object descriptions.

Type

The type of the object: EN (emission nebula), GC (globular cluster), Gx (galaxy), OC (open cluster), PN (planetary nebula), RN (reflection nebula), or SR (supernova remnant). Objects that are of multiple types are listed, for example, in the form EN/RN (emission nebula/reflection nebula) or EN/OC (emission nebula/ open cluster).

Mv

The visual magnitude of the object. For many objects, visual magnitude is not well defined, and different sources disagree. In those cases, we have attempted to choose the best available or consensus value. In cases where we were unable to find a reliable value for visual magnitude, we list the magnitude as 99.9.

Size

The extent of the object. Unless otherwise stated, all values are in arcminutes ('). For some very small objects, including many planetary nebulae and some galaxies, we list the size in arcseconds ("), which is always explicitly indicated. Values given for size are always approximate, because the visible extent of an object depends on so many factors, including the size of your telescope, the darkness of your site, your level of dark adaption, and so forth. In general, the object will appear noticeably smaller in typical amateur scopes than the size listed, which is usually a photographic size. For comparison, the apparent size of the full moon is about 30 arcminutes.

RA and Dec

The right ascension and declination of the object, J2000.0 epoch. Right ascension is specified in hours, minutes, and decimal minutes, in the form 00 40.4, which may also be read as 00h40m24s (0.4 minutes = 24 seconds). Declination is specified in degrees and minutes as a signed value in the form +41 41, where positive declination is north and negative is south.

TABLE i-1.

Featured star clusters, nebulae, and galaxies in Andromeda.

Object	Type	Mv	Size	RA	Dec °	M	B	U	D	R	Notes
NGC 205	Gx	8.9	21.9 x 10.9	00 40.4	+41 41	◉	◉				M 110; Class E5 pec; SB 13.2
NGC 221	Gx	9.0	8.7 x 6.4	00 42.7	+40 52	◉	◉	◉			M 32; Class cE2; SB 10.1
NGC 224	Gx	4.4	192.4 x 62.2	00 42.7	+41 16	◉	◉	◉			M 31; Class SA(s)b; SB 12.9
NGC 752	OC	5.7	49.0	01 57.8	+37 51			◉	◉		Cr 23; Mel 12; Class II 2 r
NGC 891	Gx	10.8	14.3 x 2.4	02 22.6	+42 21					◉	Class SA(S)b? sp; SB 14.6
NGC 7662	PN	9.2	37.0"	23 25.9	+42 32			◉		◉	Blue Snowball Nebula; Class 4+3

M, B, U, D, R

A bullet in any of these columns indicates that the object is a member of that observing list: M (Messier), B (Binocular Messier), U (Urban Observing), D (Deep Sky Binocular), or R (RASC Finest NGC Objects).

Notes

The common name of an object and/or other catalog designations and, where available, the class and surface brightness (SB) of the object.

Table i-2 is the multiple star summary table for Andromeda.

The columns in the multiple star table include the following information:

Object

The catalog designation for the multiple star, usually by Flamsteed number and/or Bayer letter.

Pair

The pair in question, usually by STF number (or Σ, which is the same thing.) If the multiple has more than two stars, additional pairs are listed on separate lines if those additional pairs are objects on either of the lists. For example, if the BC pair of 57-gamma Andromedae appeared on either of the multiple star lists covered in this book, which it does not, there would be a separate line for STF 205BC. That line would provide magnitudes, separation angle, and so forth for B as the primary and C as the companion. In this case, 57-γ is treated as a simple double star, with A as primary and BC together as the companion.

M1 and M2

The visual magnitudes of the primary and companion, respectively.

Sep

The separation between the primary and companion, listed in arcseconds (").

PA

The position angle from the primary to the companion, as described earlier in this chapter.

Year

The year of the observation for the data given. In particular, the position angle and separation may change significantly over a relatively short span for some multiples.

RA and Dec

As described above.

UO and DS

A bullet in either of these columns indicates that the object is a member of the Astronomical League Urban Observing club or the Double Star club, respectively.

Notes

The common name of the star, and/or other interesting information about it.

TABLE 01-2.

Featured multiple stars in Andromeda

Object	Pair	M1	M2	Sep	PA	Year	RA	Dec	UO	DS	Notes
57-γ	STF 205A-BC	2.3	5.0	9.7	63	2004	02 03.9	+42 20	◉	◉	Almach

A constellation chapter concludes with short sections devoted to each featured object. Those sections begin with a summary table like the one shown in Table i-3 for NGC 7662.

The top row of the summary table includes the following information:

Object catalog number

The object identifier, usually the Messier, NGC, or IC number. If the object is known under another designation, that information is included in parentheses.

Visual rating

A subjective evaluation of the appearance of the object. High-rated objects reveal detail and interesting features. Low-rated objects may be visible only as dim gray smudges. Ratings are based on observations from reasonably dark sites, using a 7X50 or 10X50 binocular for binocular objects or an 8" to 10" telescope for telescopic objects, with a narrowband or O-III filter when appropriate. We use the following scale to rank the appearance of objects.

* ★★★★ — a showpiece of the heavens
* ★★★ — shows considerable detail and/or interesting features
* ★★ — limited detail/features visible; a routine object
* ★ — unimpressive even with large aperture from a dark site

Finding difficulty

A subjective evaluation of the effort required to locate an object. High-rated objects are easy to find using a short star hop from a bright nearby star or other prominent object. Low-rated objects may require an extended star hop using dim stars. Note that difficulty is very much relative to level of experience. Beginners may consider some "easy" objects to be very difficult to locate, and "difficult" objects to be nearly impossible to find. We use the following scale to rank the difficulty of locating an object.

* ◉◉◉◉ — easy to find
* ◉◉◉ — can be found with some effort
* ◉◉ — difficult to find
* ◉ — very difficult to find

Object type

The type of the object: EN (emission nebula), GC (globular cluster), Gx (galaxy), MS (multiple star), OC (open cluster), PN (planetary nebula), RN (reflection nebula), or SR (supernova remnant). Objects that are of multiple types are listed, for example, in the form EN/RN (emission nebula/reflection nebula) or EN/OC (emission nebula/open cluster).

List membership(s)

The observing lists of which the object is a member: M (Messier), B (Binocular Messier), U (Urban Observing), D (Deep Sky Binocular), or R (RASC Finest NGC Objects). List memberships are in **bold text**. For example, **MBUDR** indicates that NGC 7662 is a member of the Urban Observing and RASC Finest NGC Objects observing lists, but not Messier, Binocular Messier, or Deep Sky Binocular.

The bottom row of the summary table includes the following information:

Chart number

For many objects, the overall constellation chart provides insufficient detail to locate the object. For these objects, we include a detailed "finder chart" that zooms in on a smaller part of the constellation. Each such chart includes one or more prominent objects from the full constellation chart to allow you to orient yourself. North is always up, and west to the right on these finder charts. The caption of each chart lists the field width and height. Most finder charts include 5° finder circles and/or 1° eyepiece circles to indicate the field of view. If your finder or eyepiece has a different field of view, it's easy to guesstimate positions using the known 1° and 5° circles.

Figure number

Most objects include a DSS image. These images are based on the POSS1 data set, and are 60 arcminutes (1°) square. When viewed under dim red illumination, they provide about the level of detail visible to an experienced observer using a very large telescope from an extremely dark site. (In other words, don't expect to see this much detail in a typical amateur telescope from a typical light-polluted observing site.)

TABLE i-3.

Object summary table

NGC 7662	★★★	◉◉◉		PN	MBUDR
Chart 01-6	Figure 01-4	m9.2, 37.0"		23h 25.9m	+42° 32'

KEY TO THE OBJECT SUMMARY TABLE

Object Catalog Number	Visual rating	Finding difficulty	Object type	List membership(s)
Chart number	DSS image of the object	Magnitude and size	Right ascension	Declination

Magnitude and size

The visual magnitude and extent of an object. Size is listed in arcminutes unless otherwise noted.

RA and Dec

The right ascension and declination (equatorial coordinates) of the object, using the J2000.0 epoch.

The object section ends with a narrative description of how to locate the object and what it looks like, based on our own observing logs or those of other observers who have contributed their own reports. Here, for example, is Robert's observing report for NGC 7662.

> NGC 7662, also called the *Blue Snowball Nebula*, is a fine planetary nebula. To locate NGC 7662, place the m4 star 19-ϰ Andromedae at the NE edge of your finder field. Once you've done this, m4 17-ι will be prominent 1.1° SSW of 19-ϰ and m6 13 Andromedae will be approximately centered in the finder, 2° W of 17-ι.
>
> NGC 7662 is visible in a low-power eyepiece as a fuzzy star 25' SSW of 13 Andromedae. At 180X and 250X in our 10" scope, NGC 7662 reveals considerable structural detail as a bright, slightly elongated, annular disk with a distinct bluish tinge and noticeable darkening toward the center. The brighter inner ring is continuous, although denser to the NE and SW. The dim outer ring is fragmented and requires averted vision to detect. An O-III filter is the best choice for this nebula, but a narrowband filter also enhances the view significantly. Although the central star is listed at m13.2, we have never been able to see that star in our 10" scope.

For most objects, the observing report is based on the appearance of the object in our 10" reflector from a moderately dark observing site. If your scope is smaller, your site is brighter, or your experience in teasing out detail from dim objects is less than ours, you'll see less detail. If your scope is larger, your site is darker, or your experience is greater, you'll see more.

Note that the amount of detail visible to you depends on many other factors. In particular, when you view dim objects such as galaxies, the transparency of the atmosphere is a key factor. On an extremely clear night, for example, we have seen more detail in the galaxy M 31 with a binocular than is visible on even a slightly hazy night with our 10" reflector. Transparency can vary dramatically from hour to hour, and one part of the sky may be very transparent while another is hazy enough to make viewing dim objects difficult.

It's also important to understand that our descriptions are based on our *best* observation for each object. We have observed many of these objects literally dozens of times, and the amount of detail visible often varies dramatically from one observation to the next. If you are using a scope of similar size to ours and you don't see what we describe, come back to the object later and view it again. Persistence pays.

DSS POSS1

The DSS (Digitized Sky Survey) is a digitized version of an all-sky photographic atlas, which was carefully scanned and digitized from images provided by several sources. The great value of DSS images for amateur astronomers is that all images are at the same scale and magnitude depth. Many planetarium programs incorporate RealSky images, which are more highly compressed versions of original DSS images. This book uses the original, full-resolution images.

For more information about DSS, visit www-gsss.stsci.edu/SkySurveys/Surveys.htm.

Observing Equipment

> The ancients observed the heavens with no equipment at all. You could do the same, but you'll get much more enjoyment from the hobby if you equip yourself properly. In this chapter, we'll tell you what you need.

Binocular

Many beginners think of a binocular as something you use for observing only if you don't yet have a telescope. Nothing could be further from the truth. If you watch experienced amateur astronomers at a star party, you'll find that most of them keep a binocular close at hand, and often use it when locating objects. (Our usual progression is chart→naked eye→binocular→Telrad→optical finder→eyepiece.)

Having even an inexpensive binocular is immensely better than having no binocular at all. A $35 Wal-Mart 7X35 binocular doesn't have the best optical or mechanical quality, but it gets the job done if that's all you can afford. If you don't yet have a binocular or you want to buy a model more suitable for astronomy, we suggest you consider the following issues:

Magnification and aperture
All binoculars are designated by two numbers, such as 7X35 or 10X50. The first number is the magnification, and the second is the aperture in millimeters. For hand-held astronomical observing, a binocular with 7X to 10X magnification is usually the best choice. (Binoculars are available with 12X to 25X or higher magnification, but anything above 10X or perhaps 12X requires tripod mounting for stability.) The aperture determines how much light the binocular gathers, proportionate to the square of the aperture. For example, a 50mm binocular has twice the aperture of a 25mm binocular, but gathers four times as much light ($2^2 = 4$). Binoculars suitable for hand-held astronomical observing are available with apertures from 35mm to 63mm, with 50mm by far the most common.

Exit pupil
The *exit pupil* of a binocular is calculated by dividing the aperture by the magnification. For example, a 10X50 binocular has a 5mm exit pupil (50 ÷ 10 = 5), as does a 7X35 binocular (35 ÷ 7 = 5), while an 8X56 binocular has a 7mm exit pupil (56 ÷ 8 = 7). The ideal exit pupil size depends on the maximum size of your fully dark-adapted entrance pupil. Young people's eyes may be able to dilate to 7mm or slightly more. Middle-aged people may be limited to 6mm or so, and older people to perhaps only 5mm. Choosing a binocular whose exit pupil is at least as large as your maximum dark-adapted entrance pupil size gives you the brightest possible image.

Let the Light Shine In

Choosing a binocular that has a larger exit pupil than your dark-adapted entrance pupil effectively stops down the binocular. For example, if your maximum entrance pupil is 5mm and you use a 7X50 binocular (7.1mm exit pupil), your eye blocks all but the central 5mm, turning that 7X50 binocular into the equivalent of a 7X35 binocular. You might just as well use a 7X35 binocular. Either that, or a 10X50 model, whose exit pupil matches your entrance pupil. All of that said, do not reject a binocular simply because its exit pupil is "too large." Robert has an entrance pupil of 6.5mm, but uses a 7X50 binocular most of the time.

Advice from Sue French

I once read a rule of thumb that says to compare binoculars, multiply the aperture by the magnification. Binoculars with a higher resulting number will show you more than those with a lower number. For the "binox" that have paraded their way through this house, I'd say that's been a fairly accurate assessment. My husband, Alan, says there's a new school of thought that says one should multiply the magnification by the square root of the aperture. As for the exit pupil, the visual acuity of the eye is very poor beyond 4mm, so you're not losing much there.

Eye relief

The *eye relief* of a binocular is the distance between the outer surface of the eyepiece lens and where your pupil needs to be placed to view the image. Standard binoculars have eye relief ranging from only a few millimeters to 20mm or more. Long eye relief—at least 17mm to 20mm—is necessary if you wear glasses while using the binocular. Short eye relief is acceptable if you use the binocular without eyeglasses or with contact lenses.

Field of view

Field of view (*FoV*) quantifies the angular range of the image visible in the binocular eyepieces. FoV is determined by the optical design of the binocular, including its focal length and the type of eyepieces used. For astronomy, a wide FoV is desirable—you can see more of the sky with a wider field—but a very wide FoV may be problematic. Increasing the FoV beyond a certain point requires optical compromises that cause blurred images at the edges of the field, distortion, short eye relief, and other problems. For 7X or 8X binoculars, something in the 6.5° to 8.5° range is reasonable. For 10X binoculars, look for something in the 5.0° to 7.0° range, and for 12X models something in the 4.5° to 6.0° range. All other things being equal, a binocular with a FoV on the low end of our recommended range will probably provide better *edge performance* (sharpness near the edges of the field of view) and eye relief than a similar model with a wider field. High-end binoculars can push the limits by using complex (and expensive) eyepiece designs that provide wider fields while maintaining optical quality and long eye relief.

Interpupilary distance

Interpupilary distance is the distance between the centers of your two pupils. Standard binoculars are adjustable to accommodate different interpupilary distances, typically within a range of 60mm to 75mm. That suffices for most people, but some women and many children have interpupilary distances too short for a standard binocular to accommodate. Unfortunately, the only answer is often compact binoculars, which are unsuitable for astronomy because of their small objective lenses.

Prism type

Binoculars use prisms to present a correct-image view, right-side up and not reversed left-to-right. Two types of prism are commonly used, *Porro prisms* and *roof prisms*. Although roof-prism models are generally more costly and are considered better for general use, Porro-prism models are superior for astronomical observing because they have higher light transmission.

Coatings

Binoculars use *anti-reflection coatings* on lenses and prisms to reduce internal reflections that would otherwise reduce light transmission and image contrast. The simplest, least expensive

---

Glass Matters

Two types of glass are used to make Porro prisms. Cheap Porro prisms are made from inferior BK-7 borosilicate flint glass. Better Porro prisms are made from superior BaK-4 barium crown glass. Although nearly any binocular that uses BaK-4 prisms advertises that fact, it's easy enough to check for yourself. Simply hold the binocular with the eyepieces several inches from your eyes, pointing at the sky or another evenly-lit light source, and look at the exit pupils. If the prisms are of BaK-4 glass, the exit pupils are round and evenly illuminated. If the prisms are of BK-7 glass, you'll see a square inside the circle, with the area inside the square brightly illuminated and the area outside the square dimmer.

Roof prism binoculars, because of their design, can use BK-7 glass without suffering the edge dimming (reduced light at the edges of the image) and vignetting (darkening at the edges of the image) that occur with BK-7 Porro prisms.

coatings are single-layer coatings, which reduce reflections significantly compared to uncoated optics. Reflection can be further reduced by using multi-layer coatings, a process known as multicoating. But good coating costs money, and good multicoating more so. To cut costs, some optics makers coat (or multicoat) only some of the surfaces. A standard terminology has arisen to describe the levels of coatings used on binoculars and other optical equipment:

Coated

Single-layer coatings have been applied to some but not all of the optical surfaces, typically only the external surfaces of the objective and eyepiece lenses. Only the cheapest binoculars—we are tempted to call them toys—are in this category.

Fully coated

Single-layer coatings have been applied to all of the optical surfaces. A fully coated binocular is the minimum acceptable for serious astronomical use. Most inexpensive binoculars sold by Wal-Mart and similar big-box retailers are in this category.

Multicoated

Multi-layer coatings have been applied to some but not all of the optical surfaces, typically only to the external surfaces of the objective lens. Usually, but not always, the other surfaces have had single-layer coatings applied. which is sometimes described as "fully coated and multicoated." Most inexpensive binoculars suitable for astronomical use are in this category.

Fully multicoated

Multi-layer coatings have been applied to all of the optical surfaces. Binoculars in this category are the best choice for astronomical observing, but are expensive. Low-end fully-multicoated models, such as the Orion UltraViews, start at $150 and go up from there.

Although there are other desirable and undesirable binocular characteristics, most of them are reflected quite accurately by the prices of various competing models. Binoculars are available in a wide range of prices. For convenience, we classify standard-size binoculars by price range as inexpensive (<$75), midrange ($75 to $250), premium ($250 to $500), and super-premium (>$500). As you might expect, it's not difficult to get a good binocular if you are willing to pay premium or super-premium prices, but even mid-priced binoculars are generally excellent.

In the sub-$250 range, each doubling of price generally buys you substantially better optical and mechanical quality, all other things being equal. For example, a $100 7X50 binocular is not twice as good as a $50 7X50 binocular, but it is likely to be more solidly built and to provide noticeably superior images. A $200 7X50 binocular provides a similar relative improvement over a $100 binocular. Once you get up to the $300 (and beyond) range, the improvements become small, if not invisible. For example, other than by examining the name plate, few people would be able to detect any difference in optical or mechanical quality between a $300 10X50 binocular and a $600 10X50 binocular.

- If your budget is very tight, nearly any inexpensive binocular is better than nothing. Avoid gimmicks, such as "instant focus", red coatings, and so on. Look for a binocular that uses BaK-4 prisms and is at least fully coated (multi-coated is better). A 7X35 model is acceptable, and in fact may be a better choice than a 7X50 or 10X50 model at the same price.

- For standard binoculars in the $75 to $250 range, most models from Bausch & Lomb, Celestron, Minolta, Nikon, Olympus, Orion, Pentax, Pro Optic, and Swift are reasonable choices. At the low end of this range, we think the 7X50 and 10X50 Orion Scenix models are the stand-out choices. At the upper end of this range, the Orion Vista binoculars in 7X50, 8X42, and 10X50 offer very high value, as do the similar but more expensive Celestron Ultima models.

- For standard binoculars in the $250 to $500 range, there are many suitable candidates. At the lower end of the range are the excellent Celestron Ultima models. At the middle and upper parts of this range, you will find many suitable models from Alderblick, Celestron, Fujinon, Nikon, Pentax, and Steiner.

- For standard binoculars in the stratospheric super-premium price range, nearly any Porro-prism model with provision for tripod mounting is an excellent choice. This is the realm of world-class optics from companies like Leitz, Fujinon, Swarovski, and Zeiss. Premium models from companies like Fujinon, Nikon, Pentax, and Steiner are also in this price class. Binoculars simply don't get any better than this.

A Multicoat of Many Colors

All coatings are not of the same quality, and, contrary to popular belief, it's impossible to judge the quality of coatings by the color of the light they reflect. Applying top-notch multicoating is a very expensive process, that if done properly can exceed the cost of making the lenses themselves. On the other hand, if all a manufacturer cares about is being able to claim that its optics are multicoated, it can slap on multicoating relatively cheaply. Optics from Zeiss, Nikon, Fujinon, or Pentax (to use just a few examples) have superb multicoating. Optics from second-tier Japanese makers, such as Vixen, have good multicoating, but not as good as that of first-tier makers. Cheap multicoated optics from Chinese makers, well, they're multicoated, but that's about the most you can say about them. Avoid any binocular with "ruby red" or other strange coatings.

Inexpensive Doesn't Have to Mean Cheap

Even some inexpensive binoculars provide surprisingly high bang-for-the-buck. When Barbara first became interested in astronomy several years ago, Robert bought her a $90 Orion Scenix 7X50 binocular, figuring that if she lost interest in astronomy it wouldn't be any great loss. Robert had been on a 20-year hiatus from active observing—college, jobs, and "real life" intervened—but was determined to get back into the hobby. At first, he planned to buy a premium Zeiss, Leitz, or Swarovski binocular, but when he saw how good Barbara's inexpensive Orion Scenix binocular was, he bought a Scenix for himself. Is the Scenix as good as premium models? No, but for a fifth to a tenth the price, it's astonishingly good.

Binoculars with apertures larger than 56mm are classified as giant binoculars. Like standard binoculars, giant and super-giant binoculars are available in various sizes and price ranges. Although we do not recommend a giant binocular as your first (or only) binocular, they have their place. In fact, some observers eschew telescopes entirely and spend all of their observing time with a good tripod-mounted giant binocular.

We classify giant binoculars as inexpensive (<$250), midrange ($250 to $1,000), premium ($1,000 to $5,000), and super-premium (>$5,000). Some inexpensive giant binoculars are surprisingly good for their price. You won't mistake them for premium units, but they are quite usable. They are generally sharp at the center of the field, but with noticeable softness in the outer 15% to 30% of the field. Inexpensive giant binoculars often show significant *chromatic aberration* (false color) on very bright objects such as Luna and bright stars, but they are not really intended for observing those types of objects. For doing what they do best—scanning Milky Way star fields and viewing open star clusters—they serve quite well. If you can afford better, you'll find that spending twice as much delivers noticeably better image quality and mechanicals. As with standard-size binoculars, there are premium and super-premium models available for those who can afford them (large Fujinon, Leitz, Takahashi, and Zeiss models sell for more than $10,000, sometimes much more.)

- If you're on a tight budget, the Chinese-made Celestron SkyMaster series is a good choice in a giant binocular. Three models are available, 15X70, 20X80, and 25X100. All use BaK-4 Porro prisms and are multi-coated. Eye relief is acceptable, at 18mm in the 15X70 model, and 15mm in the two larger models. The 15X70 model sells for well under $100, and even the 25X100 model can sometimes be found on sale for under $250. All models have provision for tripod mounting, which is necessary for these large instruments. The 15X70 model is water resistant, and the two larger models are waterproof. Image quality is mediocre, particularly at the edges, but is surprisingly good for the price. We think most occasional users will be quite pleased with a Celestron SkyMaster binocular.

- For inexpensive semi-giant binoculars, we think the stand-out choice is the Orion Mini-Giant series. These fully multi-coated Japanese-made optics are available in 8X56, 9X63, 12X63, and 15X63 models, ranging in price from $159 to $219 and with fields of view from 5.8° in the 8X model to 3.6° in the 15X model. Eye relief across the line is excellent, from 17.5mm to 26mm. The smaller models are hand-holdable, although all models have tripod sockets. Image quality is, if not quite up to the level of the premium brands, more than acceptable to most people.

We can't make other recommendations, because our experience with giant binoculars is very limited. However, we will say that Fujinon offers several giant binoculars that are extremely popular with amateur astronomers and receive uniformly excellent reviews. If we wanted to buy a premium or super-premium giant binocular, we'd look first at Fujinon models.

Nearly all of the objects in this book are visible from a dark site using a 6" telescope. Most are visible with a 4" scope, and many with a 60mm (2.4") scope. Some are even visible with the naked eye. That's not to say that small instruments are ideal for the purpose. For DSO observing in particular, the larger your scope, the better.

If you already have an astronomical telescope of any size or type, we suggest you dive right in and use it to begin observing the objects cataloged in this book. If it turns out that you need (or want) a larger scope, that will become obvious to you soon enough. By exploring the limits of your current scope, you'll get a much better idea of what your next scope should be. (The desire for a larger scope, so-called "aperture fever," is a common malady among amateur astronomers, but the truth is that few people could exhaust the potential of even a 6" telescope in a lifetime of observing.)

If you do not already have a telescope, we suggest you make haste slowly. Do some research before you buy. Among the best sources for advice about choosing and buying a telescope are the books *Star Ware* by Phil Harrington (Jossey-Bass) and our own *Astronomy Hacks* (O'Reilly). Read both of them, and decide which type and size of telescope is best for you, based on your budget and personal preferences. Once you have done so, visit your local astronomy club or attend a public observation session and get some hands-on experience with different types of telescopes. (You can find a searchable list of astronomy clubs worldwide at skyandtelescope.com/resources/organizations/).

All of that said, the vast majority of amateur astronomers choose either Dobsonian reflectors ("Dobs") or Schmidt-Cassegrain

Telescopes (SCTs) with apertures between 8" and 12" as their primary instruments. Each has advantages and disadvantages.

One major advantage of Dobsonian scopes is their very high bang-for-the-buck. In any price range, a Dob offers more aperture per dollar spent than any other type of telescope. Basic 6" Dobs are widely available for $250 to $350, and were formerly the most popular size. Nowadays, though, 8" models—which sell for only $50 to $100 more than comparable 6" models and are nearly the same size and weight—have overtaken 6" models in popularity. Even 10" Dobs ($450 - $750) and 12" Dobs ($800 - $1,200) are now considered mainstream telescopes, suitable even for beginners. Figure ii-1 shows a typical 10" "econo-dob," which happens to be our primary telescope.

Advantages of a Dob

Besides low price, basic Dobs have many other advantages that have made them the first choice of many amateur astronomers:

- Dobsonian mounts are inherently immensely stable, eliminating the "jigglies" that make inexpensive tripod mounts such a pain to use.

- Dobsonian mounts use altazimuth motions. That means they move in altitude (up and down) and azimuth (left and right), which most people, particularly beginners, find more intuitive than the equatorial mounts often used with other types of telescopes.

- The relatively fast focal ratios and short focal lengths of most Dobs allows them to display very wide true fields of view. For example, using a 2" eyepiece, our 10" f/5 Dob has a maximum possible true field of view of about 2.25°. Conversely, a 10" f/10 SCT with a 2" eyepiece has a

FIGURE ii-1.

An older-model Orion XT-10 10" Dobsonian telescope

Size Matters

The aperture of a telescope, which may be specified in inches, millimeters, or centimeters, determines how much light it gathers. Light-gathering ability varies with the square of the aperture. For example, an 8" telescope gathers nearly twice as much light as a 6" model ($8^2 = 64$ and $6^2 = 36$; $64/36 = 1.78$ times as much light). A 10" telescope ($10^2 = 100$) gathers almost three times as much light as a 6" model, which translates to seeing more than one full magnitude deeper with the 10" model.

Find It Yourself

Go-to scopes and DSCs are very popular because they find objects automatically, which allows you to devote less time to finding the objects and more time to looking at them. However, most of the Astronomical League observing clubs specifically forbid the use of go-to or DSC functions for observations that count toward completing club requirements. That's not to say that you can't use a go-to scope or one with DSCs for completing the requirements of these clubs. Simply turn off the computerized locating functions when you are observing objects "for the record."

maximum possible true field of view of only about 1.10°. Although that difference may not seem large, it means that the Dob can display four times as much sky in the eyepiece as can the 10" SCT.

- Although they are somewhat bulky, Dobsonian scopes are relatively light and portable. Many people regularly transport 10" and 12" Dobs in sub-compact cars.

- Because they use an open tube, the cool-down time required for Dobsonians to reach ambient air temperature is generally less than that required by closed-tube telescopes, including SCTs. Proper cool-down is critical for high-magnification observing, such as lunar and planetary work, but less important for DSO observing.

- Dobsonians are fast to set up and tear down. Setup requires literally a minute or less, because you need only place the base in position and set the tube on top of it. Teardown is just as fast, which matters at the end of a long observing session.

Disadvantages of a Dob

Dobs also have several disadvantages relative to SCTs:

- Dobs are inherently manual scopes, which means they do not track the apparent motion of the stars unless you add relatively costly and sometimes finicky special equipment such as a Dob Driver or equatorial platform.

- In addition to their lack of motorized tracking, Dobs are otherwise generally unsuited for astrophotography. Many Dobs use low-profile focusers, which have insufficient in-travel to allow even a CCD camera to reach focus for prime-focus imaging. Nearly all Dobs lack sufficient in-travel to allow using a standard 35mm or digital SLR camera for prime-focus imaging.

- To keep their tube lengths manageable, Dobs use relatively fast focal ratios. (The focal ratio of a scope is its focal length divided by its aperture.) An 8" Dob typically has a focal length of 48", which translates to a focal ratio of f/6; a 10" or 12" Dob typically has a focal ratio of f/5 or lower. Fast focal ratios are very hard on eyepieces, particularly inexpensive wide-field models. There's much truth in the old saying that the money you save by buying an inexpensive Dob, you'll eventually spend on expensive eyepieces. If you want an image that's sharp from edge to edge, your choices are to settle for the narrow apparent fields of view of Plössls or other inexpensive eyepieces, or to buy premium (expensive) wide-field models. (In the last few years, this situation has improved with the introduction of relatively inexpensive, high-quality, wide-field eyepieces that have good performance at fast focal ratios, such as the Orion Stratus line.)

- Although the relatively short focal lengths of Dobs make it possible for them to display wide fields of view, the flip side is that those short focal lengths make it more difficult to reach high magnification. In a typical Dob with a focal length of 1,200mm, for example, a 4mm eyepiece is needed to reach 300X magnification. Inexpensive eyepieces of such short focal lengths have tiny eye lenses and almost no eye relief, and so are very uncomfortable to use. The alternatives are to use a high-power Barlow with a longer focal length eyepiece, or to buy premium short focal length eyepieces, which offer large eye lenses and generous eye relief, but are relatively expensive.

- Dobsonians are Newtonian reflectors, which, like any telescope, must be properly *collimated* to offer their best image quality. Collimation

is the process of aligning the primary and secondary mirrors so that they share a common optical axis, and ensuring that that optical axis corresponds to the optical axis of the eyepiece. Collimation sounds more difficult than it is. Initial collimation of a Dob takes five or ten minutes, and need be done only when the scope is first assembled or when you make extensive changes to it (such as replacing the focuser). Routine collimation takes only a minute or two, but should be checked each time you set up the scope.

Advantages of an SCT

SCTs are considerably more expensive than Dobs of similar aperture. An entry-level 8" SCT on a mediocre mount, for example, may cost $1,100, and a 12" model on a good mount may set you back $4,000 or more. But in return for that higher price, SCTs have advantages all their own:

- Other than the least expensive models, all SCTs provide motor-driven tracking of the stars, which allows you to concentrate on what you're looking at rather than on keeping the object centered in the eyepiece. In particular, if you sketch, an SCT is often the ideal choice.

- Most SCT mounts provide go-to functionality. With a go-to mount, you can simply dial in the desired object on a hand controller and watch the scope move automatically to that object. (Take care, though. Very inexpensive go-to mounts are doubly-unreliable; they are prone to physical breakage and malfunctions, and they are often unsuccessful at putting the object in the eyepiece.)

- SCTs are generally well-suited to astrophotography (although only the best and most expensive mounts are suitable for long-exposure prime-focus imaging, where the main telescope mirror or objective lens projects the image directly onto the film or sensor). For casual short-exposure imaging with a CCD camera or digital SLR, nearly any SCT yields adequate results. SCTs generally provide sufficient back-focus for use even with standard 35mm film SLRs.

- Because they use a folded optical path, SCTs are physically more compact than Newtonian reflectors, including Dobs. For example, a typical 8" SCT has an effective focal length of 80", but the optical tube is actually only 18" or so long. This compactness makes an SCT tube easy to store and transport. Note, however, that an SCT mount often takes back in bulkiness and weight what you gain by using the smaller optical tube.

- The relatively long focal ratios of SCTs, typically f/10, means that even inexpensive wide-field eyepieces generally work quite well, providing sharp images from edge to edge.

- The relatively long focal lengths of SCTs mean that it's correspondingly easier to reach higher magnifications using eyepieces of moderate focal length, which typically have larger eye lenses and greater eye relief than the short focal length eyepieces necessary to reach similarly high magnifications in a Dob.

- Although, like any telescope, an SCT must be properly collimated to yield its best image quality, the slow focal ratios of SCTs are more forgiving of slight collimation errors. Also, SCTs generally hold their collimation well from one year to the next.

Disadvantages of an SCT

In addition to their higher price, SCTs have other disadvantages relative to Newtonian reflectors, including Dobs:

- Inexpensive SCT mounts are often quite shaky. Just touching the focuser knob may cause the scope to vibrate noticeably for five seconds or more. This problem is more pronounced at the higher magnifications used for Lunar and planetary observing than at the lower magnifications typically used for DSO observing, but is a factor nonetheless.

- Many SCT mounts are equatorial rather than altazimuth. Although an equatorial mount has some advantages, including the ability to track stellar motion with only one motor or slow-motion control, many beginners find the movements of an equatorial mount much less intuitive than those of an altazimuth mount.

- The relatively long focal length of typical SCTs limits their maximum possible true field of view significantly, particularly if they are used with the 1.25" visual back and diagonal that is bundled with most SCTs. (A visual back provides a threaded attachment point for a diagonal, camera, or other accessory; a diagonal uses a mirror to allow an eyepiece to be mounted at a 90° angle to the optical axis of the scope, making it easier to look through the eyepiece without contorting your neck and body.) Although it is possible to get a wider true field of view by substituting a 2" visual back, diagonal, and eyepieces, adding high-quality 2" components boosts the price of the scope by several hundred dollars.

- Although the SCT optical tube is small and easily portable, a typical SCT mount is relatively large, heavy, and awkward to transport.

- Because they use closed tubes, SCTs generally require much longer to cool down (or warm up) to ambient air temperature. Under conditions where a 10" Dob cools to ambient temperature in 45 minutes, a 10" SCT may require two hours or more. (Proper cooling of either type of scope may be expedited by adding supplemental cooling fans.)

FIGURE ii-2.

An older-model Celestron 8" SCT

- SCTs take longer to set up and tear down than Dobs, particularly if you are doing a critical polar alignment for an SCT equatorial mount. (For visual use, you can just eyeball the polar alignment; for long-exposure astrophotography, exact polar alignment is critical, and may require ten minutes or more of tweaking.)

Figure ii-2 shows a typical SCT. This one is a 1983-vintage Celestron C8 on a high-quality (Japanese) Vixen Super Polaris equatorial mount, owned by our observing buddy Paul Jones.

We really do recommend doing your homework before you buy a scope. But if you just want to get started Right Now, without doing much research, here are some models we think you'll be happy with:

Orion Telescopes and Binoculars

Orion is a retailer rather than a manufacturer. Orion offers two lines of Dobsonian telescopes in 6", 8", 10", and 12" models, which are actually made by the Chinese company Synta. XT Classic series scopes are traditional Dobsonians. IntelliScope series scopes add *DSCs* (*digital setting circles*), which can locate objects under computer control. IntelliScope scopes are not motorized, and do not have go-to functionality or tracking. You move the scope manually to point at the selected object, but the computerized hand controller points the way. We call these scopes *push-to telescopes*. With the hand controller, IntelliScope models sell for about $250 more than XT Classic scopes of the same aperture. Orion also offers relabeled Celestron 8", 9.25", and 11" SCTs on a variety of mounts. Any of the Orion Dob or SCT models is an excellent choice. (www.telescope.com)

Celestron

Celestron is best-known for its SCTs, which it offers in 8", 9.25", 11", and larger apertures on various mounts. Celestron also offers StarHopper series traditional Dobsonians in 6", 8", 10", and 12" models. Celestron telescopes are sold by numerous retailers, including Orion Telescopes and Binoculars. Any of the Celestron SCT or Dob models is an excellent choice. (www.celestron.com)

Meade

Like Celestron, Meade is best-known for its SCTs, which it offers in 8", 10", 12", and larger apertures on various mounts. In 2005, Meade introduced a new series of Dobsonian telescopes, called LightBridge scopes, which are available in 8", 10", and 12" models. Unlike most inexpensive Dobsonians, which use solid tubes, the LightBridge scopes use light, aluminum truss tubes in place of the middle section of the solid tube. When assembled, the truss structure is as rigid as a solid tube, but when it is torn down the scope can be stored in a very small space. The LightBridge models typically sell for $100 or so more than comparable solid-tube Dobs of similar aperture. Any of the Meade SCT or LightBridge Dob models is an excellent choice. (www.meade.com)

What About Refractors?

We love refractors for their pristine image quality and their suitability for astrophotography. We've used our 90mm (3.5") refractor to observe many of the objects in this book. But even their biggest fans will admit that refractors are not the best instruments for observing *faint fuzzies* (hard to see DSOs; so called because they look like fuzzballs unless you're using the Hubble telescope). The problem is that affordable, practical refractors are limited to about 4" aperture, which is marginal at best for observing many of the dimmer objects in this book. Although 6" refractors are available, they are heavy, clumsy, and require large, heavy, expensive mounts—and you still end up with only 6" aperture.

We would never be without a refractor, if only for its historical significance, but observing dim DSOs with a small refractor is like hammering nails with a screwdriver. You can do it if you must, but there are better tools available.

Essential Accessories

In addition to your binocular and/or telescope, you'll need an assortment of accessories to make your observing sessions more productive and enjoyable. In the following sections, we'll describe the accessories you really need (and a few that are just nice to have.)

Red Flashlight

Astronomers use red flashlights to avoid impairing their night vision. If you're observing Luna or the planets, you can use an ordinary white flashlight, because you needn't be dark adapted to view these bright objects. For other astronomical observing activities, you need a red flashlight.

You can make your own red flashlight by covering the lens of a standard flashlight with red film, or by buying an accessory kit for your flashlight that includes a red filter. Such makeshift solutions are not ideal, however, because even the best red filters pass a significant amount of white light, which can impair your night vision.

A better solution is to buy a red LED flashlight designed for astronomy. We phrase it that way because not all red LED flashlights are "red enough" to protect your night vision. Some are orange-red rather than the deep ruby red needed to preserve dark adaptation. (Monochrome LED flashlights emit light at one particular wavelength; 660 nanometers is ideal for astronomical use.)

Astronomy stores and on-line vendors sell a variety of red LED flashlights, made by Celestron, Rigel, Orion, and other companies. We've used many of them, and have never been completely satisfied with any of them. The least expensive models are fragile and prone to switch failures. More expensive models are more durable, but often inconvenient to use in the field. We finally found a red LED flashlight we could love when we bought our first Astrolite II, shown in Figure ii-3. The Astrolite II (and the similar Astrolite III model, which uses AAA cells instead of AA cells) features rubber armor, a 180° swivel head, a durable, recessed switch, and a clamp that allows hands-free use. The lens is textured to provide an even lighting pattern rather than the hot spots common with other LED lights. We've been using our Astrolite flashlights for more than three years now, and wouldn't consider using any other model. Astrolite flashlights sell for less than $20. (www.astrolite-led.com)

FIGURE ii-3.

Astrolite II red LED flashlights

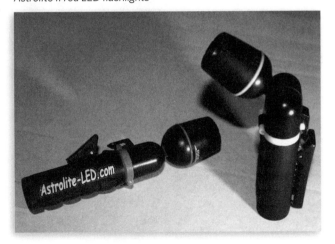

Planisphere

Although it is often thought of as a "beginner" tool, most experienced astronomers also use a *planisphere*, shown in Figure ii-4. A planisphere displays the positions of stars and other celestial objects for any date and time you "dial in". Planispheres are designed to work at a particular latitude, and are generally produced in latitude increments of 10°. The one shown is designed for use at 40° N, but is usable from 30° N to 50° N. Choose the version that's closest to your own latitude.

Planispheres are sold by many companies, including Orion, Sky Publishing, and others. Prices range from $5 to $20, depending on size and material. The cheapest models are cardboard, and are quickly ruined by dewing. Better models are constructed of heavily laminated card stock or plastic, and are much more durable. Our favorite planisphere is *David H. Levy's Guide to the Stars*, a huge (16") model that uses all-plastic construction.

FIGURE ii-4.

A planisphere

Barlow Lens

A *Barlow lens* is one of the most useful accessories you can have in your eyepiece case. A Barlow fits between the telescope's focuser and the eyepiece, where it increases the magnification provided by the eyepiece. A Barlow is designated by its *amplification factor*, which may range from 1.5X to 5X or more. A 2X Barlow, for example, effectively doubles the magnification of any eyepiece it is used with. (So-called "zoom Barlows" with variable amplification factors are available, but most of them are of low quality.)

Using a Barlow effectively doubles the number of focal lengths in your eyepiece collection. For example, if you have 25mm and 10mm eyepieces, using a 2X Barlow adds the equivalent of 12.5mm and 5mm eyepieces to your collection. Because a good Barlow costs no more than a mid-range eyepiece, buying a Barlow is an efficient way to increase the range of magnifications available to you.

Barlows are available in 1.25" models, which accept 1.25" eyepieces and fit 1.25" and 2" focusers, and in 2" models, which

---

"Apochromatic" Barlows

Ignore the marketing hype. It doesn't matter if a Barlow has two or three elements or is described as "apochromatic" (which is marketing-speak for a 3-element Barlow). What matters is the figure and polish level of the lenses and their coatings and the mechanical quality of the Barlow. There are superb 2-element Barlows, including both Tele Vue models, and very poor 3-element Barlows.

accept 1.25" and 2" eyepieces and fit 2" focusers. 1.25" models are far more popular than 2" models, because most astronomers use Barlows with mid-power and high-power eyepieces—nearly all of which are 1.25" models—to reach the high magnifications needed for Lunar and planetary observing as well as observing small DSOs.

Figure ii-5 shows a selection of high-quality 1.25" Barlows. From left to right are a Tele Vue 3X, an Orion Ultrascopic 2X, and a Tele Vue Powermate 2.5X.

Broadly speaking, there are two types of Barlows:

- Standard Barlows, such as the Orion Ultrascopic 2X and Tele Vue 3X models shown in Figure ii-5, are 5" to 6" long, and are used primarily in Newtonian reflectors, including Dobs. You can use a standard Barlow in a refractor, SCT, or other scope that uses a diagonal, either by inserting it between the telescope and diagonal or by (carefully) inserting it between the diagonal and the eyepiece. (The danger is that the long Barlow may protrude too far into the diagonal, damaging the mirror.)

- Short Barlows are about half the length of standard Barlows, and may be used in any type of scope, including Newtonian reflectors with very low-profile focusers, in which a standard Barlow may protrude into the light path.

Tele Vue makes a series of Barlow-like devices called Powermates. In effect, a Powermate is a standard 2-element Barlow with a second doublet lens added to minimize vignetting and excessive extension of eye relief when used with long focal length eyepieces. Powermates are available in 1.25" 2.5X and 5X models ($190) and 2" 2X and 4X models ($295). We don't doubt that Powermates are excellent products, but we've never been able to tell any difference in image quality between a Powermate and a high-quality standard Barlow.

Standard and short Barlows of comparable quality sell for similar prices, but there are significant optical differences. The shorter tube of a short Barlow means it must use a stronger negative lens to achieve the same level of amplification as a longer Barlow. That has three disadvantages:

Inferior image quality
Although the best short Barlows are very good indeed, the laws of optics dictate that they must be inferior optically to a full-length Barlow that uses lenses of similar quality. This inferiority most commonly manifests as lateral color (fringing), particularly near the edge of the field.

Vignetting
Because a short Barlow must bend light much more sharply than a full-length Barlow, short Barlows are subject to vignetting, particularly when used with longer focal length eyepieces.

Excessive eye relief
Although any Barlow extends the eye relief of most eyepiece designs, the stronger negative lens of a short Barlow exaggerates this effect. For example, we have no problem using our Orion Ultrascopic 30mm eyepiece with either our Orion Ultrascopic

2X Barlow or our Tele Vue 3X Barlow. But the Ultrascopic 30mm used with a short Barlow has its eye relief extended so far that we have trouble holding the exit pupil.

As you might have guessed, we're not fans of short Barlows. In fact, we don't own one. We use only full-length Barlows in our scopes, including our refractor. We freely confess, though, that many very experienced observers use and recommend short Barlows such as the Orion Shorty Plus Barlow and the Celestron Ultima Barlow (which are identical except for the brand name).

There are many cheap Barlows available, but we suggest you avoid them. A Barlow is a lifetime investment, and the difference in price between a mediocre model and an excellent one is not great. For a full-length Barlow, we recommend the $85 Orion Ultrascopic 2X model—which is often on sale for $75—and the $105 Tele Vue 3X model. (Tele Vue also makes a superb 2X Barlow, but it sells for $20 or $30 more than the Orion Ultrascopic, and we can discern no difference in image quality between them.) If you must have a short Barlow, get the $80 Celestron Ultima or the identical $70 Orion Shorty Plus, which is often on sale for $60.

Choose the amplification factor of your Barlow with your current eyepiece collection in mind, as well as any plans you have for expanding it. Avoid duplication between Barlowed and native focal lengths. For example, if you have 32mm and 16mm eyepieces, using a 2X Barlow with the 32mm effectively duplicates the 16mm. Using a 3X Barlow instead provides the equivalent of 10.7mm and 5.3mm eyepieces, both of which are useful extensions to your arsenal. Conversely, if you have the 25mm and 10mm eyepieces commonly bundled with inexpensive scopes, a 2X Barlow adds the equivalent of 12.5mm and 5mm eyepieces, again a useful expansion of your selection.

FIGURE ii-5.

Tele Vue 3X, Orion Ultrascopic 2X, and Tele Vue Powermate 2.5X Barlows

Additional Eyepieces

Many inexpensive and midrange telescopes are sold with one or two eyepieces, usually a 25mm or 26mm Plössl and perhaps a 9mm or 10mm Plössl. Although the bundled eyepieces suffice to get you started, they are generally of mediocre quality—particularly the 9mm or 10mm eyepiece—so replacing or supplementing them is a high priority for most beginning astronomers. As you begin to expand your eyepiece collection, we suggest that you keep the following points firmly in mind:

- Your first acquisition should be a top-quality Barlow, which effectively doubles your eyepiece collection.

- It's better to have two or three good eyepieces than a dozen cheap eyepieces.

- Eyepieces are a lifetime investment; you'll still be using good eyepieces long after your current scope is a distant memory.

- The eyepiece is as important as the telescope to the quality of view; a good eyepiece improves the view in even the cheapest telescope.

- Don't attempt to economize on eyepieces; buy the best quality eyepieces you can afford.

- High quality is more important for medium- and high-power eyepieces than for low-power eyepieces; if you must economize, do so on your lower power eyepiece(s).

In the following sections, we describe the important characteristics of eyepieces and offer advice about choosing eyepieces that fit your own needs and budget.

FIGURE ii-6.

Tele Vue 1.25" Radian (left) and 2"
Panoptic eyepieces

Eyepiece Characteristics

The fundamental characteristics of an eyepiece are its barrel size and its focal length.

Barrel size

Most eyepieces have 1.25" barrels. Some eyepieces, particularly low-power, wide-field models use 2" barrels. A 1.25" eyepiece can be used in a 1.25" focuser or, with an adapter, in a 2" focuser. A 2" eyepiece can be used only in a 2" focuser. The only advantage to using a 2" barrel is that the physically larger barrel allows the eyepiece to provide a wider field of view. Accordingly, 2" barrels are used almost exclusively for low-power, wide-field eyepieces. Figure ii-6 shows a typical 1.25" eyepiece on the left and a 2" eyepiece on the right.

Focal length

The focal length of an eyepiece, which is always specified in millimeters, determines how much magnification that eyepiece provides with a telescope of a particular focal length. Magnification is calculated by dividing the focal length of the telescope by the focal length of the eyepiece. For example, a 25mm eyepiece used in a telescope of 1,200mm focal length yields 48X magnification (1,200 ÷ 25 = 48).

The following characteristics are desirable in an eyepiece. Cheap eyepieces have few or none of these characteristics; expensive eyepieces have most or all of them; midrange eyepieces are somewhere in the middle.

Apparent field versus true field of view

At any given focal length, a wider apparent field of view translates to a wider true field, which determines how much sky is visible in the eyepiece. To calculate true field, divide the apparent field by the magnification. For example, in our 10" f/5 Dob, which has a focal length of 1,255mm, a 27mm Panoptic eyepiece provides about 46X magnification (1,255 ÷ 27 = 46). The apparent field of the 27mm Panoptic is 68°. Dividing that 68° apparent field by the 46X magnification yields a true field of view of about 1.46°.

Wide apparent field of view

The *apparent field of view* (*AFoV*) of an eyepiece is the angular size of the circular image it presents. At any given magnification, an eyepiece with a wide AFoV shows more of the sky than an eyepiece with a narrower AFoV. For example, Figure ii-7 shows simulated views of M 42, the Great Orion Nebula, in two eyepieces of the same focal length but with different apparent fields. Because the focal length of the two eyepieces is the same, so is the magnification, and the nebula appears to be the same size. But the eyepiece on the left has an 82° AFoV, and so shows much more of the surrounding sky than the image from the 50° eyepiece, shown on the right.

Long eye relief

Eye relief is the distance an eyepiece projects its exit pupil from the outer surface of the eyepiece eye lens. When you view with an eyepiece that has short eye relief, you have to press your eye right up against the eyepiece. Using an eyepiece with long eye relief allows you to maintain some separation between the eyepiece and your eye. If you must wear eyeglasses while observing, or if you just find longer eye relief more comfortable, look for an eyepiece with 20mm or so of eye relief. If you observe without glasses, 12mm or so of eye relief is adequate.

High mechanical and optical quality

The mechanical fit and finish of eyepieces varies dramatically. In general, all Japanese-made eyepieces and some Taiwanese models have excellent mechanical quality. Name-brand eyepieces, regardless of where they are made also have excellent fit and finish. Low-cost eyepieces, referred to generically as "Chinese eyepieces" are usually of noticeably lower quality. Labels are painted on rather than engraved, tolerances are looser, lens edges are not blackened, internal baffling is poor or absent, and so on.

Optical quality is even more important than mechanical quality. The major reason for the price difference between cheap generic eyepieces and expensive name-brand models is the level of attention given to details such as lens polish and coatings. More expensive eyepieces generally have much better lens polish and coatings, which translates to sharper images, higher contrast, and less ghosting and flaring.

Edge performance in fast scopes

In fast focal ratio scopes, the light cone converges over a shorter distance than in slower scopes. This fast-converging cone of light means that individual light rays arrive from significantly different angles, which is difficult for an eyepiece to handle without showing visible aberrations. Those aberrations are minimal at the center of the field, but become increasingly apparent as you approach the edge of the field, where the incoming light rays must be bent more sharply by the eyepiece. Edge performance is particularly problematic for eyepieces that provide a wide apparent field of view.

FIGURE ii-7.

The Great Orion Nebula (M 42) with an 82° eyepiece (left) and a 50° eyepiece

In an f/10 or slower scope, for example, nearly any well made wide-field eyepiece provides a sharp, pleasing image across most or all of its field. In an f/5 or faster scope, only top-notch eyepieces of modern design—such as Pentaxes, Naglers, Radians, and Panoptics—are capable of providing a good image across a wide apparent field of view.

If you have a slow focal ratio scope, even older wide-field designs such as Erfles and Königs work reasonably well (although the modern premium eyepieces are still better). If you have a fast focal ratio scope, you basically have three choices: (1) pay the price for premium wide-field eyepieces, (2) limit yourself to Plössls and similar older designs, which have narrower apparent fields, but are sharp to the edge even in fast scopes, or (3) buy inexpensive wide-field eyepieces and resign yourself to very poor edge performance.

Choosing Eyepieces

Choosing an optimum eyepiece collection depends on many factors, including your budget, the focal ratio of your scope(s), the types of objects you prefer to observe, and so on. If you use a Barlow, which we recommend for those on a budget, you can get by with two or perhaps three eyepieces. If you prefer not to use a Barlow, you may eventually have five, six, or even more eyepieces in your accessory case.

We categorize eyepieces by their magnification class, shown in Table ii-1. The actual eyepiece focal lengths in each class differ according to the focal ratio of your scope, but the magnifications remain consistent. For example, used in a 10" f/5 Dob, a 25mm eyepiece provides a 5mm exit pupil (25 ÷ 5 = 5) at 50X, which makes it a low-power eyepiece. Conversely, in a 10" f/10 SCT, that same 25mm eyepiece provides a 2.5mm exit pupil at 100X, and is therefore a medium- to high-power eyepiece.

The slow focal ratio of SCTs makes it difficult to achieve very low or even low magnifications with mainstream eyepieces. For example, although there are a few exceptions such as the 56mm Tele Vue Plössl, the longest focal length mainstream eyepieces are in the 40mm to 42mm range, which barely qualifies as low-power in an f/10 scope. (Of course, low magnification per se is never really the goal; we want a wide field of view, which is one result of using lower magnification.) Conversely, the fast focal ratio of most Dobs makes it difficult to achieve very high magnifications with mainstream eyepieces, although a Barlow can always be used to increase magnification in that situation.

When you choose eyepieces, we recommend you keep the following in mind:

- Maximum visual acuity for most people occurs with an exit pupil in the 2.0mm to 3.0mm range. Visual acuity begins to decrease rapidly with exit pupils smaller than 1.0mm. Below 0.7mm, visual acuity is significantly degraded, and, for many people, "floaters" and other artifacts begin to interfere.

- If you are near- or far-sighted, you can use the focuser to accommodate your vision without wearing glasses or contacts. If you have astigmatism or other visual aberrations, they are likely to be minimal with 2.5mm or smaller exit pupil, so you may be able to use high-power eyepieces without your glasses or contacts.

- In most locations, seeing quality (atmospheric instability) limits maximum useful magnification to 300X or less (possibly much less), regardless of aperture, so there is little point to buying eyepieces that provide higher magnification than your seeing supports. On rare nights with excellent seeing, use a Barlow with your standard eyepieces to reach very high magnifications.

- Large DSOs, including many emission/reflection nebulae and open clusters and some galaxies, are best viewed at very low to medium magnification.

- Most DSOs are best viewed at medium to high magnification.

- Many globular clusters and most planetary nebulae as well as Luna and the planets are best viewed at high to very high magnification.

So, which eyepieces should you buy? We suggest you use the following guidelines:

Finder eyepiece

Regardless of the type of scope you use or your observing habits, you'll need a low-power, wide-field *finder eyepiece*. As the name indicates, a finder eyepiece is used primarily for locating objects, but is also useful for viewing very large DSOs and star fields. Don't underestimate the importance of a finder eyepiece; it will probably spend more time in your focuser than any of your other eyepieces. The good news, though, is that low power is more forgiving than high power, so if you need to economize, the finder eyepiece is the place to do it. If you're just getting started and have a bundled 25mm or 26mm Plössl eyepiece, that will serve the purpose until you can get something better. If you do not have a finder eyepiece, we recommend the following, depending on the type of focuser you have:

1.25" focuser

On a tight budget, regardless of the focal ratio of your scope, choose an inexpensive 32mm Plössl, such as the Orion Sirius 32mm (~$40). If you can afford something in the $120 range, choose a 30mm or 35mm Orion Ultrascopic. If price doesn't matter, choose the $310 Tele Vue 24mm Panoptic.

2" focuser

On a tight budget, for an f/6 or slower scope, choose the

TABLE ii-1. *Magnification classes by exit pupil*

MAGNIFICATION CLASS	EXIT PUPIL	4"	6"	8"	10"	12"
Very Low	7.0 – 5.5 mm	14X – 18X	21X – 27X	28X – 36X	36X – 45X	43X – 55X
Low	5.5 – 4.0 mm	18X – 25X	27X – 38X	36X – 50X	45X – 63X	55X – 75X
Medium	4.0 – 2.5 mm	25X – 40X	38X – 60X	50X – 80X	63X – 100X	75X – 120X
High	2.5 – 1.0 mm	40X – 100X	60X – 150X	80X – 200X	100X – 250X	120X – 300X
Very High	1.0 – 0.7 mm	100X – 143X	150X – 215X	200X – 286X	250X – 357X	300X – 429X

Eyepiece overlaps and gaps

Smaller apertures and slower focal ratios both compress the range of useful eyepiece focal lengths. For example, with an 8" f/10 SCT, the 2.0mm to 3.0mm "workhorse" range translates roughly to 20mm (~100X) and 32mm (~64X) eyepieces. If that SCT has a 1.25" focuser, the longer 32mm workhorse eyepiece will probably also be your finder eyepiece. The shortest generally useful eyepiece for that scope will be in the 7mm (~290X) range. That means the useful range of eyepiece focal lengths for that scope is 32:7 or 4.6:1. Even if you have a 2" focuser, there is little point to using an eyepiece longer than 40mm to 42mm, which expands the range only to 6:1. Contrast that with the useful range on an f/5 scope, which runs from 35mm down to about 3.5mm, for a 10:1 ratio.

Small apertures reduce the range of useful eyepiece focal lengths by eliminating the middle range. By that, we mean there's little point to operating at anything between the lowest practical magnification and very high magnification, because "very high" magnification with a small aperture translates to relatively moderate absolute magnification. For example, with our 90mm f/11 refractor, which has a 1.25" focuser, we use a 30mm Ultrascopic (33X, 1.6° true field) as our wide-field "finder" eyepiece. The next reasonable step up in magnification is our 10mm Pentax (100X, 0.9mm exit pupil), because that puts us in the right general range for observing most DSOs. For Lunar/planetary magnifications, we use eyepiece focal lengths of 7mm (140X, 0.6mm exit pupil) or 5mm (200X, 0.5mm exit pupil, which is *really* pushing a 90mm scope). So, although our actual range is 30mm to 5mm, or 6:1, the middle of that range is unused.

$65 42mm Guan Sheng Optics SuperView, which provides the widest possible true field in a 2" focuser. If you can afford something in the $225 range, choose the University Optics 40mm MK-80 or, better yet, a used Pentax 40mm XL. In the premium price range, the $345 Tele Vue 27mm Panoptic, the $360 Vixen 42mm Lanthanum SuperWide, the $380 Tele Vue 35mm Panoptic, the $510 41mm Tele Vue Panoptic, the $500 Pentax 30mm or 40mm XW, and the $640 Tele Vue 31mm Nagler are all excellent choices, as they should be for the price.

For a scope faster than f/6, we think the best budget choice is the $65 30mm GSO SuperView. At higher budget levels, our choices are the same as listed in the preceding paragraph, except that we would not choose a focal length that yielded an exit pupil larger than 6.5mm to 7mm.

Workhorse eyepiece(s)
For your workhorse eyepiece, the one you'll use most of the time when you're not using your finder eyepiece, choose an eyepiece that provides an exit pupil between 2.0mm and 3.0mm. Used alone, this eyepiece provides medium- to medium-high power, which you will use for observing most DSOs. Used with a good 2X Barlow, this eyepiece provides a 1.0mm to 1.5mm exit pupil, with magnification into the high range, and is suitable for observing smaller DSOs.

Many people prefer to have two eyepieces in this range, one at either end of the exit pupil range stated. For example, our first workhorse eyepiece is a 14mm Pentax (2.7mm exit pupil at 90X in our f/5 scope), and our second workhorse eyepiece is a 10mm Pentax (2.0mm exit pupil at 125X).

Lunar/planetary eyepiece(s)
For observing Luna and the planets, as well as for small DSOs such as many planetary nebulae and globular clusters, you'll want an eyepiece or eyepieces that provide high to very high magnification. These can be actual eyepieces, or you can use a Barlow with your workhorse eyepieces to achieve higher magnification. For example, in our 10" f/5 Dob, we generally use 7mm (180X) and 5mm (250X) focal lengths, which we can achieve by using eyepieces having those native focal lengths, or by using a 2X Barlow with our 14mm and 10mm Pentaxes.

The best choices for workhorse and Lunar/planetary eyepieces are strongly influenced by your budget. Many amateurs are content with generic Plössls or University Optics Orthos, which sell in the $40 to $60 range. Many others prefer Tele Vue Plössls or Orion Ultrascopics (also sold under the Antares and Parks Gold Series brands), which sell for $75 to $120. All of these eyepieces provide good to excellent image quality, but are hampered by relatively small apparent fields and (particularly in shorter focal lengths) relatively short eye relief. If you prefer wider apparent fields, longer eye relief, or both, here are several eyepiece lines to consider:

Vixen Lanthanum LV
Vixen Lanthanum LV eyepieces sell in the $130 range, and can be thought of as Plössls that happen to have 20mm of eye relief regardless of their focal lengths. Focal lengths of 2.5, 4, 5, 6, 7, 9, 10, 12, 15, 18, 20, 25, and 40mm are available in 1.25" models, with 30mm ($200) and 50mm models available for 2" focusers. Apparent fields are 50° across most of the line, dropping to 45° in the shortest focal lengths, and jumping to 60° in the 30mm 2" model. Sharpness and contrast are excellent, although many owners report that light transmission is somewhat lower than with most eyepieces. Image quality is excellent down to f/5.

Orion Epic ED-2
Orion Epic ED-2 eyepieces sell in the $70 range, and are, in effect, Chinese clones of the Vixen Lanthanum LV eyepieces. Focal lengths of 3.7, 5.1, 7.5, 9.5, 12.3, 14, 18, 22, and 25mm are available for 1.25" focusers. Apparent fields are 55° across the line, with 20mm of eye relief in all models. Fit and finish are inferior to the Vixen Lanthanum LV models, as are the coatings. Some ghosting and flare are visible when bright objects are in or near the field, and contrast is inferior to that of the Vixen LVs or a good Plössl. Edge performance is decent down to f/6, but some softness is evident at f/5 and faster focal ratios. All of that said, if you require long eye relief and are on a tight budget, these eyepieces are a good choice.

Burgess/TMB Planetary Series

Burgess/TMB Planetary Series 1.25" eyepieces sell for $99 each, are available in 2.5mm, 3.2mm, 4mm, 5mm, 7mm, 8mm, and 9mm focal lengths, and provide 60° apparent fields with 10mm to 14mm of eye relief, much more than the eye relief of Plössls or Orthos of similar focal length. Despite their midrange price, these eyepieces are considered by many to be true premium eyepieces, with optical performance approaching that of the $250 Tele Vue Radians, or nearly so. Although we've never had the opportunity to test one, based on reports from observers whose opinions we respect, we'd consider these eyepieces to be excellent choices even in scopes as fast as f/5.

Vixen Lanthanum LVW SuperWide

Vixen Lanthanum LVW SuperWide eyepieces sell in the $250 range, and are available in focal lengths of 3.5, 5, 8, 13, 17, and 22mm for 1.25" focusers, and 42mm ($360) for 2" focusers. Apparent fields are 65° across the line (72° for the 42mm model), with eye relief of 20mm in all models. These are world-class premium eyepieces. Image quality is excellent down to f/5.

Orion Stratus

Orion Stratus eyepieces were introduced in 2005. These wide-field, long eye-relief eyepieces sell in the $130 range—although they are sometimes offered on sale for less than $100—and are available in focal lengths of 3.5, 5, 8, 13, 17, and 21mm for 1.25" focusers. Apparent fields are 68° across the line, with eye relief of 20mm in all models. Stratus eyepieces are, in effect, Chinese clones of the Vixen Lanthanum LVW SuperWide eyepieces. Their fit and finish are very good, although not quite up to Japanese standards. Their coatings and polish are also very good, although again they don't quite meet the world-class standards of the Vixen, Tele Vue, and Pentax models. Still, the Stratus eyepieces offer 80% to 90% the performance of the Vixen LVWs at half the price, and provide good image quality down to f/5. If the true premium eyepiece lines are a bit rich for your pocketbook, we think you'll be very happy with the mid-priced Stratus eyepieces.

Pentax XW

Pentax XW eyepieces are, in our opinion and that of many others, the finest eyepieces available. The $310 1.25" models are available in focal lengths of 3.5, 5, 7, 10, 14, and 20mm. The $500 2" models are available in focal lengths of 30mm and 40mm. Apparent fields are 70° across the line, with eye relief of 20mm in all models. Fit and finish is top-notch, and the lens polish and coatings are simply unbeatable. Image quality is superb down to f/4.

Tele Vue Radian

Tele Vue Radian eyepieces sell for $250, and are considered by many to be the best medium- to very high power eyepieces available. We prefer the Pentax XWs for their wider apparent fields and because we find them more comfortable to use, but we certainly have no argument with those who prefer Radians. Radians are available in focal lengths of 3, 4, 5, 6, 8, 10, 12, 14, and 18mm, all for 1.25" focusers. Apparent fields are 60° across the line, with eye relief of 20mm in all models. Fit and finish is top-notch, and the lens polish and coatings are excellent. Image quality is superb down to f/4.

Tele Vue Nagler

Tele Vue Nagler eyepieces are considered by many to be the best of the best. Naglers are available in focal lengths of 2.5, 3.5, 5, 7, 9, 11, 12, 13, and 16mm to fit 1.25" focusers, and in focal lengths of 17, 20, 22, 26, and 31mm to fit 2" focusers, with prices ranging from $190 to $640. Apparent fields are 82° across the line, with eye relief ranging from 8mm to 19mm. Fit and finish is top-notch, and the lens polish and coatings are excellent. Image quality is superb down to f/4.

Nebula Filters

A *nebula filter* selectively passes specific wavelengths of light emitted by emission nebulae and planetary nebulae and blocks other wavelengths. Because these nebulae emit light almost exclusively at these wavelengths, a nebula filter darkens the background substantially while passing almost all of the light emitted by the nebula, enhancing the contrast and detail visible in nebulae, even from a dark-sky site. The results can be striking. In some cases, a nebula that is only dimly visible without a filter in a large-aperture telescope may be visible to the naked eye with a filter. More commonly, a nebula filter expands the visible extent of a nebula and allow you to see finer detail. For an in-depth discussion of the technical aspects of nebula filters, read *Choosing a Nebula Filter* by Greg A. Perry, Ph.D. (members.cox. net/greg-perry/filters.html).

Three types of nebula filters are commonly used by amateur astronomers:

Narrowband filter

A *narrowband filter* is the most generally useful nebula filter. It provides a moderate to dramatic contrast improvement for most emission nebulae, and is also helpful for many planetary nebulae. A narrowband filter passes the bluish hydrogen-beta (H-beta or H-β) line at 486 nanometers (nm) and the blue-greenish doubly-ionized oxygen (Oxygen-III or O-III) lines at 496 and 501 nm, and may also pass the blue-greenish cyanogen (CN) lines at 511 and 514 nm and the red hydrogen-alpha (H-alpha or H-α) line at 656 nm. A narrowband filter is useful on more objects than any other type of nebula filter. It is likely to help and unlikely to hurt the view of any emission or planetary nebula.

Nebula Filters are Only for Nebulae

Although nebula filters are useful—indeed sometimes almost essential—for viewing emission nebulae and planetary nebulae, they are actually counterproductive for other types of objects. Using a nebula filter on a star cluster or galaxy simply dims the view without offering much contrast enhancement, because stars emit light across the entire visible spectrum. Reflection nebulae shine from reflected starlight, so nebula filters are generally not useful for observing them.

Filter Sizes

Nebula filters are available in several sizes. The most common are those that fit standard threads on 1.25" and 2" eyepieces and also those that fit the visual backs of Schmidt-Cassegrain Telescopes (SCTs). Meade goes its own way, using a non-standard thread on their 1.25" and 2" eyepieces and filters, and Questar uses still a different thread. A few filters are available for 0.96" eyepieces, which are common in Japan and with some old telescopes but rare elsewhere. Filters for 1.25" eyepieces are available in by far the widest variety. The selection for 2" eyepieces is somewhat more limited, and those for SCT visual backs much more limited.

If you use both 1.25" and 2" eyepieces, consider buying only 2" filters. Although a 2" filter cannot be attached directly to a 1.25" eyepiece, there are several common workarounds. First many 1.25"/2" focuser adapters provide threads for mounting a 2" filter, which can then be used with any 1.25" eyepiece you insert into the adapter. Second, if you have a Newtonian (or Dobsonian) reflector, you can install a *filter slide*, which attaches to the inside of the tube between the bottom of the focuser and the secondary mirror. With a filter slide, you can quickly position any of several filters under the focuser simply by moving the slide. Third, there are external filter slide adapters that integrate with your focuser. Finally, of course, you can simply hold a 2" filter between your eye and the eye lens of the eyepiece, assuming the eyepiece provides sufficient eye relief (swapping the filter in and out is called *blinking*, and is a useful trick for bringing elusive objects to your brain's attention).

Nebula Filters with Smaller Scopes

You may hear it said that nebula filters are useless in telescopes smaller than 6" or 8", because they dim the view too much. We have not found that to be true. For example, we once attempted to view the North America Nebula with an 80mm short-tube refractor. Without a filter, no nebulosity was visible. With an O-III nebula filter, the view was very dim, it is true. But we were able to see the North America Nebula quite clearly with the filter, and in an 80mm scope.

The most popular narrowband filters are the *Orion Ultrablock* and the *Lumicon UHC* (Ultra-High Contrast) models, although similar models are offered by Celestron, Tele Vue, Thousand Oaks, Astronomiks, and others. If you can afford only one nebula filter, a narrowband model is the one to buy. Although there are minor differences, the performance of narrowband filters is quite similar regardless of brand name, so we suggest you buy on price. We use an Orion Ultrablock.

Oxygen-III (O-III) filter

An *Oxygen-III filter*, usually called an *O-III filter*, has a sharper cutoff than a narrowband filter, typically passing only the 496 and 501 nm O-III lines. Although an O-III filter may be helpful for some emission nebulae—and may in fact be the best choice for some emission nebulae—the O-III filter serves best for most planetary nebulae. (Not all planetary nebulae are best viewed with an O-III filter; a few planetaries actually look better if you use a narrowband filter.)

The most popular O-III filters are the *Orion O-III* and the *Lumicon O-III* models, although similar models are offered by several other companies. If you already have a narrowband filter and there's still room in the budget, an O-III filter should be your next acquisition. Although there are minor differences, brand name makes little real difference, so we suggest you buy on price. We use a Thousand Oaks O-III.

Hydrogen-beta filter

The *Hydrogen-beta filter*, also called an *H-beta filter*, *H-β filter* or *Hydrogen-β filter*, is often referred to as the *Horsehead Nebula filter*, only partly in jest. The H-beta filter is extremely specialized. It passes only the H-beta line at 486 nm, and is useful for only a handful of objects, of which the Horsehead Nebula is by far the most renowned. But for those few objects, the H-beta filter is a magic bullet. It's not overstating the case much to say that without an H-beta filter the Horsehead is elusive visually even in large scopes, while with an H-Beta filter the Horsehead becomes possible—although challenging—from a dark site with scopes as small as 8" to 12", and relatively easy in large scopes. We do not own an H-beta filter, although we admit to borrowing one from time to time.

Unit-power Finder

If you enjoy tracking down DSOs manually, installing a *unit-power finder* (often mistakenly called a zero-power finder) is the best single upgrade you can make to your scope. A unit-power finder provides no magnification. Instead, it allows you to view the night sky naked-eye with a superimposed dim red target-locating pattern. By orienting that pattern geometrically relative to the background stars, you can locate most deep-sky objects in a fraction of the time needed if you use an optical finder alone. Figure ii-8 shows a popular unit-power finder model, called a Telrad.

A unit-power finder is simplicity itself. No computers, no fancy optics. Just a dim red bulls-eye pattern projected on the night sky. Figure ii-9 shows a simulation of using a Telrad unit-power finder to point the scope at the Great Orion Nebula, M 42.

The two most popular unit-power bullseye finders are the Rigel QuikFinder (www.rigelsys.com) and the Telrad (no web site), both of which sell for under $40. Both products have strong advocates in the amateur astronomy community, and any Telrad versus QuikFinder discussion soon degenerates into a religious debate.

We prefer the larger bulls-eye pattern of the Telrad—0.5°, 2°, and 4° circles versus only 0.5° and 2° circles with the QuikFinder— and its absence of parallax (the apparent movement of the target object relative to the aiming pattern, depending on how your eye is placed relative to the finder), but many astronomers prefer the QuikFinder for its smaller size and lighter weight. Whatever your preference, do install one or the other. Until you try it, you won't believe how much easier it is to find objects using a bulls-eye finder.

Avoid red-dot finders

Rather than a bulls-eye pattern, a red-dot finder projects a simple (you guessed it) red dot. A red-dot finder allows you to point your scope quickly and accurately. The problem is, that's all it does. Because it doesn't provide a bulls-eye pattern, it offers no help in locating objects geometrically.

FIGURE ii-8.

A Telrad unit-power finder

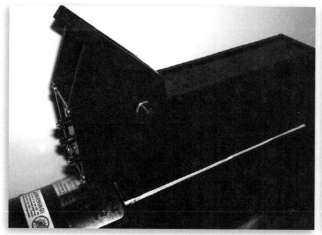

FIGURE ii-9.

Using a Telrad unit-power finder to put M 42 in the eyepiece

Observing Chairs

We're always surprised at how many astronomers observe while standing at their scopes. Whatever type and size of scope you have, you'll see much more if you observe while seated comfortably. In our experience, observing while seated is the equivalent using the next size larger scope. For example, if you observe while seated at an 8" scope, you'll see about as much image detail as you would observing while standing at a 10" scope.

We've seen observing chairs that run the gamut from an inverted 5-gallon plastic bucket to folding lawn chairs to wooden stools to expensive adjustable observing chairs built specifically for observing, like the one shown in Figure ii-10. The ideal observing chair for you depends on your budget, the type of scope you use, where you use the scope, and other factors. It's important to choose a chair of the proper height to put your eye level near the eyepiece without requiring excessive stretching or bending.

For many astronomers, the ideal observing chair is a drummer's throne, which can be purchased at any musician supply store. The cheapest models sell for under $25, but those are generally fragile and awkward to adjust. Better models cost $50 to $75, are much sturdier, and can be adjusted quickly over a wide range of heights by unlocking a retaining mechanism and lifting or lowering the seat to the desired height.

Using a telescope is not the only observing activity for which a chair is helpful. It's helpful to have a second chair at your observing table so that you can sit comfortably while you study charts, record your observations, and so on. And a chair or similar aid can be immensely useful for binocular observing because it supports you as you observe, helping you avoid the shakes and allowing you to observe an object longer without strain. We use an ordinary folding lawn chair for that purpose, but some binocular astronomers prefer to observe while lying in a partially-inflated child's swimming pool, using the sides of the pool to support their arms.

FIGURE ii-10.

An adjustable observing chair

Planetarium Software

Planetarium software is an almost indispensable adjunct to traditional printed charts. Unlike static charts, planetarium software can provide a simulation of the night sky as viewed from any specified location, in real-time or for some past or future date. Astronomers use planetarium software at home to plan future sessions, and in the field to provide updated charts of how the sky appears as of that moment and to provide detailed data about the objects they are observing.

The major advantage of planetarium software relative to printed charts is flexibility. With printed charts, including those in this book, what you see is what you get. With planetarium software, you can specify the field of view, the level of detail, the types of objects to be plotted, the limiting magnitude, and so on. You can zoom, rotate, and flip charts on-screen to correspond to the view in your finder or telescope, and then print them for use during the observing session.

There are scores of planetarium programs available for Windows, Mac OS X, and Linux. Many of them, including some of the better ones, are free. Others cost money. We use half a dozen planetarium programs for their different strengths and features, but our primary planetarium software is MegaStar (www.willbell.com), shown in Figure ii-11. Although MegaStar costs $130, it provides more features and flexibility than any other planetarium software we have used, including some programs that cost twice as much. In fact, we used MegaStar to plot the charts in this book.

If you'd like to try a zero-cost planetarium program that offers about 80% of the features and functionality of MegaStar, download the extraordinary Cartes du Ciel (Sky Charts) v2.76, written by Swiss astronomer Patrick Chevalley. Cartes du Ciel is attractive, powerful, and immensely flexible. The complete program is a 15 MB download (www.stargazing.net/astropc/), and includes several of the most important star and nebulae

---

Observation Planning Software

Most planetarium programs, including MegaStar and Cartes du Ciel, are weak in the area of observation planning. For example, you might want to plan an observing session by generating a list of objects of a particular type or subgroup that are visible on a particular evening. Planetarium software generally provides only limited tools for this purpose.

Fortunately, there are specialized observation planning programs to address this need, including the $40 Deepsky Astronomy Software (www.deepskysoftware.net) and the $100 SkyTools 2 (www.skyhound.com/skytools.html). Although both support other features, their real purpose is to generate customized observing lists, and both do an excellent job at this primary task. Most serious observers use one or the other, and many use both.

FIGURE ii-11.

Megastar, showing the belt region of Orion

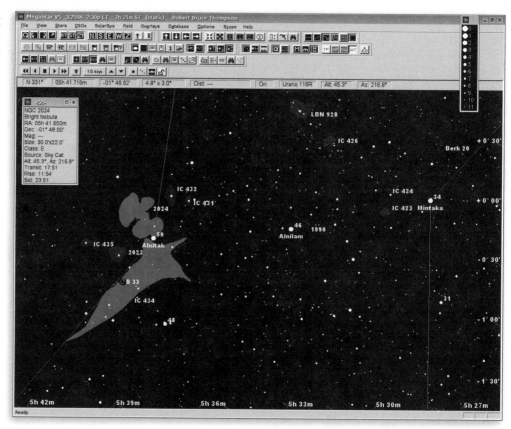

catalogs, as well as data for the planets, comets, and asteroids, which can be updated on-line. Cartes du Ciel also makes it easy to install supplemental catalogs, of which dozens are available and many are indexed and searchable. Figure ii-12 shows Cartes du Ciel displaying a section of the southern summer sky with the Messier Catalog activated. Version 3.0 is in development as we write this, and will be available for both Windows and Linux.

Cartes du Ciel also provides the advanced features you'd expect in a heavyweight planetarium program—such as controlling computerized telescopes, CCD integration, updating Solar system ephemerides on-line (including the moons of Mars, Jupiter, Saturn, and Uranus), using custom horizon maps, and so on. In short, Cartes du Ciel probably does nearly everything you need to do. If you do eventually outgrow it, there's always MegaStar.

Lunar Observing Software

Most planetarium programs provide few tools for Lunar observers. For example, although Cartes du Ciel shows a nicely detailed image of Luna, including an accurate terminator, that's as far as it goes. Fortunately, the authors of Cartes du Ciel wrote a second program, Virtual Moon Atlas (VMA), to fill that hole. Full-time Lunar observers will still want a copy of Rukl's classic book, *Atlas of the Moon*, but even dedicated part-time Lunar observers will probably be happy with VMA. If VMA doesn't quite do the job for you, have a look at the $40 commercial product Lunar Phase Pro (LPP), available at www.nightskyobserver.com/LunarPhaseCD. In general, LPP has Lunar mapping features similar to those of VMA, but superior observation planning features.

Planetarium Software for PDAs

Many astronomers run planetarium software in the field on their primary notebook computers. Some purchase an inexpensive or used notebook computer for that purpose. But if you don't have a notebook computer, or prefer not to risk it in the field, you can still run planetarium software during observing sessions if you have a PDA. Although there are several PDA-based planetarium programs, our favorite is Planetarium (www.aho.ch/pilotplanets/). You can download a free demo version to test the software. The full version costs only $24.

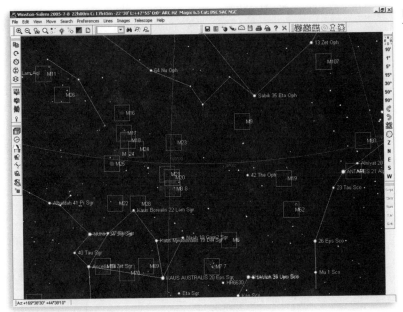

FIGURE ii-12.

Cartes du Ciel v 2.76, showing Messier objects in the southern summer sky

Printed Charts

Despite the proliferation of computerized go-to telescopes and planetarium software for notebook computers and PDAs, printed star atlases remain, if not essential, at least extremely useful. Many astronomers, for example, use star atlases to plan their observing sessions because they prefer the "big picture" of a printed atlas to the constraints of a relatively small notebook or PDA screen. Many also use printed charts during observing sessions, preferring their familiarity and reliability. (You'll never have to cut short an observing session because the battery in your printed atlas dies.)

There are numerous star atlases to choose from, and those atlases vary greatly in level of detail, scale, physical size, price, and other important considerations. Which atlas or atlases are best for your needs depends on numerous factors, including:

- The aperture of your scope.
- How dark your observing site is.
- The types of objects you observe.

- The limiting stellar magnitude of the atlas.
- The number, type, and limiting magnitude of DSOs included.
- Portability, dew resistance, and price.

For example, a star atlas that is ideal for someone who observes bright star clusters from an urban location with a 3.5" refractor is poorly suited for someone who observes dim galaxies from a dark rural site with a 20" Dobsonian reflector, and vice versa. The urban observer can't use the rural observer's atlas, because it shows thousands of objects that are too dim to be visible in a small scope from a light-polluted urban site. It's impossible to see the forest for the trees. Conversely, the urban observer's atlas is useless for the rural observer, because it doesn't provide adequate detail to locate the dim galaxies that are the target of the rural observer.

Like many astronomers, we have both large and small instruments and observe from both urban and rural sites, so one star atlas is insufficient for our needs. The following sections describe the atlases we use, and our reasons for choosing them.

Top of the Charts

Okay, we admit it. We're star chart junkies. We carry more than a dozen different charts and atlases in our chart case. But the truth is you don't really *need* any of the star charts or atlases we discuss in this section. The large-scale and detailed charts in this book are sufficient for finding and observing all of the more than 300 objects covered in this book, including all of the Messier objects, RASC Finest NGC objects, and AL Double Star Club objects.

So why should you buy a separate set of star charts? Because hundreds of deep-sky objects are visible in amateur scopes that aren't included in the charts in this book, and you may want to observe and log some of these other objects while you're observing the objects that are covered here. For example, while you're observing the 21 objects we cover in Virgo, all of which are Messier or RASC objects, you'll probably see many other objects, most of which will be members of the Astronomical League Herschel 400 list. You might just as well observe and log them while you're working Virgo, but you can't do that unless you know what they are.

Mag 6 atlases, also called naked-eye star atlases, chart stars down to magnitude 6.5 or so. No amateur astronomer should be without a Mag 6 atlas. They're useful for impromptu naked-eye or binocular observing sessions, and for times when you simply don't feel like hauling out the "serious" charts.

Mag 6 atlases show only a few hundred of the brightest deep-sky objects (DSOs), including (usually) all of the Messier Objects and the brightest non-Messier NGC objects, such as the Double Cluster, as well as prominent double stars. Mag 6 atlases use a small image scale, typically mapping the entire night sky on a dozen or so small charts. This small scale limits the number of objects that can be shown and the amount of label detail for each.

For example, Figure ii-13 shows the cluttered belt and sword region of the constellation Orion as shown on the Mag 6 Orion DeepMap 600 chart. Only the brightest stars and DSOs appear on this chart, and the small scale means that it can be difficult to determine which label refers to which object. This is not a slam on the Orion DeepMap 600; in fact, it's one of the better Mag 6 charts.

When you want just an overview of bright stars and objects rather than a detailed (and cluttered) view, a Mag 6 atlas is just what you need. We always have a Mag 6 atlas handy for that reason. Here are the Mag 6 star atlases we use and recommend:

Orion DeepMap 600

The Orion DeepMap 600 ($15; Orion Telescope and Binocular Center) uses maps drawn by the incomparable uranographer (cartographer of the heavens) Wil Tirion and is unique among star atlases. Rather than loose individual maps or bound map pages, DeepMap 600 is a single large page that folds up like a standard roadmap. It is printed on water-resistant, tear-resistant plastic stock, and is quite durable. The front of DM600 covers the night sky from declination -60° to +70°, with the north circumpolar region appearing as a separate map on the rear side. (The south circumpolar region is not covered.) DM600 charts more than 500 of the brightest DSOs, including the Messier Objects, and about 100 of the brightest and most interesting double and variable stars. Its portability and durability make it an ideal part of a portable kit. With just a binocular and the DM600, you're ready for an impromptu observing session any time.

Bright Star Atlas

The Bright Star Atlas ($10; Willmann-Bell; 2001; ISBN 0943396271) features objects selected by noted astronomer Brian Skiff and maps by the ubiquitous Mr. Tirion. BSA divides the night sky into ten charts, showing stars down to magnitude 6.5, open and globular star clusters to magnitude 7.0, galaxies to magnitude 10.0, and double stars and nebulae visible in small scopes. Facing each full-page map is a full-page table of the objects plotted on that map and their characteristics. At only 32 pages, BSA contains little of the supplementary information included in book-size atlases. It limits itself to providing first-rate

FIGURE ii-13.

The belt and sword of Orion, as shown by the Orion DeepMap 600

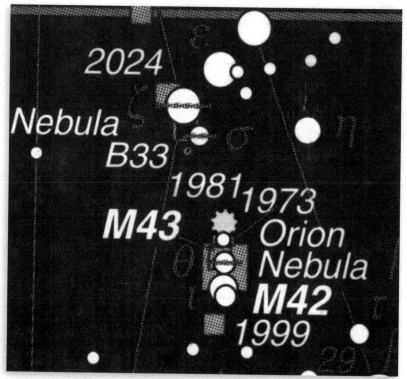

Mag 6 maps and tabular information to support the maps. As such, BSA is an ideal companion for field observing sessions with the naked eye, binocular, or a small telescope.

Norton's Star Atlas

Norton's Star Atlas ($30; Pi Press; 2003; ISBN: 0131451642), edited by renowned astronomer Ian Ridpath, is probably the most popular of the book-style Mag 6 atlases. It maps the entire night sky with seven double-page maps that cover declination -60° to +60°, with the north and south circumpolar regions from declination ±60° to ±90° each allocated a double-page map. The scope of objects covered is similar to BSA, including the tabular lists of objects. Unlike BSA, which provides only star maps and lists of objects, Norton's Star Atlas devotes a significant percentage of its 200+ pages to general information of interest to amateur observers. That makes it clumsier to use in the field, but an excellent desk reference when a Mag 6 atlas suffices.

Pocket Sky Atlas

If you want a portable atlas that goes deeper than the Mag 6 atlases, check out the Pocket Sky Atlas by Roger W. Sinnott (Sky Publishing; 2006; ISBN: 978-1931559317). At 6 by 9 inches, it's a bit large to fit most pockets, but it's certainly small enough to stick in the glove box or eyepiece case. The Pocket Sky Atlas contains 80 charts in a spiral binding that makes it very convenient to use when a full-size set of charts is awkward. PSA maps almost 31,000 stars down to magnitude 7.6—bright enough to be prominent in a standard finder—and about 1,500 deep-sky objects, including 675 galaxies down to magnitude 11.5 and all of the Astronomical League Herschel 400 and Caldwell objects. The charts show constellation borders and outlines, which makes it easy to locate objects. Detailed charts in the back cover the Pleiades, Orion's Sword, the Virgo Galaxy Cluster, and (for southerly observers) the Larger Magellanic Cloud. Figure ii-14 shows Orion's belt and sword.

The PSA is the atlas we find ourselves using most often during observing sessions. The scale is large enough and the amount of detail sufficient to use it right at the eyepiece as we're tracking down yet another Herschel 400 object to add to our bag. Those few times when we need more detail, it's easy enough to walk over to the chart table and look up the object in our Sky Atlas 2000.0 or an observing buddy's Uranometria, described shortly.

SkyAtlas 2000.0

The next step up from the Pocket Sky Atlas is Sky Atlas 2000.0 (Sky Publishing; 1999; ISBN numbers vary by version), which is considered by many amateur astronomers to be the gold standard among "serious" star atlases. Sky Atlas 2000.0 (SA2K) covers the entire sky with 26 charts drawn by, you guessed it, Wil Tirion. SA2K charts 81,312 single, multiple, and variable stars of magnitude 8.5 and brighter—which is to say about nine times as many stars as a Mag 6 atlas—and about 2,700 deep-sky objects, including most of the Herschel 2,500 list. SA2K also includes supplemental detailed charts for the celestial poles, the Coma-Virgo galaxy cluster, and other cluttered areas. It also includes a transparent coordinate grid overlay that provides fields of view for finders, a Telrad, and eyepieces.

SA2K provides more detail than the PSA and much more detail than a Mag 6 atlas. For example, Figure ii-15 shows the same region around Orion's belt and sword as shown in the preceding figures. Comparing the figures, it's obvious that the SA2K chart is not only much larger scale, but includes many more objects and more detail about each object.

Many of the objects charted by SA2K are beyond the reach of a binocular or small telescope, even from a dark site. We have heard it said that a skilled observer with a 4" scope at a very dark site can observe all of the objects charted by SA2K. That may be true, but our regular observing sites aren't that dark (and, perhaps, we're not that skilled.) We think SA2K is a perfect match for a typical intermediate to advanced amateur astronomer who uses a 6" to 10" telescope from what passes nowadays for a dark site.

If SA2K sounds good to you, the next decision is which version to buy. SA2K is available in six versions:

- Field Version ($30; 13.5" X 18.5"; white stars on black sky; loose sheets)

- Laminated Field Version ($70; 13.5" X 18.5"; white stars on black sky; spiral bound)

- Desk Version ($30; 13.5" X 18.5"; black stars on white sky; loose sheets)

- Laminated Desk Version ($70; 13.5" X 18.5"; black stars on white sky; spiral bound)

- Deluxe Version ($50; 16" X 21"; black stars on white sky with features highlighted in color; loose sheets)

- Laminated Deluxe Version ($120; 16" X 21"; black stars on white sky with features highlighted in color; spiral bound)

Both Field Versions use white stars on a black sky, mimicking the appearance of the night sky. The rationale of using these mostly black charts in the field is that they minimize glare and make it easier to remain dark adapted. In reality, if you use a proper

FIGURE ii-14.

The belt and sword of Orion, as shown by Pocket Sky Atlas

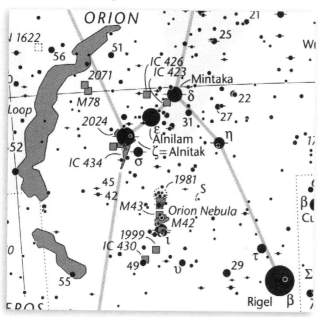

FIGURE ii-15.

The belt and sword of Orion, as shown by Sky Atlas 2000.0 Deluxe

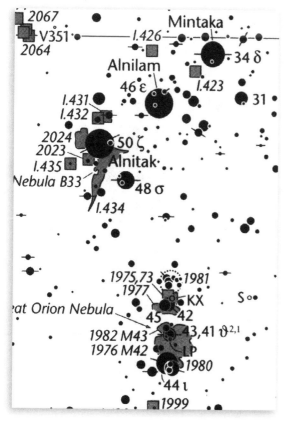

red LED flashlight for viewing charts, your dark adaptation is unaffected no matter how bright the red light. The downside to the Field Version is that the black background makes it impossible to annotate the charts or to draw in constellation lines, which are not printed on the charts. The two Desk Versions are identical to the Field Versions, except the Desk Versions use black stars on a white background, which we greatly prefer.

The Field and Desk Versions are available either as loose paper charts, at $30 list price, or as laminated, spiral-bound books, at $70 list price. Despite the higher price, many amateur astronomers buy the laminated versions because they are immune to dew, which can quickly wreck unprotected paper charts. We dislike the stiffness of the laminated versions, and prefer to use a standard version and takes steps to protect it from dew. The easiest way to do that is simply to cover the charts when they're not actually being used. Covering them with a towel will protect them adequately under any but the worst dewing conditions. Alternatively, Sky & Telescope sells a $28 chart carrier that is specifically designed to carry and protect any version of SA2K. The chart carrier includes a protective plastic sheet that allows you to view any particular chart while keeping it dry.

Unlike the Field and Desk versions, which use monochrome 13.5" X 18.5" charts, the Deluxe versions use full-color 16" X 21" charts. In addition to making them prettier, we think the use of color adds significantly to the usability of the charts. (The charts are equally readable under white or red light.) The $50 Deluxe Version is spiral-bound into a 12" X 16" book, with each chart folded once. The $120 Laminated Deluxe Version is also spiral-bound, but without folding, yielding an oversize 16" X 21" book that we consider too awkward for use in the field.

On balance, we think the unlaminated Deluxe Version is the best choice for most people. Despite extensive use in the field, our copy remains in good condition, without any dew damage.

The only missing piece in Sky Atlas 2000.0 is an index of the objects it includes. That lack is remedied by the Sky Atlas 2000.0 Companion ($30; Sky Publishing; 2000; ISBN: 0933346956) by Robert A. Strong and Roger W. Sinnott. Its nearly 300 pages list and describe each of the 2,700 DSOs charted by SA2K and cross-reference each object by chart number. We don't use this book, because we generally work standard catalogs such as the Herschel 400, which list objects by declination and right ascension, making it easy to locate the object in the SA2K charts. But if you want to locate objects in SA2K using only their NGC or other standard designations, the SA2K Companion is essential as an index to the charts.

If Sky Atlas 2000.0 is the gold standard in serious star charts, then Uranometria 2000.0 must be the platinum standard. As useful as SA2K is, it doesn't go deep enough for many serious astronomers, particularly those who observe with larger instruments from very dark sites. If your scope is at least 8" to 10" and you observe from a reasonably dark site, you'll see many prominent objects that don't appear in SA2K. Welcome to the world of serious DSO observing.

Dedicated DSO observers consider Uranometria 2000.0 their bible. Uranometria 2000.0 (U2K) charts 280,000+ stars down to magnitude 9.75 and 30,000+ deep-sky objects. That's more than three times as many stars as SA2K and more than ten times as many DSOs. In a region where SA2K charts two or three galaxies, for example, U2K may chart 20 or 30. U2K also includes 26 even more detailed charts of cluttered regions, with a limiting stellar magnitude of about 11.0.

With the exception of the Millennium Star Atlas, which after being out of print for years is now available in a 3-volume, $150 paperback version, U2K is by far the deepest mainstream printed star atlas available. And, although MSA goes deeper than U2K for stars, U2K goes much deeper than MSA for DSOs. The only close competition for U2K is the Herald-Bobroff AstroAtlas (www.heraldbobroff.com). Many observers, including some whose opinions we respect, like the HBA, but we consider it too cluttered for use in the field. It may be an acquired taste.

Figure ii-16 shows the same region around Orion's belt and sword in U2K as is shown in the preceding Figure for SA2K. At first glance, the level of detail may seem similar, but closer examination shows that U2K provides significantly more detail than SA2K. Look, for example, at the region between Alnitak and Alnilam, where many more stars are visible on the U2K chart. Also examine the region of the Great Orion Nebula, where U2K again provides immensely more detail.

Uranometria 2000.0, by Wil Tirion, Barry Rappaport, and Will Remaklus, is packaged as a three-volume set of hardback books. Uranometria 2000.0 Deep Sky Atlas Volume 1 ($50; Willmann-Bell; 2001; ISBN: 0943396719) covers the northern hemisphere from declination +90° to -6°. Uranometria 2000.0 Deep Sky Atlas Volume 2 ($50; Willmann-Bell; 2001; ISBN: 0943396727) covers the southern hemisphere from declination -90° to +6°. Both of these volumes are necessary for complete coverage of the night sky. Uranometria 2000.0 Deep Sky Field Guide Volume 3 ($60; Willmann-Bell; 2001; ISBN: 0943396735), by Murray Cragin and Emil Bonanno, lists and indexes every DSO charted by

FIGURE ii-16.

The belt and sword of Orion, as shown by Uranometria 2000.0

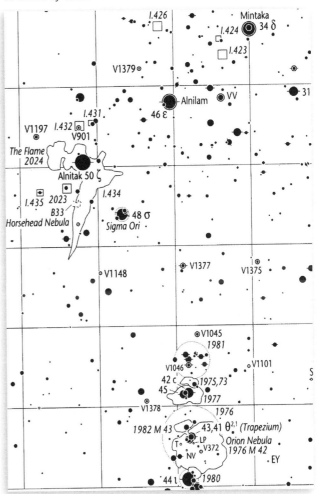

the two atlas volumes. Each of the 220 double-page charts in the atlases has a corresponding table in the Field Guide that lists co-ordinates, dimensions, classification, and notes for every object plotted on that chart. The Field Guide also includes a full index that lists every object and is cross-referenced to the appropriate chart number. All volumes are printed on high-quality paper. Although they are not dew-resistant, taking minimal precautions such as closing them when not in use will keep a copy of U2K in good shape through years of field use.

Choosing Charts

The real decision in choosing printed charts is where to stop. Even the deepest printed charts cannot plot every object visible in large scopes from dark sites, so at some point most dedicated DSO observers find themselves abandoning printed charts for planetarium software, which has no such limits.

For most observers, we think the Pocket Sky Atlas or Sky Atlas 2000.0 is the ideal compromise between depth, cost, and usability. We observe primarily with a 10" scope from reasonably dark sites, and find that SA2K is a good fit in terms of objects plotted versus objects visible with our equipment. When we observe from a darker site or with a larger instrument, we use our notebook computer running planetarium software to chart the truly faint fuzzies or beg a few minutes with an observing buddy's copy of Uranometria 2000.0.

If you observe mostly faint DSOs from a dark location with an 8" or larger scope, and particularly if you don't have a notebook computer you're willing to risk in the field, don't bother buying SA2K. You'll be much happier with Uranometria 2000.0.

Miscellaneous Accessories

In addition to your major accessories, you'll find it helpful to have various miscellaneous accessories. Here's our basic checklist:

- Cellular phone, with emergency numbers for your observing site.
- First-aid kit, insect repellent, and similar safety and comfort supplies.
- For cold-weather sessions, warm clothing, blankets, chemical hand warmers, propane heater, etc.
- Lens-cleaning supplies, tool kit, and essential spare parts.
- Spare batteries for your cell phone, unit-power finder, flashlights, drive motors, etc.
- Folding lawn chairs and table.
- Folding chart table.
- Collimating eyepiece and/or laser collimator.
- Black "pirate" eye patch, for protecting night vision.

- Sketch book and/or writing paper, with spare pens and pencils.
- An observing notebook with log sheets or a digital voice recorder, for recording observations.
- Wrist watch or clock, for noting the time of each observation.
- White flashlight, for clean up after the observing session; never use it *during* an observing session.
- Trash bags, for carting away waste.
- Cooler, packed with munchies and drinks.

We have an SUV dedicated to observing trips, which is always packed and ready to go. If we didn't have a dedicated vehicle available, we'd pack our accessories into boxes and duffle bags to keep them organized and tidy. It's also a good idea to keep a checklist of your accessories, to make sure you don't forget anything important.

Eyepiece/Accessory Case

Amateur astronomers accumulate an incredible amount of gear. Eyepieces, Barlows, flashlights, filters, diagonals, collimation tools, adapters, spare batteries, lens cleaning kits—the list just goes on and on. As you continue to accumulate accessories, you'll need some means of keeping those accessories safe and organized. The best way to do that is to buy or make an eyepiece/accessory case (or cases; we use two).

Many amateur astronomers make their own accessory cases by modifying fishing tackle boxes, drink coolers, tool boxes or similar enclosures with dividers, partitions, and other organizing methods. We prefer to use aluminum tool cases with pluckable foam, which are sold by Home Depot and similar home-supply stores for $20 or so. These cases are large enough to hold half a dozen or more eyepieces, a Barlow or two, collimating eyepiece and laser collimator, Telrad, filters, and numerous other small accessories. Figure ii-17 shows one of our cases, populated with eyepieces and other accessories.

Freeze the Foam

Some people dislike pluckable foam, and prefer to use standard foam with custom cut-outs to fit their accessories. The problem is, standard foam is very difficult to cut neatly. One of our readers pointed out an easy solution. Simply put the foam in the freezer for a couple hours, which makes it rigid enough to cut neatly with a sharp knife or even a hole saw.

FIGURE ii-17.

One of our eyepiece/ accessory cases

Where (and Where Not) to Buy Stuff

The best places to buy astronomy equipment are specialty astronomy stores and on-line astronomy vendors. You'll generally pay less and get better quality gear if you restrict your purchases to these sources. Although we do not formally endorse vendors—don't blame us if you buy from one of our preferred vendors and your goldfish dies—here are several sources of astronomy gear that we believe are reliable, based on our own experiences and those reported by our readers:

* Anacortes Telescope and Wild Bird (www.buytelescopes.com)

* Apogee, Inc. (www.apogeeinc.com)

* Astronomics (www.astronomics.com)

* Eagle Optics (www.eagleoptics.com)

* Hands-On Optics (www.handsonoptics.com)

* Helix Observing Accessories (www.helix-mfg.com)

* High Point Scientific (highpointscientific.com)

* Island Eyepiece and Telescope Ltd. (www.islandeyepiece.com)

* Kendrick Astro Instruments (www.kendrick-ai.com)

* Oceanside Photo and Telescope (www.optcorp.com)

* O'Neil Photo and Optical (www.oneilphoto.on.ca)

* Orion Telescope and Binocular Center (www.telescope.com)

* ScopeStuff.com (www.scopestuff.com)

* ScopeTronix Astronomy Products (www.scopetronix.com)

* University Optics (www.universityoptics.com)

Adorama, B&H, and the other New York City camera stores often have excellent prices on astronomy gear, but you must know exactly what you want. Don't expect skilled advice from them. Their shipping charges are often very high, so make sure to get a total price before you order. We prefer to support specialty astronomy retailers. They may charge a few bucks more, but they are run by astronomers for astronomers, and their expert advice is usually worth the small extra cost, particularly if you're not entirely sure what you're doing.

Brian Jepson Notes

Some local shops, particularly those that have birdwatching equipment, can probably order stuff from Orion. That's where I got my scope, and because my scope arrived damaged, it was a massive convenience to not have to ship it back myself.

For More Information

The best first source of information about astronomy gear is your local astronomy club, where you can get hands-on experience with a lot of different equipment, as well as unbiased advice from club members. In addition, we recommend the following printed and online sources as good starting points:

- Our own book, **Astronomy Hacks** (O'Reilly; 2005; ISBN: 978-0596100605), provides much more detailed advice about choosing, maintaining, and modifying astronomy gear including binoculars, telescopes, eyepieces, and other accessories, as well as advice about using that equipment to best advantage.

- Phil Harrington's book, **Star Ware** (Wiley; 2007; ISBN: 978-0471750635)

- **Sky & Telescope** magazine and **Astronomy** magazine

- Phil Harrington's Talking Telescopes forum (groups.yahoo.com/group/telescopes)

- Phil Harrington's Starry Nights forum (groups.yahoo.com/group/starrynights/)

- Gene Baraff's Skyquest Telescopes forum (groups.yahoo.com/group/skyquest-telescopes)

- Rod Mollise's SCT Users' forum (groups.yahoo.com/group/sct-user)

01
Andromeda, The Princess

NAME: Andromeda (an-DROM-eh-duh)

SEASON: Autumn

CULMINATION: 9:00 p.m., late November

ABBREVIATION: And

GENITIVE: Andromedae (an-DROM-eh-dye)

NEIGHBORS: Ari, Cas, Lac, Peg, Per, Psc

BINOCULAR OBJECTS: NGC 205 (M 110), NGC 221 (M 32), NGC 224 (M 31), NGC 752

URBAN OBJECTS: NGC 221 (M 32), NGC 224, (M 31), NGC 752, NGC 7662, 57-γ (STF 205)

Andromeda is a large constellation, ranking 19th in size among the 88 constellations. It covers 722 square degrees of the celestial sphere, or about 1.8%. Andromeda is one of the ancient constellations. In Greek mythology, Princess Andromeda was the daughter of King Cepheus and Queen Cassiopeia of Æthiopia. The vain Cassiopeia bragged publicly that she was more beautiful than the Nereids, the sea nymph goddesses renowned for their beauty. Bad idea. Poseidon, hearing the angry protests of the proud Nereids, sent floods and Cetus the sea monster to wreak havoc on the land and people of Æthiopia.

In desperation, King Cepheus consulted the oracle of Ammon, which told him that the only way to appease Poseidon was to volunteer his daughter Andromeda to be devoured by Cetus. Cepheus chained Andromeda to a rock near the sea, to await Cetus. Andromeda, resigned to her fate, watched Cetus approaching. Purely by luck, the Hero Perseus, on his way home after slaying the Gorgon, happened to be sailing past and noticed the sea monster about to devour the pretty girl. Perseus, of course, confronted and slew the sea monster, freed Andromeda from her chains, and married her. All in a day's work, for a Hero.

Andromeda returned home with Perseus. They went on to have six sons and a daughter and to found the kingdom of Mycenae and the Perseidae dynasty. Upon the death of Andromeda, the goddess Pallas Athena placed her among the stars as the

TABLE 01-1.

Featured star clusters, nebulae, and galaxies in Andromeda

Object	Type	Mv	Size	RA	Dec	M	B	U	D	R	Notes
NGC 205	Gx	8.9	21.9 x 10.9	00 40.4	+41 41	◉	◉				M 110; Class E5 pec; SB 13.2
NGC 221	Gx	9.0	8.7 x 6.4	00 42.7	+40 52	◉	◉	◉			M 32; Class cE2; SB 10.1
NGC 224	Gx	4.4	192.4 x 62.2	00 42.7	+41 16	◉	◉	◉			M 31; Class SA(s)b; SB 12.9
NGC 752	OC	5.7	49.0	01 57.8	+37 51			◉	◉		Cr 23; Mel 12; Class II 2 r
NGC 891	Gx	10.8	14.3 x 2.4	02 22.6	+42 21					◉	Class SA(S)b? sp; SB 14.6
NGC 7662	PN	9.2	37.0"	23 25.9	+42 32			◉		◉	Blue Snowball Nebula; Class 4+3

TABLE 01-2.

Featured multiple stars in Andromeda

Object	Pair	M1	M2	Sep	PA	Year	RA	Dec	UO	DS	Notes
57-γ	STF 205A-BC	2.3	5.0	9.7	63	2004	02 03.9	+42 20	◉	◉	Almach

constellation Andromeda, with Perseus, Cassiopeia, and Cepheus always nearby.

Andromeda is a mid-northerly constellation that lies far above the plane of our own Milky Way galaxy. When you look toward Andromeda, you are looking through a thinly-populated part of our own galaxy and toward intergalactic space. Accordingly, you won't find bright nebulae or globular clusters in Andromeda, because those objects are almost exclusively denizens of the galactic plane. You will find numerous galaxies—four of which (look for the Type Gx in Table 01-1) are among our featured objects, including the magnificent Great Andromeda Galaxy—the open cluster NGC 752, and the fine, bluish planetary nebula NGC 7662.

Andromeda is easy to find, located between the prominent W-shape of Cassiopeia to the north and the Great Square of Pegasus to the southwest. The brightest stars in Andromeda form a long, narrow V-shape, with its apex at Alpheratz, the northeastern corner star of the Great Square of Pegasus. For observers at mid-northern latitudes, Andromeda is well-placed for evening viewing from late summer, when it rises after dusk, through mid-winter, when it sets a couple hours after dark. Andromeda culminates at zenith at 9:00 p.m. in late November.

CHART 01-1.

The constellation Andromeda (field width 50°)

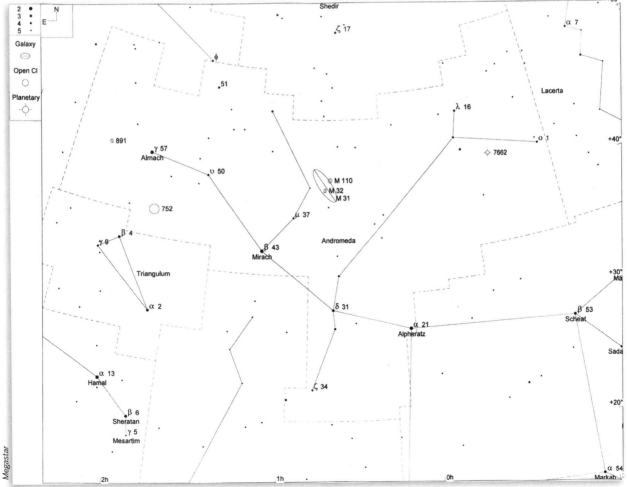

Clusters, Nebulae, and Galaxies

M 31 (NGC 224)	★★★★	✧✧✧		GX	MBUDR
Chart 01-2	Figure 01-1	m4.4, 192.4' x 62.2'	00h 42.7m		+41° 16'

M 32 (NGC 221)	★★	✧✧✧		GX	MBUDR
Chart 01-2	Figure 01-1	m9.0, 8.7' x 6.4'	00h 42.7m		+40° 52'

M 110 (NGC 205)	★★	✧✧✧		GX	MBUDR
Chart 01-2	Figure 01-1	m8.9, 21.9' x 10.9'	00h 40.4m		+41° 41'

Along with our own Milky Way galaxy, the galaxies M 31, M 32, and M 110 are members of the *Local Group* of galaxies. M 31, the famous Great Andromeda Galaxy, is the nearest major galaxy to our own Milky Way galaxy, and is in many respects a twin sister to it. Recent data suggest that M 31 is somewhat larger than our Milky Way, but is much less densely populated, with only about half the mass. At a distance of about 2.9 million light years, M 31 has the distinction of being the most distant object that is easily visible without optical assistance.

M 31 stands out to the naked eye from a moderately dark site, and is clearly visible with the slightest optical aid even from light-polluted urban sites. Although M 32 is a much more difficult binocular object than M 31, its relatively high surface brightness (10.1) makes it visible with a 50mm or larger binocular from a dark site as a fuzzy star about 25' S of the bright core of M 31. M 110, located about 37' NW of the core of M 31 and with lower surface brightness of 13.2, is extremely challenging but may be possible with a tripod-mounted 60mm or larger binocular from a dark site.

M 31 begins to reveal detail in a 3.5" or larger telescope, and shows nuanced detail in typical amateur instruments of 6", 8", or 10" aperture. All three galaxies are visible in the field of a 1° eyepiece, although M 32 and M 110 show little detail. M 31 shows a mottled appearance, with two dark lanes clearly visible. At 90X to 125X, the core of M 31 takes on the grainy appearance of a globular cluster that is not quite resolvable into individual stars. M 32 is visible with direct vision as a prominent circular nebulosity with gradual brightening to a star-like core, located just S of the core of M 31, but embedded within the nebulosity of M 31. M 110 is visible as a faint, oval nebulosity on the NW side of M 31, opposite M 32, and near the edge of the visible extent of M 31.

FIGURE 01-1.

Figure 01-1. NGC 224 (M 31), NGC 221 (M 32, bottom center), and NGC 205 (M 110, top right) (60' field width)

Image reproduced from Digitized Sky Survey courtesy Palomar Observatory and Space Telescope Science Institute

The Andromeda Galaxy cluster is easy to find. From m2 Alpheratz (21-α), which is the NE star in the Great Square of Pegasus, hop 6.9° ENE (east northeast) to m3 31-δ and then 7.9° NE to m2 Mirach (43-β). Alternatively, you can locate Mirach directly by using the westernmost triangle of Cassiopeia—of which m2 Shedir is the apex—as a pointer to Mirach. Follow that pointer 21.4° SSE (about two fists south southeast) to Mirach, which is the brightest star in the near vicinity.

Once you have located Mirach, move the finder NW, placing Mirach on the edge of the field. The m4 star 37-μ will be prominent in the finder. Continue moving the finder on the same line until 37-μ reaches the edge of the field, and M 31 will be visible in the finder with the m5 star 35-ν prominent about 1.5° E.

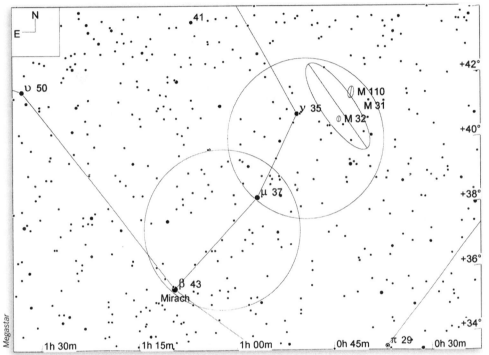

CHART 01-2.

M 31, M 32, and M 110 (15° field width; 5° finder circles; LM 9.0)

NGC 752	★★★	❖❖❖	OC	MBUDR
Chart 01-3	Figure 01-2	m5.7, 49.0'	01h 57.8m	+37° 51'

NGC 752 is a moderately rich, detached, loosely scattered open cluster that is best viewed with a large binocular or at low magnification in a telescope. The cluster is easy to locate with a 50mm binocular or finder. With Almach and m4 50-υ, it forms nearly an equilateral triangle with 5° sides, with the center of the cluster located about 2.1° dead W of m5 58-Andromedae and 40' NE of the prominent m6 pair 57-Andromedae. In a 50mm binocular or finder, the cluster is visible as a dozen or so m9 stars, with a very faint nebulosity of unresolved dimmer stars. In our 10" scope at 42X, 60+ stars are visible, most m10 and m11.

FIGURE 01-2.

NGC 752 (60' field width)

Image reproduced from Digitized Sky Survey courtesy Palomar Observatory and Space Telescope Science Institute

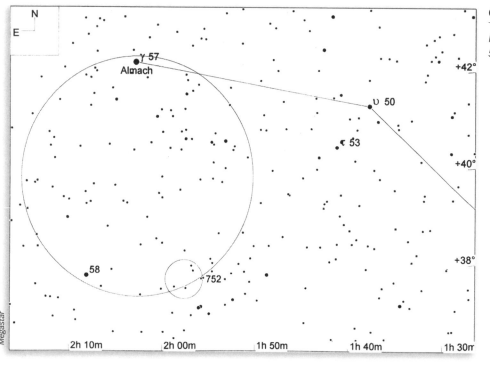

CHART 01-3.

NGC 752 (10° field width; 5° finder circle; LM 9.0)

NGC 891	★★	☉☉☉	Gx	MBUDR
Chart 01-4	Figure 01-3	m10.8, 14.3' x 2.4'	02h 22.6m	+42° 21'

NGC 891 is a small, faint galaxy located 3.4° dead E of Almach. Although NGC 891 is invisible in a 50mm finder, it is relatively easy to locate geometrically. NGC 891 forms a triangle with Almach and m5 60 Andromedae, and is located 1° dead N of a prominent m6 field star. NGC 891 is also located just ESE of an asterism of m8 stars that are visible in a 50mm finder and resemble the "ice cream cone" pattern of the constellation Cepheus. The 14.6 surface brightness of NGC 891 means it is much fainter visually than its 10.8 visual magnitude suggests. In our 10" telescope at 125X, NGC 891 is visible as a faint, slender streak of light extending about 6' NNE-SSW.

FIGURE 01-3.

NGC 891 (60' field width)

Image reproduced from Digitized Sky Survey courtesy Palomar Observatory and Space Telescope Science Institute

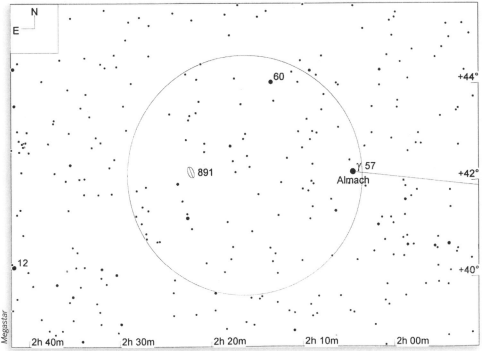

CHART 01-4.

NGC 891 (10° field width; 5° finder circle; LM 9.0)

NGC 7662	★★★	✦✦✦	PN	MBU**DR**
Chart 01-5	Figure 01-4	m9.2, 37.0"	23h 25.9m	+42° 32'

NGC 7662, also called the *Blue Snowball Nebula*, is a fine planetary nebula. To locate NGC 7662, place the m4 star 19-ϰ Andromedae at the NE edge of your finder field. m4 17-ɩ is prominent 1.1° SSW of 19-ϰ and m6 13 Andromedae shows prominently, 2° W of 17-ɩ. NGC 7662 is visible in a low-power eyepiece as a fuzzy star 25' SSW of 13 Andromedae. At 180X and 250X in our 10" scope, NGC 7662 reveals considerable structural detail as a bright, slightly elongated, annular disk with a distinct bluish tinge and noticeable darkening toward the center. The brighter inner ring is continuous, although denser to the NE and SW. The dim outer ring is fragmented and requires averted vision to detect. An O-III filter is the best choice for this nebula, but a narrowband filter also enhances the view significantly. Although the central star is listed at m13.2, we have never been able to see that star in our 10" scope.

FIGURE 01-4.

NGC 7662 (DSS image, 60' field width)

Image reproduced from Digitized Sky Survey courtesy Palomar Observatory and Space Telescope Science Institute

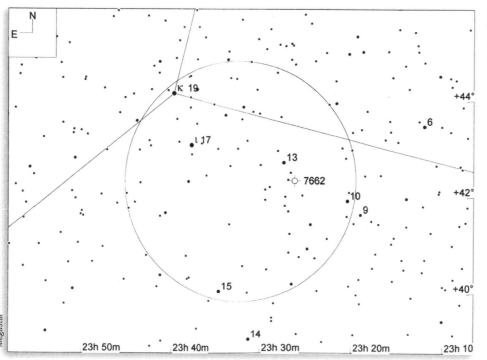

CHART 01-5.

NGC 7662 (10° field width; 5° finder circle; LM 9.0)

Multiple Stars

57-γ (STF 205A-BC)	★★★	٥٥٥٥	MS	UD
Chart 01-1		m2.3/5.0, 9.7", PA 63° (2004)	02h 03.9m	+42° 20'

ALMACH, 57-γ Andromedae, is a beautiful multiple star system, with a yellow m2 primary and a blue-green m5 secondary located just under 10" ENE of the primary. This AB pair is easy to split, even with a small telescope. At only 0.5" separation, the BC pair is a different matter. We have never split BC (m5.0/6.3) in our 10" scope, despite using stupid high magnification (500X+) on nights of excellent seeing. At most, we thought we saw a (very) slight easterly elongation in B, and that was only by using averted imagination.

Almach is easy to find, because it is the only prominent naked-eye star in the vicinity about 7° dead N of Triangulum. Alternatively, you can use Cassiopeia as a pointer. The eastern triangle of the Cassiopeia, with Ruchbah at its apex, points directly to Almach, about 19° to the SSE.

02

Aquarius, The Water Bearer

NAME: Aquarius (uh-KWAIR-ee-us)

SEASON: Summer

CULMINATION: midnight, 26 August

ABBREVIATION: Aqr

GENITIVE: Aquarii (uh-KWAIR-ee-ih)

NEIGHBORS: Aql, Cap, Cet, Del, Equ, Peg, PsA, Psc, Scl

BINOCULAR OBJECTS: NGC 6981 (M72), NGC 7089 (M2)

URBAN OBJECTS: NGC 7009, NGC 7089 (M2)

Aquarius, the Water Bearer, is a large constellation, ranking 10th in size among the 88 constellations. It covers 980 square degrees of the celestial sphere, or about 2.3%. Despite its size, Aquarius is not a prominent constellation. Its brightest stars, Sadalmelik, 34-Alpha (α) Aquarii, and Sadalsuud, 22-Beta (β) Aquarii, are only third magnitude. The most prominent star pattern in Aquarius is the Y-shaped *asterism* (a group of stars that forms a distinct pattern, but is not a constellation; for example, the Big Dipper is an asterism that is a part of the constellation Ursa Major) formed by the triangle of 48-Gamma (γ), 62-Eta (η), and 52-Pi (π) Aquarii, with 55-Zeta (ζ) at its center. This pattern has been known at least since Babylonian times as "The Water Jar."

Aquarius is the eleventh sign of the Zodiac. It is an ancient constellation, recognized as such by the Babylonians, and it was probably ancient even then. Aquarius lies in an area of the sky that has been associated since ancient times with water, and is surrounded by other watery constellations including Pisces (the Fishes), Pisces Austrinus (the Southern Fish), Capricornus (the Sea Goat), Delphinus (the Dolphin), and Cetus (the Sea Monster).

Among the Babylonians, who suffered periodic devastating floods, Aquarius was feared. To the ancient Greeks, Egyptians, and Arabs, who lived in dry climates, Aquarius represented a generally benevolent god, who brought water when it was needed to nourish their crops. But the Greeks also identified Aquarius with the legend of Deucalion and his wife, Pyrrha, who built a large ship and stocked it with provisions against a coming flood. After nine days and nights afloat in the flood waters, their ship grounded on Mount Parnassus. The parallels with the later Biblical myth of Noah are unmistakable.

In modern times, Aquarius gained prominence during the late 60s when the counter-culture proclaimed the dawning of the Age of Aquarius. They were a bit premature, however. Astrology defines a Zodiacal age as the period during which the Sun appears in that constellation at the Vernal Equinox, or the first day of spring. The Sun will not appear in Aquarius at the Vernal Equinox until about 600 years from now.

TABLE 02-1.

Featured star clusters, nebulae, and galaxies in Aquarius

Object	Type	Mv	Size	RA	Dec	M	B	U	D	R	Notes
NGC 6981	GC	9.2	6.6	20 53.5	-12 32	◉	◉				M 72; Class IX
NGC 6994	OC	8.9	2.8	20 58.9	-12 38	◉					M 73; Cr 426; Class IV 1 p
NGC 7009	PN	8.3	70.0"	21 04.2	-11 22			◉		◉	Saturn Nebula; Class 4+6
NGC 7089	GC	6.6	16.0	21 33.5	-00 49	◉	◉	◉			M 2; Class II
NGC 7293	PN	7.5	16.0	22 29.6	-20 50					◉	Helix Nebula; Class 4+3

Aquarius lies far from the galactic plane, and so lacks open clusters and diffuse nebulae. Many galaxies lie in Aquarius, but all are dim and unsatisfying targets for any but the largest amateur telescopes. Aquarius is home to two prominent Messier globular clusters, M 2 and M 72, and two notable planetary nebulae, NGC 7009 (the Saturn Nebula) and NGC 7293 (the Helix Nebula). Aquarius also contains a third Messier object, M 73, which vies with M 40 for the title of Least Impressive Messier Object. M 73 is usually considered to be one of Messier's few "mistakes" because it appears to be a random asterism of four stars without nebulosity. Recent data suggest, however, that Messier may have the last laugh. A substantial body of opinion now holds that M 73 is indeed an actual open cluster, albeit a very small, dim, and unimpressive one.

Aquarius is easy to find, as it lies just north of the prominent first-magnitude star Fomalhaut in Piscis Austrinus and surrounds the north and east borders of Capricornus. For observers at mid-northern latitudes, Aquarius is well-placed for evening viewing from early summer, when it rises after dusk, through mid-winter, when it sets a couple hours after dark.

CHART 02-1.

The constellation Aquarius (field width 50°)

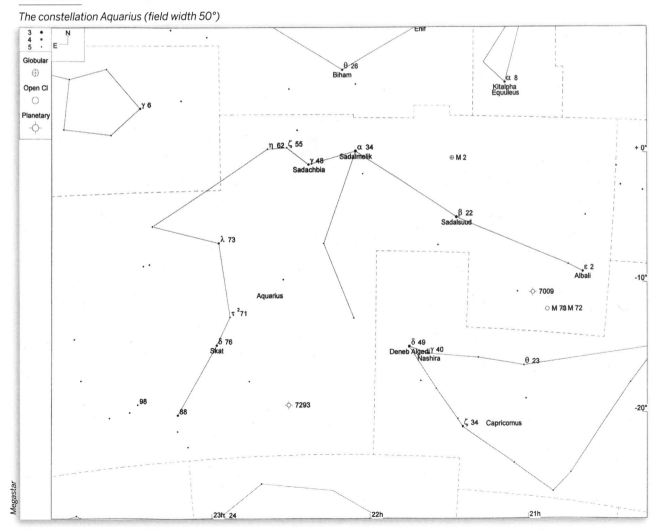

TABLE 02-2.

Featured multiple stars in Aquarius

Object	Pair	M1	M2	Sep	PA	Year	RA	Dec	UO	DS	Notes
55-zeta	STF 2909	3.7	3.8	2.0	190	1997	22 28.8	-00 01		◉	
94	STF 2998	5.2	5.2	12.4	351	1994	23 19.1	-13 28		◉	

Clusters, Nebulae, and Galaxies

M 72 (NGC 6981)	★★★	✹✹✹	GC	**MBUDR**
Chart 02-2	Figure 02-1	m9.2, 6.6'	20h 53.5m	-12° 32'

M 72 is notable among the Messier globular clusters as being among the loosest, most distant, and most inherently bright. M 72 is assigned a Shapley-Sawyer Concentration Class of IX. Among the Messier globulars, only M 56, at Class X, and M 71, at Class X or XI, are looser. At about 53,000 light years distant, M 72 lies well beyond the galactic center, so far from us that only its extremely high inherent brightness allows us to see it with small and medium-size telescopes.

M 72 is faintly visible with averted vision in a 50mm binocular or finder scope as a fuzzy m9/10 star. In a small scope or at low magnification in a larger scope, M 72 is obviously an extended object, but resembles a tiny comet or planetary nebula. In a larger scope at higher magnification, M 72 reveals itself as a globular cluster, notable for its even appearance, with a bright central core and only very slight dimming toward the edges of the object. At 180X in our 10" scope, a few stars are resolvable near the extreme edges, but even 350X leaves the halo and core stubbornly unresolved.

To locate M 72, locate the m4 star 2-epsilon (ε) Aqr (Albali), which lies about 8° ENE (east northeast) of the prominent m3 pair Algedi/Dabih at the NW corner of Capricornus, and is the brightest star in the immediate vicinity. Place Albali on the NW edge of your finder field, and look for an m6 star that appears prominently in the SW quadrant of the field. M 72 lies 41' dead E of that star, and is visible in a 50mm finder with averted vision as a fuzzy m9/10 star. M 73 is invisible in the finder, but appears as a distinct grouping of stars in a low-power, wide-field eyepiece.

FIGURE 02-1.

NGC 6981 (M 72) (60' field width)

Image reproduced from Digitized Sky Survey courtesy Anglo-Australian Observatory and Space Telescope Science Institute

CHART 02-2.

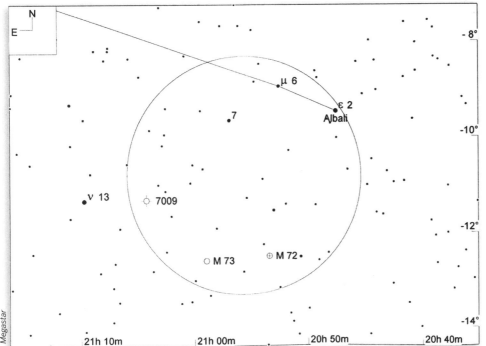

M 73 (NGC 6994)	★	✪✪✪	OC	MBUDR
Chart 02-2, 02-3	Figure 02-2	m8.9, 2.8'	20h 58.9m	-12° 38'

M 73 (NGC 6994) is a tiny, dim, open cluster. M 73 comprises a 1' triangle of three m11/12 stars, with a fourth m13 star lying close to the NNW. For many years, M 73 was considered to be a "Messier Mistake" and was even omitted from many listings of the Messier catalog. Until recently, M 73 was thought to be an asterism, a random collection of stars that were not gravitationally bound rather than a true open cluster. However, statistical analysis shows that the chances of such a grouping occurring randomly are less than one in four, so M 73 is now acknowledged by most astronomers as a true open cluster, albeit a very small, poor, and sparse one.

Despite its lack of prominence, M 73 is relatively easy to locate once you have found M 72. M 73 lies 1.3° E of M 72, in an area devoid of prominent stars. If you center your optical finder on M 72, M73 lies about half a finder field E and just slightly S. Near the ESE (east southeast) edge of the finder field, two m6/7 stars appear prominently, separated by about 0.5°. M 73 lies about the same distance NNW on that line.

FIGURE 02-2.

NGC 6994 (M 73) (60' field width)

Image reproduced from Digitized Sky Survey courtesy Anglo-Australian Observatory and Space Telescope Science Institute

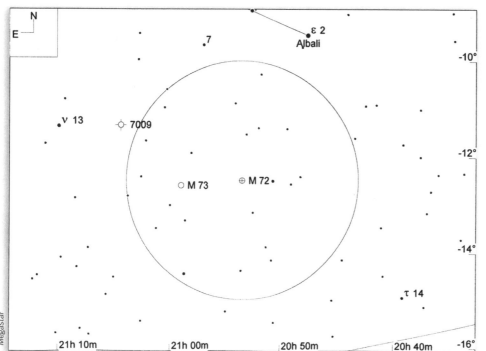

CHART 02-3.

M 73 detail (10° field width; 5° finder circle; LM 8.5)

NGC 7009	★★★	✹✹✹	PN	MBUDR
Chart 02-2, 02-4	Figure 02-3	m8.3, 70.0"	21h 04.2m	-11° 22'

In 1782, William Herschel discovered and logged the planetary nebula NGC 7009, one of the first of many planetary nebulae he would discover. In 1850, Lord Rosse named this object the Saturn Nebula, noting that its *ansae* (projections) bore a distinct resemblance to the rings of the planet Saturn. NGC 7009 is an excellent example of how this class of objects came to be named. Under anything less than high magnification, it resembles the distant gas-giant planets Uranus and Neptune, including its distinct bluish-green coloration.

In our 10" scope at 90X with direct vision, NGC 7009 is a beautiful, bright, blue-green planetary nebula of about 25" extent, with slight east-west elongation and noticeable central brightening. NGC 7009 responds well to a narrowband filter, and very well to an O-III filter. Averted vision at 180X shows very faint ansae extending a few arcseconds on the E and W sides. Although the central star is listed in some catalogs as m12.7, we are unable to detect it in our 10" scope.

To locate NGC 7009, move your finder scope ESE (east southeast) from 2-Epsilon (ε) Aqr (Albali) until the m4.5 star 13-Nu (ν) Aqr comes into view. 13-Nu (ν) is by far the brightest star in the near vicinity, and shows very prominently in the finder. NGC 7009 lies 1.3° dead W of 13-Nu (ν), and is visible as a fuzzy star in a low-power wide-field eyepiece. If necessary, you can verify NGC 7009 at low power by blinking the object with your narrowband or O-III filter.

FIGURE 02-3.

NGC 7009 (60' field width)

Image reproduced from Digitized Sky Survey courtesy Anglo-Australian Observatory and Space Telescope Science Institute

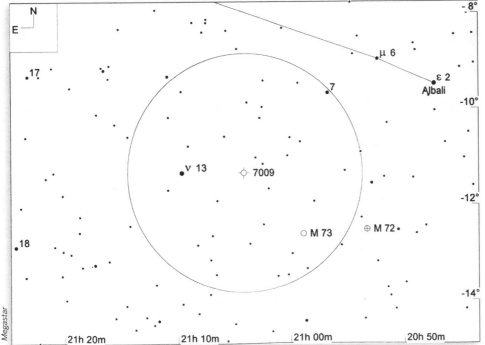

CHART 02-4.

NGC 7009 detail (10° field width; 5° finder circle; LM 9.0)

Megastar

M 2 (NGC 7089)	★★★★	⊙⊙⊙⊙	GC	MBUDR
Chart 02-5	Figure 02-4	m6.6, 16.0	21h 33.5m	-00° 49'

The globular cluster M 2 (NGC 7089) was first logged by Giovanni Domenico Maraldi in 1746. In 1760, Charles Messier independently rediscovered the cluster and logged it as the second object in his catalog. M 2 is not the largest or brightest of the Messier globular clusters, but nonetheless is a perennial favorite of many observers.

As its Shapley-Sawyer Concentration Class II indicates, M 2 is one of the most concentrated globular clusters easily visible in amateur instruments. Although M 2 is cataloged as having a 16' extent, about half the size of the full Moon, that size is derived from long-exposure astrophotographs. Visually, M 2 has an apparent extent of 5' or 6', with the dim outer halo brightening dramatically to a small central core. With a visual magnitude of 6.6, M 2 has been reported by some observers to be visible to the naked eye, but only on exceedingly clear nights from completely dark sites. The relatively small size of the object and its relatively low magnitude combine to yield high surface brightness, which makes the cluster prominent with even the slightest optical aid.

At 42X in our 10" scope, M 2 stands out prominently as a diffuse, unresolved, bright nebulosity in a field devoid of bright stars. At 125X, the outer halo is resolved into individual stars, and at 250X the cluster is resolved nearly to the core. Although some observers report fully resolving M 2 in scopes as small as 8", we have been able to do so in nothing smaller than a 17.5" scope.

M 2 is easy to find, lying about 4.7° N of m2.9 22-beta (β) Aqr (Sadalsuud). To locate M 2, place Sadalsuud at the S edge of your finder field with the prominent m6 pair 20/21-Aquarii near the W edge of the field, and look for a prominent nebulosity at the N edge of the field.

FIGURE 02-4.

NGC 7089 (M 2) (60' field width)

Image reproduced from Digitized Sky Survey courtesy Anglo-Australian Observatory and Space Telescope Science Institute

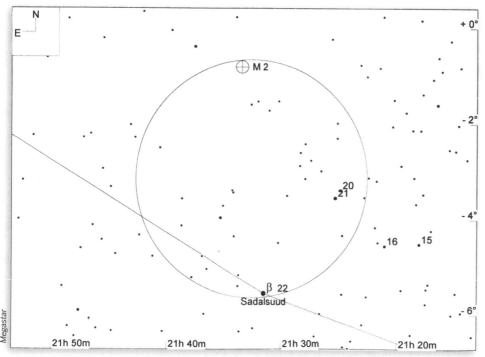

CHART 02-5.

NGC 7089 (M 2) (10° field width; 5° finder circle; LM 9.0)

NGC 7293	★★	⊙⊙⊙	PN	MBUDR
Chart 02-6		m7.5, 16.0	22h 29.6m	-20° 50'

NGC 7293, also known as the Helix Nebula, was discovered by Karl Ludwig Harding in 1824. NGC 7293 is probably closer to us than any other planetary nebula. Its distance has been estimated at between 85 and 600 light years, with the best current estimates clustering around 400 light years. Its proximity contributes to its very large apparent size, which is generally stated as 16', or more than half the size of the full Moon. This gigantic size means that, despite its bright visual magnitude of 7.5, NGC 7293 has extremely low surface brightness and is therefore a very difficult object to see. (The Helix name is a modern one, derived from the appearance of the nebula in long-exposure photographs; there is no hint visually of a helical structure.)

NGC 7293 is best viewed with a binocular, ideally a 70mm or larger model, or at very low magnification in a telescope. The nebula responds well to a narrowband filter, and very well to an O-III filter. At 42X in our 10" scope with averted vision and an O-III filter, NGC 7293 is visible as faint wisps, brightest to the NNE (north northeast) and SSW (south southwest), with very faint nebulosity filling in most of the annulus. The central core is noticeably darker than the ring structure. The central star is visible with averted vision and no filter. Under excellent viewing conditions, NGC 7293 resembles a much larger, much fainter version of the Ring Nebula (M 57) in Lyra.

NGC 7293 is relatively easy to find. Put the m3 star 76-delta (δ) Aqr (Skat) at the NE edge of your finder field, and locate the m5 stars 66-Aqr and 68-Aqr near the SW edge of the field. Move the finder to put 66- and 68-Aqr near the ENE edge of the finder field, and locate the m5 star 59-upsilon (υ) Aqr. NGC 7293 lies 1.2° W of 59-upsilon (υ) Aqr, where it is faintly visible in a low-power eyepiece.

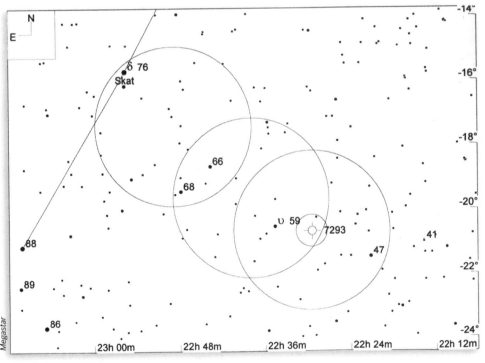

CHART 02-6.

NGC 7293 (15° field width; 5° finder circles; 1° eyepiece circle; LM 9.0)

Multiple Stars

55-zeta (STF 2909)	★★★	✧✧✧✧	MS	UD
Chart 02-7		m3.7/3.8, 2.0", PA 190°	22h 28.8m	-00° 01'

At 180X in our 10" scope, 55-zeta (ζ) Aquarii (STF 2909) is a pretty, evenly-matched close pair of yellowish stars. This pair is challenging in smaller scopes. In our 90mm (3.5") refractor at 200X, 55-zeta appears elongated, but is not cleanly split and has no hint of color. 55-zeta is easy to locate, as it is the center of the "Water Jar" asterism.

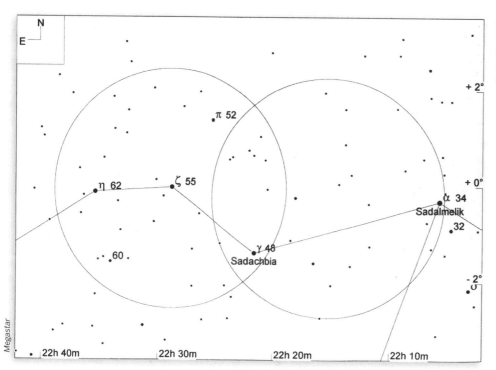

CHART 02-7.

55-Zeta Aquarii (STF 2909) (10° field width; 5° finder circles; LM 9.0)

94 (STF 2998)	★★★★	⟡⟡⟡	MS	UD
Chart 02-8		m5.2/5.2, 12.4", PA 351°	23h 19.1m	-13° 27'

94-Aquarii (STF 2998) is a beautiful, evenly-matched, widely-separated pair that offers a striking color contrast in any size scope. In our 10" scope at 90X, the pair appears reddish-white and deep green. In our 90mm (3.5") refractor at 100X, the pair appears yellow-orange and a pale lime green.

To locate 94-Aquarii, place 90-phi (φ) Aquarii at the N edge of your finder field and look for the prominent arc of three m4/5 stars—91-ψ^1, 93-ψ^2, and 95-ψ^3 Aquarii—in the S half of the finder field. Move the finder until that arc of stars is near the N edge of the finder field, and look for m5 94-Aquarii near the S edge of the finder field.

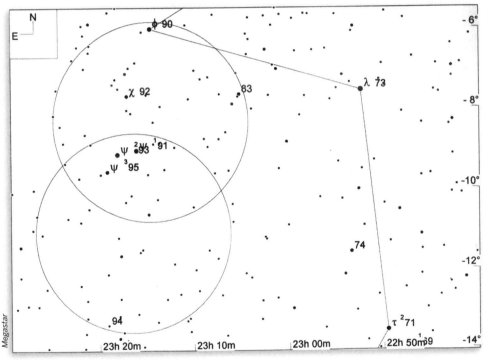

CHART 02-8.

94-Aquarii (STF 2998) (12° field width; 5° finder circles; LM 9.0)

03

Aquila, The Eagle

> **NAME:** Aquila (ACK-will-uh)
>
> **SEASON:** Summer
>
> **CULMINATION:** midnight, 12 July
>
> **ABBREVIATION:** Aql
>
> **GENITIVE:** Aquilae (ACK-will-eye)
>
> **NEIGHBORS:** Aqr, Cap, Del, Her, Oph, Sct, Ser, Sge, Sgr
>
> **BINOCULAR OBJECTS:** NGC 6709
>
> **URBAN OBJECTS:** NGC 6709

Aquila is a medium-size constellation, ranking 22nd of 88. It covers 652 square degrees of the celestial sphere, or about 1.6%. Aquila is an ancient constellation, old even to the Homeric Greeks. Aquila has been recognized as a constellation at least since the time of the Sumerians, between 3,000 and 5,500 years ago. Aquila has almost universally been recognized as a bird, usually an eagle, across cultures and across time.

To the Greeks, Aquila the Eagle was Zeus's pet and companion. Zeus was enraged when the Titan Prometheus stole fire and gave it to humankind. Zeus chained Prometheus to a rock, and sent Aquila to torment Prometheus with his fearsome beak and talons. Hercules came to the defense of Prometheus, slaying Aquila with his mighty bow. Saddened by the loss of his companion, Zeus raised Aquila to the heavens, where he would forever soar.

The *lucida* (brightest object) in Aquila is the star Altair, also designated alpha (α) Aquilae. Altair, which translates from the Arabic as "the flying eagle," is the southern apex of the famed Summer Triangle, with Vega, alpha (α) Lyrae, at the northwestern apex, and Deneb, alpha (α) Cygni, at the northeastern apex.

Aquila is on the celestial equator, with its northwestern half embedded in the Milky Way. Despite this, Aquila is poor in open clusters and other objects associated with the Milky Way. Other than the prominent open cluster NGC 6709 and the striking planetary nebula NGC 6781, Aquila is bereft of showpiece objects.

Aquila is easy to find, as it lies about halfway along the line between the prominent constellations Cygnus and Sagittarius. For observers at mid-northern latitudes, Aquila is well-placed for evening viewing from early summer, when it rises after dusk, through late autumn, when it sets a couple hours after dark.

TABLE 03-1.

Featured star clusters, nebulae, and galaxies in Aquila

Object	Type	Mv	Size	RA	Dec	M	B	U	D	R	Notes
NGC 6709	OC	6.7	13.0	18 51.5	+10 20			◉	◉		Cr 392; Mel 214; Class IV 2 m
NGC 6781	PN	11.8	1.8	19 18.5	+06 32			◉			Class 3+3

TABLE 03-2.

Featured multiple stars in Aquila

Object	Pair	M1	M2	Sep	PA	Year	RA	Dec	UO	DS	Notes
57-Aquilae	STF 2594	5.7	6.5	35.6	170	1991	19 54.6	-08 14		◉	
-	STF 2404	6.4	7.4	3.6	181	1991	18 50.8	+10 59		◉	

CHART 03-1.

The constellation Aquila (field width 45.0°)

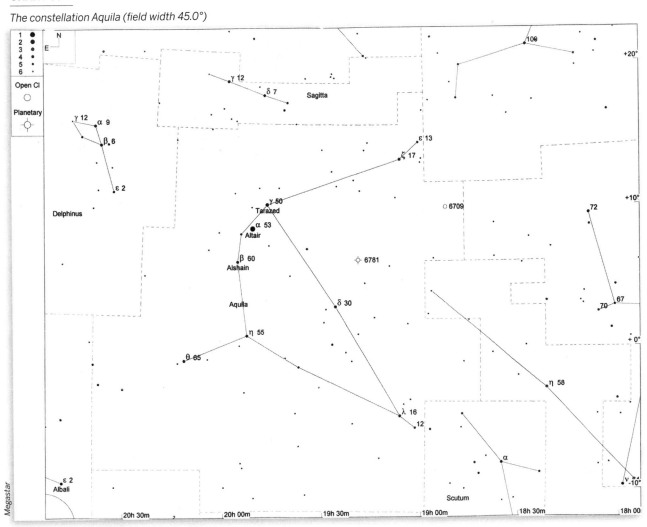

Clusters, Nebulae, and Galaxies

NGC 6709	★★★	✦✦✦✦	OC	MBUDR
Chart 03-2	Figure 03-1	m6.7, 13.0'	18h 51.5m	+10° 20'

The open cluster NGC 6709 is visible as a bright hazy patch in a 50mm finder or binocular, and resolves into dozens of stars in 4" and larger telescopes. In our 10" scope at 90X, NGC 6709 shows a prominent pair of m9/10 stars near the W edge of the cluster, one blue-white and one yellowish. An m9 star lies near the center of the cluster, with a chain of half a dozen m10/11 stars running from its W to NNW (north northwest), and 40+ m11/13 stars filling out the cluster. The member stars are gathered in clumps and chains, interspersed with gaps.

NGC 6709 is easy to find. Place m3 17-zeta (ζ) Aql and m4 13-epsilon (ε) Aql on the NE edge of your finder field. These two stars form a prominent triangle with m5/6 10- and 11-Aql, with the apex pointing toward the nebulous glow of NGC 6709 to the SSW (south southwest). With a 5.5° or wider finder field, NGC 6709 is located near the SSW edge of the finder field. With our 5.0° 9X50 RACI (right angle correct image) finder, the object is located just outside the finder field, so we have to pivot our finder SSW until the object comes into view.

As long as you have NGC 6709 in your eyepiece, you might as well also locate and log the double star STF 2404, described at the end of this chapter.

FIGURE 03-1.

NGC 6709 (60' field width)

Image reproduced from Digitized Sky Survey courtesy Palomar Observatory and Space Telescope Science Institute

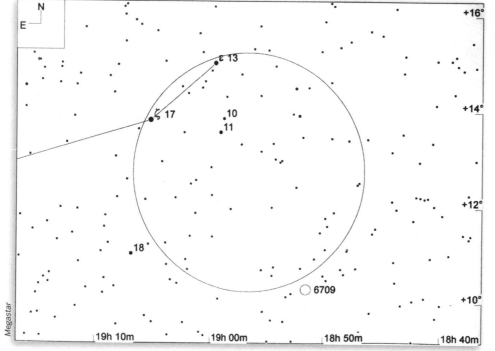

CHART 03-2.

NGC 6709 (10° field width, 5.0° finder field, LM 9.0)

Megastar

NGC 6781	★★★	◐◐◐	PN	MBUDR
Chart 03-3	Figure 03-2	m11.8, 1.8'	19h 18.5m	+06° 32'

First logged by William Herschel in 1788, NGC 6781 is by far the brightest and most impressive of the planetary nebulae in Aquila. In our 10" scope, it is obviously non-stellar even at 42X. At 125X with an O-III or narrowband filter, NGC 6781 shows a large, round, conspicuous nebulosity. With averted vision, an annulus is visible from the W edge all the way around to the E edge, but fades to invisibility from NE through NW.

To locate NGC 6781, begin with your finder in the position shown in Chart 03-3, with 38-mu (μ) and 30-delta (δ) near the edges of the finder field. Pivot the finder WSW until 30-delta (δ) is near the SE edge of the field, m6 22-Aql appears near the center of the field, and m5 21-Aql appears near the SSW edge of the field, as shown in Chart 03-3. Near the N edge of the finder field, look for an m7 star, which is prominent among the dimmer field stars and has a close m9 companion. Center your crosshairs on that m7 star, and then move the crosshairs about 0.5° E. NGC 6781 appears nearly centered in your low-power eyepiece, where it is visible without filtration as a moderately bright nebulosity. Use your O-III or narrowband filter with averted vision to reveal further detail, including the annular ring structure.

FIGURE 03-2.

NGC 6781 (60' field width)

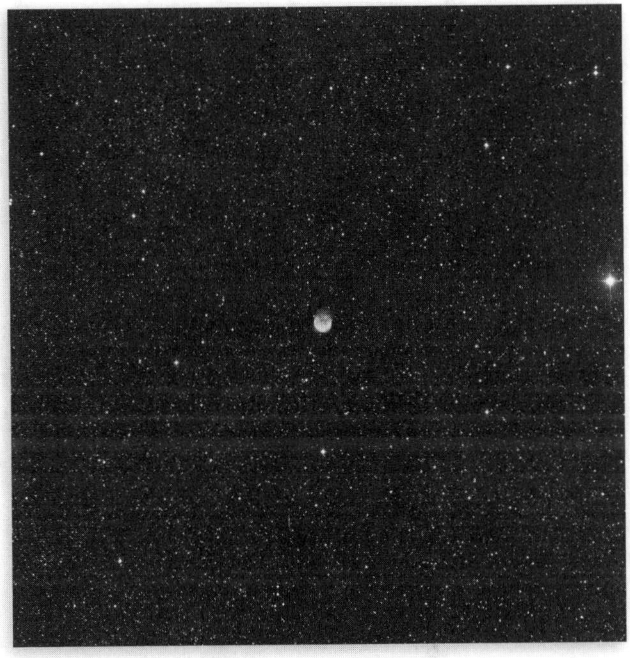

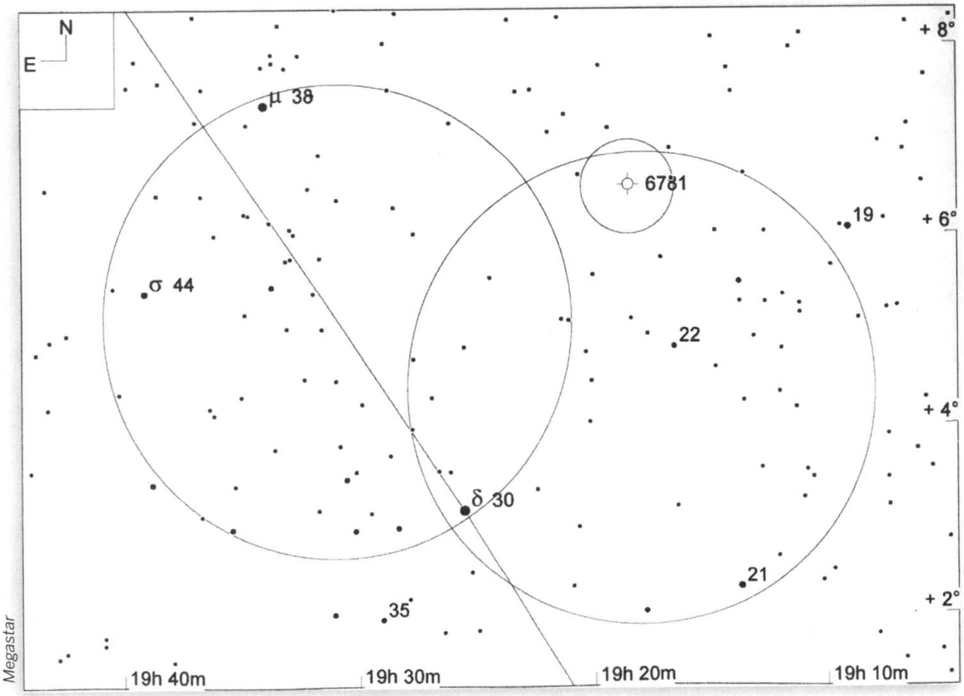

CHART 03-3.

NGC 6781 (10° field, 5.0° finder fields, 1.0° eyepiece field, LM 9.0)

Multiple Stars

57-Aql (STF 2594)	★★	✧✧✧		MS	UD
Chart 03-4		m5.7/6.5, 35.6", PA 170° (1991)		19h 54.6m	-08° 14'

57-Aql (STF 2594) is a closely matched pair of white stars, wide enough to be split easily even at low power. 56-Aql, about 22' to the SSW and visible in the same low-power eyepiece field, is also double. 56-Aql is considerably more difficult than 57-Aql, though. Although, at 46", the separation of the 57-Aql pair is wider than 56-Aql, the m12.3 companion of 56-Aql is easy to lose in the glare of the m5.8 primary.

Although 57-Aql lies in Aquila, the easiest way to find it is to start from the bright star 6-alpha^2 (α^2) Capricorni (Algedi), which lies in the NW corner of Capricornus. Place Algedi on the E edge of your finder field and look for the m6 star 63-Sgr, which appears prominently near the SW edge of the field. Move the finder W until 63-Sgr lies on the SE edge of the finder field and m5 51-Aquilae comes into view on the NW edge of the field. Move the finder N until 51-Aql is at the SW edge of the field and look for the m6 pair 56- and 57-Aquilae, which appear prominently near the center of the field.

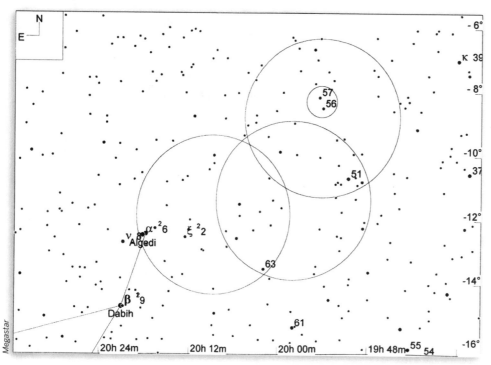

CHART 03-4.

57-Aquilae (STF 2594) (10° field, 5.0° finder fields, 1.0° eyepiece field, LM 9.0)

STF 2404	★	◊◊◊◊		MS	UD
Chart 03-2, 03-5		m6.4/7.4, 3.6", PA 181° (1991)		18h 50.8m	+10° 59'

STF 2404 is an unremarkable double star, located about 40' N of NGC 6709 (see Chart 03-2).

STF 2404 is easy to find, because it's located in the same low-power eyepiece field as NGC 6709, described at the beginning of this chapter. To locate STF 2404, begin by finding NGC 6709, as shown on Chart 03-2. Place NGC 6709 at the S edge of your low-power eyepiece field, and look for a prominent (m6) star in the NW quadrant of the field, as shown in Chart 03-5. Center that star in the field, and switch to an eyepiece that provides 90X or higher magnification to split the double.

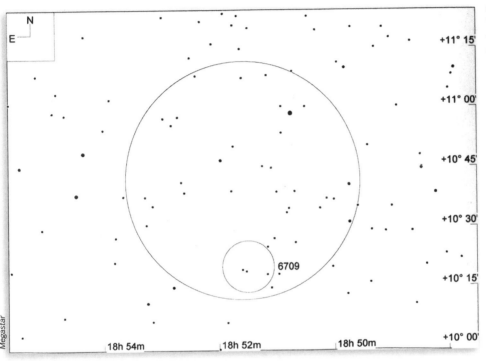

CHART 03-5.

STF 2404 (2° field, 1.0° eyepiece field, LM 11.0)

04
Aries, The Ram

NAME:	Aries (AIR-eez)
SEASON:	Autumn
CULMINATION:	midnight, 20 October
ABBREVIATION:	Ari
GENITIVE:	Arietis (AIR-ee-ET-us)
NEIGHBORS:	Cet, Per, Psc, Tau, Tri
BINOCULAR OBJECTS:	none
URBAN OBJECTS:	5-gamma (STF 180)

Aries is an inconspicuous constellation, but its position west of the Pleiades in Taurus, and south of Triangulum, makes it easy to locate. Aries is a medium-size constellation, ranking 39th of the 88 constellations. It covers 441 square degrees of the celestial sphere, or about 1.1%. Aries is an ancient constellation, well-known to the Greeks and Romans of antiquity.

Because it is located on the ecliptic, Aries is one of the twelve zodiacal constellations. In the time of the ancient Greeks, the Sun lay in Aries at the Vernal Equinox, the first day of Spring, so Aries was considered the first of the zodiacal constellations. As a result of *precession* (changes over time in the direction that Earth's rotational axis points in the celestial sphere) over the last few thousand years, the Sun now lies in Pisces at the Vernal Equinox, but Aries is still considered the first of the zodiacal constellations.

In Greek mythology, it was the golden fleece of Aries the Ram that was sought by Jason and the Argonauts. King Athemus of Thessaly took the young and beautiful Ino as his second wife. Ino, jealous of Athemus's first wife, Nephele, as well as of her son, Phryxus, and her daughter, Helle, plotted to dispose of the competition. Ino engineered a crop failure and suborned the messenger sent by Athemus to an oracle for advice about how to deal with the problem. Bribed by Ino, the messenger reported to Athemus that the oracle advised sacrificing Phryxus to appease the gods and avoid famine.

Regretfully, Athemus decided to take the oracle's advice and kill his son. But Nephele appealed to Zeus, who sent Aries the Ram to rescue Phryxus and his sister Helle in the nick of time. Unfortunately, as the royal children rode Aries to safety, Helle fell off and plunged to her death, drowning in the straits that separate Europe and Asia, which are still known as the Hellespont. Phryxus arrived safely in Colchis, the land of King Aeetes. Phryxus, not known for remembering his friends, promptly slew Aries as a sacrifice to Zeus and hung the fleece in a sacred grove, where it turned golden. The fleece was guarded by a fierce dragon, and remained safe until the arrival of Jason, who stole it.

The most prominent stars in Aries—Hamal, Sheratan, and Mesarthim—lie in the far western part of the constellation, where they form the head of the Ram. Aries contains only three featured objects, a galaxy and two double stars. Fortunately, all of these objects are easy to find and easy to see.

TABLE 04-1.

Featured star clusters, nebulae, and galaxies in Aries

Object	Type	Mv	Size	RA	Dec	M	B	U	D	R	Notes
NGC 772	Gx	11.1	7.2 x 4.2	01 59.3	+19 01					◉	Class SA(s)b; SB 13.6

TABLE 04-2.

Featured multiple stars in Aries

Object	Pair	M1	M2	Sep	PA	Year	RA	Dec	UO	DS	Notes
5-Gamma	STF 180	3.9	3.9	7.8	1	1994	01 53.5	+19 18	◉	◉	Mesartim
9-Lambda	n/a	4.8	7.3	36	50	n/a	01 57.9	+23 36		◉	SAO75051 / 75054

CHART 04-1.

The constellation Aries (field width 25°)

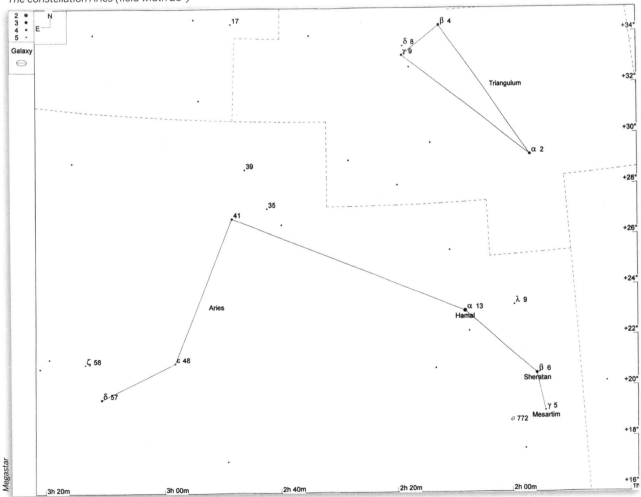

Clusters, Nebulae, and Galaxies

NGC 772	★★	❋❋❋	GX	MBUDR
Chart 04-2	Figure 04-1	m11.1, 7.2' x 4.2'	01h 59.3m	+19° 01'

NGC 772 is a small, relatively bright galaxy. At 42X in our 10" scope, NGC 772 is visible as a small, very faint, hazy patch surrounding a brighter core. At 125X with averted vision, the bright core is visible with some hint of a stellar nucleus. Very faint nebulosity with some hint of mottling is visible extending elliptically about 2' SE-NW.

To locate NGC 772, place Sheratan and Mesarthim in your finder, as shown in Chart 04-2. The m5 star 8-iota (ι) Ari appears prominently in the SW quadrant of the field, with m6 15-Ari near the eastern edge. Bump the finder about 15' dead W to center the crosshairs on NGC 772.

FIGURE 04-1.

NGC 772 (60' field width)

Image reproduced from Digitized Sky Survey courtesy Palomar Observatory and Space Telescope Science Institute

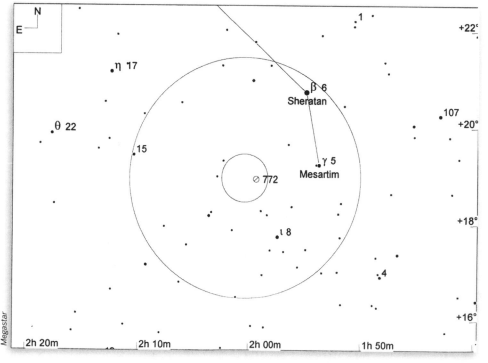

CHART 04-2.

NGC 0772 (10° field width, 5.0° finder field, 1.0° eyepiece field, LM 9.0)

Multiple Stars

5-gamma (STF 180)	★★★	✦✦✦✦	MS	UD
Chart 04-1		m3.9/3.9, 7.8", PA 1° (1994)	01h 53.5m	+19° 18'

The naked-eye star Mesarthim, also known as 5-gamma (γ) Arietis or STF 180, is a beautiful, evenly-matched, close pair of blue-white stars. At 125X in our 10" scope, the pair splits cleanly, floating in a rich field of m8-12 stars.

9-lambda	★★	✦✦✦✦	MS	UD
Chart 04-3		m4.8/7.3, 36", PA 50°	01h 57.9m	+23° 36'

9-lambda (λ) Arietis is a wide, easy pair with a yellowish m4.8 primary and a bluish m7.3 companion. The pair splits easily at 42X in our 10" scope. A wider m9/11 pair lies just to the E.

To locate 9-lambda (λ) Ari, place Hamal and Sheratan in the finder, oriented as shown in Chart 04-3. At m4.8, 9-lambda (λ) Ari shines brightly just N of center in the finder, with m6 7-Ari about 0.5° dead W.

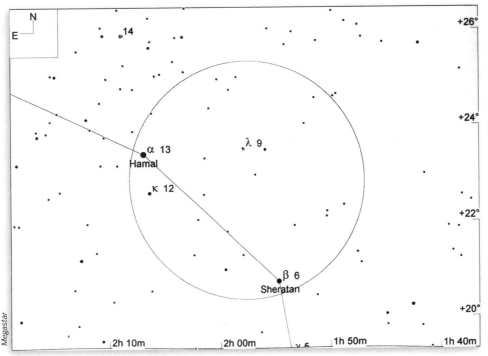

CHART 04-3.

9-Lambda (λ) Arietis (10° field width, 5.0° finder field, LM 9.0)

05

Auriga, The Charioteer

NAME:	Auriga (uh-RYE-guh; ore-RYE-guh)
SEASON:	Late autumn
CULMINATION:	midnight, 9 December
ABBREVIATION:	Aur
GENITIVE:	Aurigae (uh-RYE-GUY; ore-RYE-GUY)
NEIGHBORS:	Cam, Gem, Lyn, Per, Tau
BINOCULAR OBJECTS:	NGC 1893, NGC 1907, NGC 1912 (M 38), NGC 1960 (M 36), NGC 2099 (M 37), NGC 2281
URBAN OBJECTS:	NGC 1912 (M 38), NGC 1960 (M 36), NGC 2099 (M 37), NGC 2281, 37-theta

Auriga is a prominent, large constellation, ranking 21st in size among the 88 constellations. It covers 657 square degrees of the celestial sphere, or about 1.6%. Auriga is an ancient constellation, well-known to the Greeks and Romans of antiquity, although they did not associate Auriga with a chariot or charioteer. The chariot association arose much earlier, at least as early as the Babylonians, and perhaps as early as the Sumerians. The Greeks and Romans instead saw Auriga as a shepherd, leading or carrying a she-goat and her kids, represented by the stars 8-zeta (ζ) Aurigae and 10-eta (η) Aurigae. The brightest star in Auriga was known to the Greeks as Amaltheia, the name of the she-goat that suckled the infant Zeus. The current proper name of that star, Capella, is of Roman origin, and means "she-goat" in Latin.

The most prominent stars in Auriga, including 0th-magnitude Capella, form an unmistakable pentagon that lies embedded in a rich area of the winter Milky Way. Opposite Capella at the far southern corner of the pentagon lies the bright 2nd-magnitude star Alnath, which for historical reasons has the distinction of belonging to two constellations. In addition to being designated gamma (γ) Aurigae, Alnath is also known as 112-beta (β) Tauri.

TABLE 05-1.

Featured star clusters, nebulae, and galaxies in Auriga

Object	Type	Mv	Size	RA	Dec	M	B	U	D	R	Notes
NGC 2099	OC	5.6	23.0	05 52.3	+32 33	◉	◉	◉			M 37; Cr 75; Class II 1 r or I 2 r
NGC 1960	OC	6.0	12.0	05 36.3	+34 08	◉	◉	◉			M 36; Cr 71; Class I 3 r
NGC 1931	OC/RN/EN	10.1	6.0	05 31.4	+34 15					◉	Cr 68; Stock 9; Class I 3 p n
NGC 1912	OC	6.4	21.0	05 28.7	+35 51	◉	◉	◉			M 38; Cr 67; Class II 2 r
NGC 1907	OC	8.2	6.0	05 28.1	+35 20				◉		Cr 66; Mel 35; Class I 1 m n
NGC 1893	OC	7.5	11.0	05 22.8	+33 25				◉		Cr 63; Mel 33; Class II 3 r n
NGC 2281	OC	5.4	14.0	06 48.3	+41 05				◉	◉	Cr 116; Mel 51; Class I 3 m

TABLE 05-2.

Featured multiple stars in Auriga

Object	Pair	M1	M2	Sep	PA	Year	RA	Dec	UO	DS	Notes
37-theta	STF 545	2.7	9.2	130.7	350	1924	05 59.7	+37 13	◉		

Auriga contains many relatively bright stars, well distributed, which serve as guideposts for locating the DSOs in Auriga. Due in no small part to its location in the winter Milky Way, Auriga is home to many magnificent open clusters, including the superb Messier open clusters M 36, M 37, and M 38. Auriga is at the anti-center of the Milky Way, which means it lies in the plane of our galaxy, but is more distant from the galactic center than we are. When we look at Auriga, we are looking out toward the edge of our galaxy, so the Milky Way banding that is so prominent in constellations like Cygnus and Sagittarius—which lie between us and the galactic center—is much less prominent.

The bright pentagon of Auriga is easy to find, anchored as it is by the 0th-magnitude star Capella and lying just north of and halfway along a line between the bright pair Castor and Pollux in Gemini and the Pleiades (M 45) in Taurus. For observers at mid-northern latitudes, Auriga is well placed for evening observing between mid-autumn, when it rises soon after dusk, to mid-spring, when it sets a couple hours after dark.

CHART 05-1.

The constellation Auriga (field width 50°)

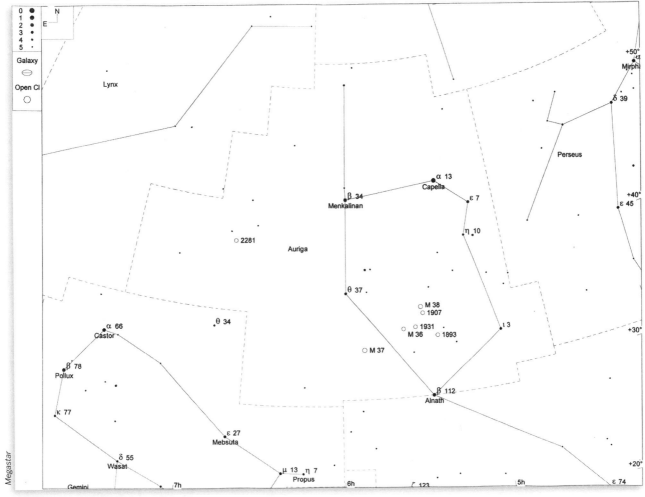

Clusters, Nebulae, and Galaxies

We begin with a sextet of open clusters that we personally call the Heart of Auriga, shown in Chart 05-2. Other than the belt and sword region of Orion, which is without equal, we consider the Heart of Auriga the most impressive single region in the winter sky. This region includes the magnificent Messier clusters M 36, M 37, and M 38, all of which are prominent in a 50mm finder scope. In fact, if your binocular or finder scope has a 6° or wider field of view, it's possible to view all three Messier clusters at the same time.

Two of the three smaller, dimmer open clusters in this region—NGC 1893 and NGC 1907—are also relatively easy binocular objects. Only NGC 1931 is beyond the reach of a standard binocular. But merely because five of these objects can be viewed with a binocular doesn't mean that's the only way you should view them. All six of these objects reveal a wealth of detail in even small telescopes.

CHART 05-2.

The Heart of Auriga (15° field width, 5.0° finder fields, 1.0° eyepiece field, LM 9.0)

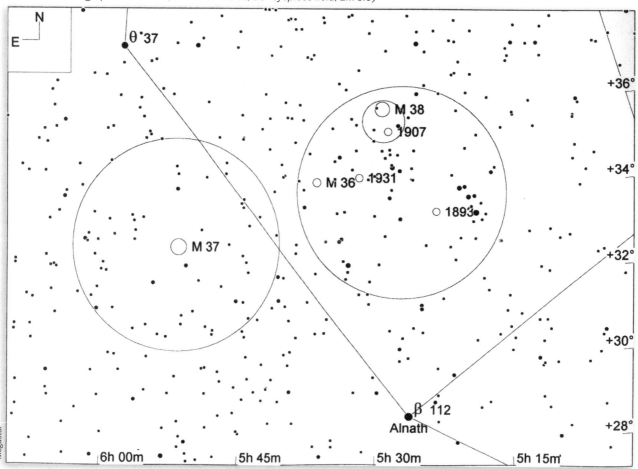

M 37 (NGC 2099)	★★★★	✦✦✦✦	OC	MBUDR
Chart 05-2	Figure 05-1	m5.6, 23.0'	05h 52.3m	+32° 33'

Begin with M 37, which we consider to be the finest open cluster in the celestial sphere. In a 50mm finder or binocular, M 37 is visible as a large, bright nebulosity with no individual stars visible with direct vision. M 37 is variously cataloged as a Trumpler Class I 2 r or II 1 r cluster (we think the latter classification is proper). At 42X in our 10" scope, M 37 stands out prominently from the surrounding star field. At 90X, M 37 is simply magnificent, a detached, rich cluster with 100+ stars visible ranging from m9 to m13. The center of the cluster is anchored by two m9 stars, with a thick chain of m10/11 stars running west-east, tapering off E of center.

M 37 is very easy to find. Imagine a line running SW from 37-theta (θ) Aurigae, at the SE corner of the pentagon to 112-beta (β) Tauri, at the S corner. Place your finder not quite halfway along that line, a bit closer to 37-theta (θ) Aurigae, and move it about half a field SE. M 37 shows prominently in the finder as a large, bright, hazy patch. Center that patch in your crosshairs, and you're off to the races.

FIGURE 05-1.

NGC 2099 (M 37) (60' field width)

Image reproduced from Digitized Sky Survey courtesy Palomar Observatory and Space Telescope Science Institute

M 36 (NGC 1960)	★★★★	✦✦✦✦	OC	MBUDR
Chart 05-2	Figure 05-2	m6.0, 12.0'	05h 36.3m	+34° 08'

M 36 is about half the size of M 37, a bit dimmer, and shows more variation in brightness among the member stars. Although it is variously cataloged as a Trumpler Class I 3 r or II 3 m cluster, M 36 actually appears in typical amateur scopes as a class I 3 p (poor) cluster, with only about 40 stars visible from m9 to m13. Don't let that put you off, though. Despite the relatively small number of stars visible, M 36 is a very impressive open cluster.

In a 50mm finder or binocular, M 36 appears as a medium-size bright nebulosity with no individual stars visible with direct vision. At 90X in our 10" scope, M 36 stands out prominently, detached from the surrounding star field. Perhaps because it is much sparser than M 37, M 36 appears less concentrated to us, despite its Class I rating. In contrast to the relatively even distribution of stars in M 37, the stars in M 36 are grouped in knots and clumps.

To locate M 36, simply move your finder half a field or so NNW (north northwest) from M 37 until M 36 slides into view.

FIGURE 05-2.

NGC 1960 (M 36) (60' field width)

Image reproduced from Digitized Sky Survey courtesy Palomar Observatory and Space Telescope Science Institute

NGC 1931	★★	✸✸✸✸	OC/RN/EN	MBUDR
Chart 05-2	Figure 05-3	m10.1, 6.0'	05h 31.4m	+34° 15'

NGC 1931 is cataloged as an open cluster, Trumpler Class I 3 p n, but it also incorporates an emission/reflection nebula. At 42X in our 10" scope, NGC 1931 is visible as a small, moderately bright, circular patch that appears nebular. At 125X, a tight group of three relatively bright stars is visible, with several much dimmer stars embedded in the nebulosity. A narrowband filter reveals no additional extent or detail in the nebulosity.

NGC 1931 is easy to find, because it is located 1° dead W of M 36, and visible in the same low-power eyepiece field.

FIGURE 05-3.

NGC 1931 (60' field width)

Image reproduced from Digitized Sky Survey courtesy Palomar Observatory and Space Telescope Science Institute

M 38 (NGC 1912)	★★★★	✸✸✸✸	OC	MBUDR
Chart 05-2	Figure 05-4	m6.4, 21.0'	05h 28.7m	+35° 51'

M 38 is the second most impressive of the Messier Auriga clusters, after M 37. M 38 is cataloged as Trumpler Class II 2 r or III 2 r, reflecting its lack of central concentration. In a 50mm finder or binocular, M 38 appears as a large, bright nebulosity with no individual stars visible with direct vision. At 90X in our 10" scope, M 38 stands out prominently, detached from the surrounding star field, with numerous m9/10 stars embedded in a scatter of 100+ m11 and dimmer stars.

To locate M 38 from M 36, simply center your finder on M 36 and look for M 38 in the NW quadrant of the finder field.

FIGURE 05-4.

NGC 1912 (M 38) (60' field width)

Image reproduced from Digitized Sky Survey courtesy Palomar Observatory and Space Telescope Science Institute

NGC 1907	★★★	✺✺✺✺	OC	MBUDR
Chart 05-2	Figure 05-5	m8.2, 6.0'	05h 28.1m	+35° 20'

NGC 1907 is a fine, compact Trumpler Class I 1 m n open cluster located about 30' SSW (south southwest) of M 38, and visible in the same low-power eyepiece field. In a 50mm finder or binocular, NGC 1907 is visible as a small, moderately bright nebulosity. At 42X in our 10" scope, NGC 1907 is visible as a nebulous patch with a dozen or so m9/10 stars resolved. At 90X, the nebulosity persists, but with 25+ m9 through m12 stars resolved. A prominent m9 field star lies just NE of the cluster, and a closely-spaced m9.5 double on the SSE edge of the cluster. With M 38 and NGC 1907 on opposite sides of the eyepiece field, the contrast between the two clusters is striking. M 38 appears loose and scattered compared to the tightly concentrated, almost globular appearance of NGC 1907. (If you like the contrasting appearance of these two clusters, compare these two clusters with M 35 and NGC 2158 in Gemini.)

NGC 1907 is easily found, lying just half a degree SSW of M 38.

FIGURE 05-5.

NGC 1907 (60' field width)

Image reproduced from Digitized Sky Survey courtesy Palomar Observatory and Space Telescope Science Institute

NGC 1893	★★★	✺✺✺✺	OC	MBUDR
Chart 05-2	Figure 05-6	m7.5, 11.0'	05h 22.8m	+33° 25'

NGC 1893, cataloged as Trumpler Class II 3 r n or II 2 m n, is a medium-size, moderately bright open cluster located in a rich Milky Way star field. In a 50mm finder or binocular, the cluster is visible as a faint, small nebulous patch. At 90X in our 10" scope, the cluster nestles in a prominent 11' equilateral triangle of m9 stars, with another m9 star 4' NE. About 30 m9 to m12 stars form an oval, elongated N-S. NGC 1893 is embedded in the faint emission nebula IC 410. At 90X, no hint of IC 410 is visible without filtration. Using an O-III filter and averted imagination, it's just possible to glimpse a hint of IC 410 nebulosity just outside the NW edge of the cluster.

NGC 1893 is easy to find. Move your finder SSW from M 38 until it is on the NNW edge of the finder field, or just past it. A prominent 1.5° NE-SW chain of four m5 stars appears in the W half of the finder field. NGC 1893 is visible as a dim, hazy patch 1° dead E of 16-Aur, the middle star in that chain.

FIGURE 05-6.

NGC 1893 (60' field width)

Image reproduced from Digitized Sky Survey courtesy Palomar Observatory and Space Telescope Science Institute

NGC 2281	★★★	◉◉◉	OC	MBUDR
Chart 05-3	Figure 05-7	m5.4, 14.0'	06h 48.3m	+41° 05'

NGC 2281 is a loose, bright, medium-size open cluster, cataloged as Trumpler Class I 3 m. In a 50mm finder or binocular, NGC 2281 is visible as a moderately bright hazy patch with one m8 star resolved at the center of the cluster. At 90X with averted vision in our 10" scope, a chain of a dozen m9/10 stars is visible extending to the E and NE of center, with about 20 more m11 through m13 filling out the cluster. The W and S regions of the cluster are very sparse. An m9 star lies on the W edge of the cluster, and another m9 star on the S edge.

NGC 2281 might have been a difficult object to locate, except that it is visible in a 50mm finder and is surrounded by numerous m5 stars, all of which are assigned the Bayer designation of psi (ψ). To locate NGC 2281, place Menkalinen on the WNW edge of the finder field, with the prominent m6 pair 38- and 39-Aurigae on the SW edge. Pan the finder ESE (east southeast) a full field or so, looking for the prominent m5 psi (ψ) group of stars. Center m5 50-psi^2 (ψ^2) Aurigae in the finder field, and locate m5 58-psi^7 (ψ^7) Aur near the SE edge of the finder field. NGC 2281 lies about 1° SSW of 58-psi^7 (ψ^7) Aur, and shows weakly in a 50mm finder.

FIGURE 05-7.

NGC 2281 (60' field width)

Image reproduced from Digitized Sky Survey courtesy Palomar Observatory and Space Telescope Science Institute

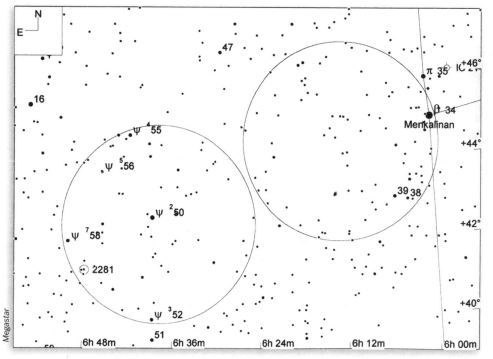

CHART 05-3.

NGC 2281 (12° field width, 5.0° finder fields, LM 9.0)

Multiple Stars

37-theta (STF 545)	★	◊◊◊◊	MS	UD
Chart 05-1		m2.7/9.2, 130.7", PA 350°	05h 59.7m	+37° 13'

37-theta (θ) Aurigae, the star that marks the eastern corner of the pentagon, is actually a physical double, but that is not the object cataloged in the Astronomical League Urban Observing club list. The physical double comprises an m2.7 primary with a much dimmer companion located only 3.7" NW. The great disparity in brightness makes this an extremely difficult split in our 10" scope. At 125X on a night of excellent seeing, 37-theta appears slightly elongated to the NW. At 250X, it is possible to achieve a clean split.

The double star listed by the Urban Observing club is STF 545, which comprises the m2.7 primary 37-theta (θ) Aurigae and an m9.2 companion lying 2.2' N by NNW. This pair is easily split at low magnification in even the smallest scope.

06

Boötes, The Herdsman

NAME: Boötes (boe-OAT-ees)	
SEASON: Spring	
CULMINATION: midnight, 30 April	
ABBREVIATION: Boo	
GENITIVE: Boötis (boe-OAT-is)	
NEIGHBORS: CrB, Com, CVn, Dra, Her, Ser, UMa, Vir	
BINOCULAR OBJECTS: none	
URBAN OBJECTS: none	

Boötes is a large constellation, ranking 13th in size among the 88 constellations. It covers 907 square degrees of the celestial sphere, or about 2.2%. Despite its size, Boötes is not prominent. The constellation is anchored by 0th magnitude Arcturus, but has only a handful of other stars brighter than 4th magnitude.

Boötes is an ancient constellation that has always been closely associated with neighboring Ursa Major in one way or another. The ancient Greeks named this constellation Boötes ("oxen driver") because they saw the Big Dipper asterism as an ox-drawn cart with Boötes holding the reins. To the ancient Romans, Boötes was the son of Jove and Callisto. They imagined (and named) Ursa Major as the Great Bear, with Boötes pursuing that bear around the heavens.

Boötes lies far from the Milky Way, and so lacks the open clusters and nebulae associated with the Milky Way. When we look at Boötes, we are looking away from the plane of the Milky Way and into intergalactic space, which is populated almost exclusively with dim external galaxies. The only featured DSO in Boötes is the

TABLE 06-1.

Featured star clusters, nebulae, and galaxies in Boötes

Object	Type	Mv	Size	RA	Dec	M	B	U	D	R	Notes
NGC 5466	GC	9.2	9.0	14 05.4	+28 32					◉	Class XII

TABLE 06-2.

Featured multiple stars in Boötes

Object	Pair	M1	M2	Sep	PA	Year	RA	Dec	UO	DS	Notes
17-kappa	STF 1821	4.5	6.6	14.0	237	1998	14 13.5	+51 47		◉	
21-iota	STF 26	4.8	12.6	86.7	194	1925	14 16.2	+51 22		◉	
29-pi	STF 1864	4.9	10.4	128.0	163	1995	14 40.7	+16 25		◉	
36-epsilon	STF 1877	2.7	12.0	175.5	256	1988	14 45.0	+27 04		◉	Izar
37-xi	STF 1888	4.7	12.6	282.7	100	1932	14 51.4	+19 06		◉	
49-delta	STF 27	3.5	7.8	103.7	78	1998	15 15.5	+33 19		◉	
51-mu	STF 28	4.3	6.5	109.1	171	1996	15 24.5	+37 23		◉	Alkalurops

globular cluster NGC 5466, which is notable for being a Shapley-Sawyer Concentration Class XII cluster—the loosest, sparest category—and resembles a tight open cluster rather than a typical globular cluster.

Boötes is easy to find, anchored by the 0th-magnitude star Arcturus. To locate Arcturus, extend the curve of the handle of the Big Dipper to "arc to Arcturus" about 30° to the SSE (south southeast). The prominent stars in Boötes form an unmistakable kite-shaped pattern to the NNW (north northwest) of Arcturus. For observers at mid-northern latitudes, Boötes is well-placed for evening observing between early spring, when it rises soon after dusk, to mid-summer, when it sets a couple hours after dark.

CHART 06-1.

The constellation Boötes (field width 45°; North to right)

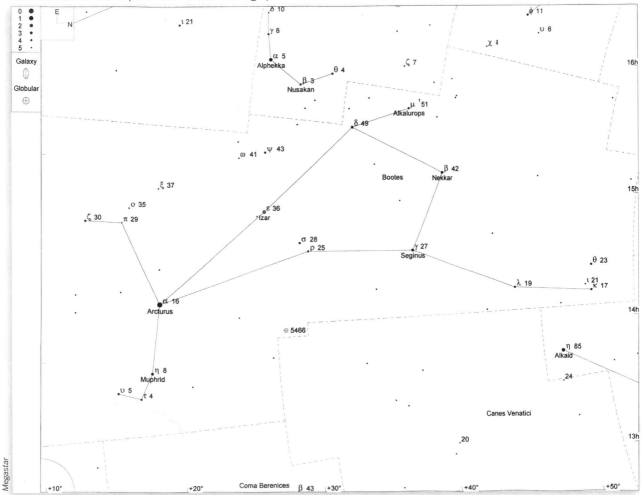

Clusters, Nebulae, and Galaxies

NGC 5466	★★	✸✸✸	GC	MBUDR
Chart 06-2	Figure 06-1	m9.2, 9.0'	14h 05.4m	+28° 32'

NGC 5466 is a large, dim, very loose globular cluster. Despite its cataloged visual magnitude of 9.2, the large extent of this object means it has very low surface brightness, and accordingly can be very difficult to see if any light pollution is present. At 42X in our 10" scope, NGC 5466 is invisible to us, even with averted vision. At 125X and 180X, the cluster becomes visible with averted vision as a large, faint, hazy patch with a dozen or so extremely faint stars resolvable. Visually, NGC 5466 more resembles a dim, sparse, loose open cluster than a typical globular cluster.

To locate NGC 5466, put the m4 stars 25-rho (ϱ) Boötis and 28-sigma (σ) Boötis on the NE edge of the finder field. Move the finder dead W about a full field, until m5 9-Boötis comes into view near the WSW (west southwest) edge of the field. Look near the center of the finder for an m7 star, which appears prominently in a field of dimmer stars. NGC 5466 is located 20' WNW (west northwest) of that star.

FIGURE 06-1.

NGC 5466 (60' field width)

Image reproduced from Digitized Sky Survey courtesy Palomar Observatory and Space Telescope Science Institute

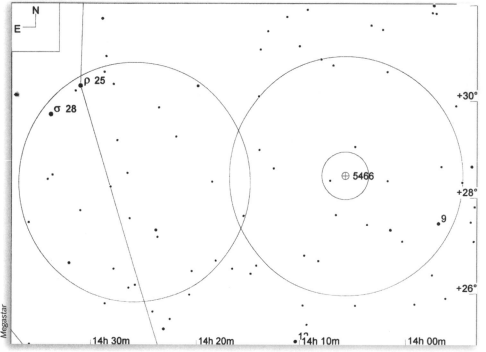

CHART 06-2.

5466 (10° field width, 5.0° finder fields, 1.0° eyepiece field, LM 9.0)

Multiple Stars

All of the featured multiple stars in Boötes are visible to the naked eye from a reasonably dark location, and so are easy to find.

17-kappa (κ) (STF 1821) ★★★	❂❂❂❂		MS	UD
Chart 06-1	m4.5/6.6, 14", PA 237° (1998)		14h 13.5m	+51° 47'

Located about 5° ENE (east northeast) of Alkaid, the last star in the handle of the Big Dipper, 17-kappa (κ) Boötis (STF 1821) is a bright, wide pair that is easily split even at medium power in a small telescope. At 90X in our 10" scope, the primary is a warmish white and the companion shows a slight blue tinge.

21-iota (ι) (STF 26) ★★	❂❂❂❂		MS	UD
Chart 06-1	m4.8/12.6, 86.7", PA 194° (1925)		14h 16.2m	+51° 22'

Located in the same medium-power eyepiece field as 17-kappa (κ) Boötis, 21-iota (ι) (STF 26) is actually a triple star. The m4.8 primary is pure white. The m8.6 secondary lies 38.6" NE at PA 36°, and has a distinct yellowish tinge. The m12.6 tertiary—which, with the primary, actually comprises the STF 26 pair—lies about 1.5' almost dead S of the primary. The primary and secondary are easily split with a 50mm binocular. The very dim tertiary is invisible in our 90mm refractor at 200X, but is detectable with some difficulty at 250X in our 10" scope. Although the tertiary is cataloged as m12.6, it appears more like m13.5 to us.

29-pi (π) (STF 1864) ★★	❂❂❂❂		MS	UD
Chart 06-1	m4.9/10.4, 128.0", PA 163° (1995)		14h 40.7m	+16° 25'

Lying 6.5° ESE (east southeast) of Arcturus, 29-pi (π) (STF 1864) is another triple star system. At 90X in our 10" scope, the m4.9 primary, π^1, is bright white with a slight blue tinge. The m5.8 secondary, π^2, is of similar color, and lies 6.1" ESE at PA 109°. The m10.4 tertiary—which, with the primary, actually comprises the STF 1864 pair—lies 128" SSE at PA 163°.

36-epsilon (ε) (STF 1877) ★★★	❂❂❂❂		MS	UD
Chart 06-1	m2.7/12.0, 175.5", PA 256° (1988)		14h 45.0m	+27° 04'

Izar, 36-epsilon (ε) (STF 1877), is another triple star system. At 125X in our 10" scope, the m2.7 primary, ε^1, is distinctly yellow. The m5.1 secondary, ε^2, is blue-white, and lies 2.8" NNW at PA 340°. The m12.0 tertiary—which, with the primary, actually comprises the STF 1877 pair—lies about 2.9' WSW at PA 256°. The tertiary seems dimmer than its cataloged magnitude of 12.0. We estimate it at m13.

37-xi (ξ) (STF 1888) ★★★	❂❂❂❂		MS	UD
Chart 06-1	m4.7/12.6, 282.7", PA 100° (1932)		14h 51.4m	+19° 06'

37-xi (ξ) (STF 1888) is a quadruple star. The AB pair are a yellow m4.7 primary and a red-orange m7.0 companion, separated by 6.5" at PA 316°. Although the tertiary (D) star is cataloged at m12.6 with 282.7" separation at PA 100°, the star at that location appears much, much brighter to us, perhaps m8.5. At 250X in our 10" scope, the m13.6 quaternary (C) star is visible with difficulty, lying about 1' NNW of the primary.

49-delta (δ) (STF 27)	★★	୧୧୧୧	MS	UD
Chart 06-1		m3.5/7.8, 103.7", PA 78° (1998)	15h 15.5m	+33° 19'

49-delta (δ) (STF 27) is a bright, wide pair, easily split even with a 50mm finder or binocular. The m3.5 primary is distinctly yellowish. The m7.8 companion lies 103.7" ENE at PA 78° and appears a warm white.

51-mu (μ) (STF 28)	★★★	୧୧୧୧	MS	UD
Chart 06-1		m4.3/6.5, 109.1", PA 171° (1996)	15h 24.5m	+37° 23'

Alkalurops, 51-mu (μ) (STF 28), is a fine triple star. At 90X in our 10" scope, the m4.3 primary, μ^1, is yellow-white. The m6.5 secondary, μ^2, is of similar color, and lies 109.1" S at PA 171°. At 125X, μ^2 shows elongation, revealing itself as a close double (cataloged as STF 1938). STF 1938 comprises a pair of yellowish stars of nearly equal brightness separated by about 2". We are able to split that pair cleanly at 250X.

07

Camelopardalis, The Giraffe

NAME: Camelopardalis (CAM-eh-low-PAR-duh-lis)

SEASON: Winter

CULMINATION: midnight, 23 December

ABBREVIATION: Cam

GENITIVE: Camelopardalis (CAM-eh-low-PAR-duh-lis)

NEIGHBORS: Aur, Cas, Cep, Dra, Lyn, Per, UMa, UMi

BINOCULAR OBJECTS: Stock 23, Kemble 1, NGC 2403

URBAN OBJECTS: Stock 23

Camelopardalis is a large, circumpolar constellation that covers an area of the sky that is devoid of bright stars. Camelopardalis ranks 18th in size among the 88 constellations, covering 757 square degrees of the celestial sphere, or about 1.8%. Despite its size, Camelopardalis is among the least prominent of the constellations. Its brightest star, 10-beta (β), is only magnitude 4.03, and the constellation contains only two other stars that are magnitude 4.5 or brighter.

Camelopardalis is one of the so-called "modern" constellations, and therefore has no legends or myths associated with it. The ancient Greeks and Romans, who felt no compulsion to include all areas of the celestial sphere in their constellations, ignored the area that is now Camelopardalis. It was not until 1624 that the German astronomer Jakob Bartsch published a book that gave this constellation its current name.

TABLE 07-1.

Featured star clusters, nebulae, and galaxies in Camelopardalis

Object	Type	Mv	Size	RA	Dec	M	B	U	D	R	Notes
Stock 23	OC	6.2	14.0	03 16.3	+60 02			◉	◉		Class II 3 p n
NGC 1502/ Kemble 1	OC/ AST	6.9/4	7.0/180	03 54.0	+63 21				◉		Cr 45; Class I 3 m / Kemble's Cascade
NGC 1501	PN	13.3	52"	04 07.0	+60 55					◉	Oyster Nebula; Camel's Eye Nebula; Class 3
NGC 2403	Gx	8.9	22.1 x 12.4	07 36.9	+65 36			◉	◉		Class Sc; SB 12.8
NGC 2655	Gx	11.0	6.6 x 4.8	08 55.6	+78 13				◉		Class SAB(s)0/a

TABLE 07-2.

Featured multiple stars in Camelopardalis

Object	Pair	M1	M2	Sep	PA	Year	RA	Dec	UO	DS	Notes
1	STF 550	5.8	6.9	10.3	308	1991	04 32.0	+53 55		◉	
32	STF 1694	5.4	5.9	21.5	329	1997	12 49.1	+83 25		◉	Laftwet

And that name has remained in some dispute. Although the IAU officially named the constellation Camelopardalis in 1933, it had previously long been known as Camelopardis, which name still appears in many astronomy books and other documents, including recent ones. Among amateur astronomers, the name Camelopardis is at least as popular as the official name, perhaps because many people find it easier to pronounce without the extra syllable. Whichever name you use, the genitive is always Camelopardalis.

Newbie astronomers often assume from the name, reasonably enough, that Camelopardalis must represent a camel. More experienced astronomers explain that Camelopardalis is named for the giraffe, which the Greeks called the leopard camel. But in fact Bartsch named Camelopardalis not just for a camel, but for the specific camel mentioned in Genesis 24:61, which carried Rebekah to meet her bridegroom Isaac, and many older sources represent the constellation as a camel rather than as a giraffe.

Despite its lack of prominence, Camelopardalis contains several interesting objects, including two bright open clusters, the famous chain of stars called Kemble's Cascade, a nice planetary nebula, and two bright galaxies. One of those galaxies, NGC 2403, is so bright that we're surprised that Messier missed it. In the telescopes of his time, it would have appeared very like a comet, and so would almost certainly have been added to his list had he observed it.

CHART 07-1.

The constellation Camelopardalis (field width 65°)

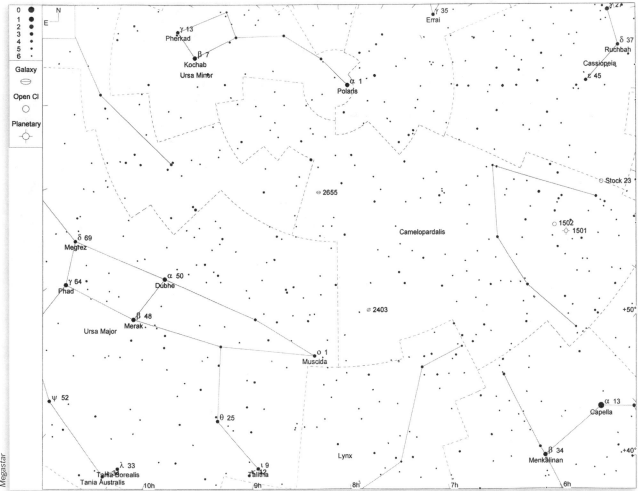

Most or all of Camelopardalis is circumpolar for observers at mid-northerly latitudes, which means that objects in the northern part of the constellation can be observed on any evening year-round. Objects in the southern part of the constellation are visible during evening hours from early autumn until early spring.

Clusters, Nebulae, and Galaxies

Stock 23	★★	✷✷✷	OC	MBUDR
Chart 07-2	Figure 07-1	m6.2, 14.0'	03 16h.3m	+60° 02'

Stock 23 is a moderately large, loose, moderately bright open cluster, cataloged as Trumpler Class II 3 p n.

With our 7X50 binoculars, Stock 23 is visible as a group of four m7/8 stars in a trapezoidal pattern about 5' on a side that shows prominently in a field devoid of bright stars, and resembles the keystone of Hercules. A fifth star of about m9 on the N edge of the pattern causes the trapezoid to take on the "house" or "ice cream cone" appearance of the constellation Cepheus, with the peak pointing NW. With averted vision, we're able to glimpse three additional m10 stars, but no hint of the reported nebulosity. An m9 star, which is actually a close double, lies about 10' SW of the center of the cluster, but is not a member.

From our urban site in our 10" scope at 90x, Stock 23 is visible as an obvious open cluster that stands out from the background star field. With averted vision, about 25 stars are visible down to m12, but no nebulosity is visible. A chain of three m10 stars extends W from the m7 central star.

Although Stock 23 lies in Camelopardalis on its border with Cassiopeia, the easiest way to locate it is to begin at the naked-eye stars 15-eta (η) Perseii and 23-gamma (γ) Perseii. Put 15-eta (η) on the W edge of your finder field and 23-gamma (γ) at the S edge, as shown in Chart 07-2. Move the finder NNE until the prominent m4 star SAO 24054 comes into view on the NE edge of the finder field, with an m5 star about 1.1° S of it. Stock 23 lies 1.6° dead W of SAO 24054, with the pattern of four stars visible in the finder as a tiny cluster.

FIGURE 07-1.

Stock 23 (60' field width)

Image reproduced from Digitized Sky Survey courtesy Palomar Observatory and Space Telescope Science Institute

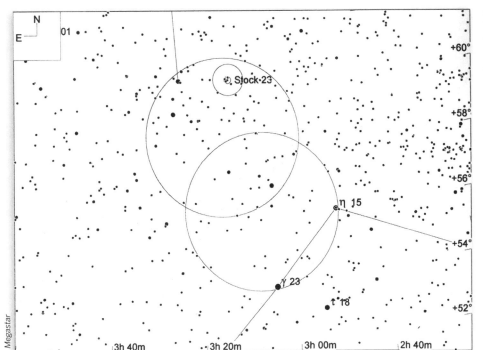

CHART 07-2.

Stock 23 (15° field width, 5.0° finder fields, 1.0° eyepiece field, LM 9.0)

Megastar

NGC 1502/Kemble 1	★★★	✹✹✹	OC/AST	MBUDR
Chart 07-3	Figure 07-2	m6.9/4, 7.0'/180'	3h 54.0m	+63° 21'

Named by Walter Scott Houston in honor of its discoverer, Canadian astronomer Fr. Lucien J. Kemble, Kemble's Cascade is a beautiful 2.5° chain of stars that extends NW from the open cluster NGC 1502. Kemble's Cascade is a chance arrangement of stars (an asterism) rather than a true cluster. Kemble's Cascade is best viewed with a binocular or other wide-field instrument.

With our 7X50 binoculars, Kemble's Cascade is visible as a chain of about 15 m8/9 stars extending NW from the m6 open cluster NGC 1502, which appears as a fuzzy m6 star. An m5 star features prominently near the middle of the chain, and an arc of three m5/6 stars lies just W of the NW end of the chain. The SE end of the chain is anchored by an m7 star that lies about 19' SW of NGC 1502.

With our 4.5" (114mm) f/4 wide-field reflector at 15X (3.5° field of view), NGC 1502 is resolved and Kemble's Cascade is spectacular. About 50 stars are visible in the Cascade, with the chain filled in and bulked out with numerous m10/11 stars. The brighter end of the chain that extends S from NGC 1502 is joined by a separate branch that extends SE from the fork at NGC 1502.

The easiest way to find Kemble's Cascade and NGC 1502 is to locate the prominent m4 star SAO 24054 as described for Stock 23 in the preceding section. Put SAO 24054 on the SW edge of your finder field. Kemble's Cascade is visible on the NE edge of the finder field.

FIGURE 07-2.

NGC 1502 and the southeastern segment of Kemble's Cascade (60' field width)

Image reproduced from Digitized Sky Survey courtesy Palomar Observatory and Space Telescope Science Institute

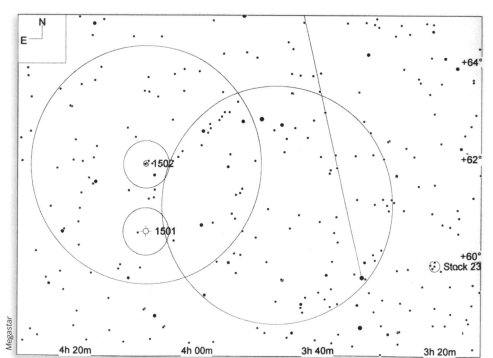

Megastar

NGC 1501	★★★	⊙⊙⊙	PN	MBUDR
Chart 07-4	Figure 07-3	m13.3, 52"	04h 07.0m	+60° 55'

NGC 1501, also known as the Oyster Nebula or Camel's Eye Nebula, is a moderately large, moderately bright planetary nebula, first logged by William Herschel in 1787. Visually, NGC 1501 resembles a smaller, dimmer, rounder version of the the Ring Nebula (M 57) in Lyra. Despite its relatively high magnitude, this object has fairly high surface brightness and is visible with direct vision even at low magnification. At 180X in our 10" scope, NGC 1501 appears as a small, bluish-green smoke ring extended slightly east-west, with some mottling visible with averted vision. The m14.3 central star flashes in and out at 180X, but is visible steadily at 250X. Unusually for a planetary nebula, this object responds equally well to a broadband or O-III filter. Although the object is clearly visible without a nebula filter, the filter enhances contrast dramatically and reveals additional subtle detail.

NGC 1501 is easily located by making an eyepiece starhop from NGC 1502, described in the preceding section. Place NGC 1502 at the N edge of your wide-field eyepiece, and continue to move the scope S as shown in Chart 07-4 until an m7 star appears prominently at the S edge of the field. NGC is visible as a fuzzy star 10' W of that star, and can be confirmed by blinking with your narrowband or O-III filter.

FIGURE 07-3.

NGC 1501 (60' field width)

Image reproduced from Digitized Sky Survey courtesy Palomar Observatory and Space Telescope Science Institute

CHART 07-4.

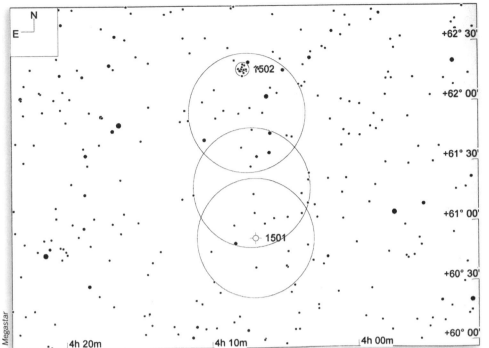

NGC 2403	★★	⚬⚬⚬	GX	MBUDR
Chart 07-5, 07-6	Figure 07-4	m8.9, 22.1' x 12.4'	07h 36.9m	+65° 36'

NGC 2403 is a face-on spiral galaxy that is an outlying member of the M 81/82 group. Bright for a galaxy at magnitude 8.9, NGC 2403 also has relatively high surface brightness at 12.8. It is visible with direct vision in our 7X50 binoculars and 9X50 finder as a moderately small hazy patch. Averted vision increases the apparent extent of the object, but reveals no additional detail. At 90X in our 10" scope, NGC 2403 is an oval halo of about 10' x 5' with slight central brightening and very faint mottling. No stellar core is visible.

To locate NGC 2403, put the m3 naked-eye star 1-omicron (o) UMa (Muscida) on the SSE (south southeast) edge of your finder field and look for an m6 star that appears prominently near the center of the field, with another m6 star prominent near the W edge of the field. Move the finder NW until the central star lies on the SE edge of the finder field, and look for another m6 star on the NW edge of the finder field. NGC 2403 lies 1° W of that star, and is easily visible in a 50mm finder.

FIGURE 07-4.

NGC 2403 (60' field width)

Image reproduced from Digitized Sky Survey courtesy Palomar Observatory and Space Telescope Science Institute

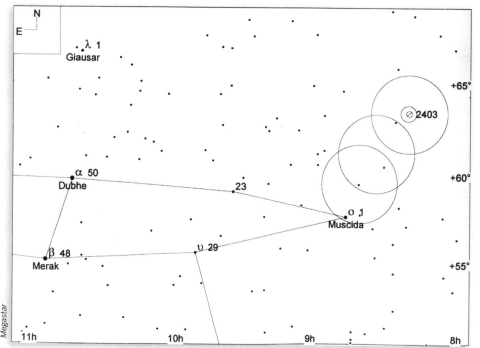

CHART 07-5.

NGC 2403 overview (30° field width, 5.0° finder

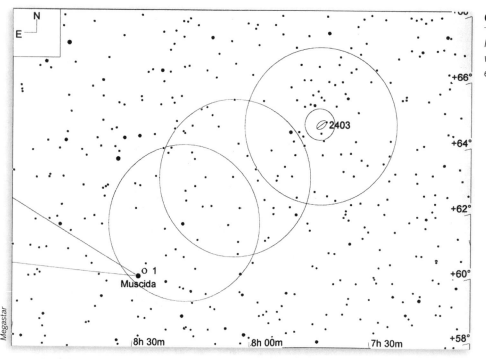

CHART 07-6.

NGC 2403 detail (15° field width, 5.0° finder fields, 1.0° eyepiece field, LM 9.0)

Blinking Planetary Nebulae

Even with moderately high magnification, small planetary nebulae may appear nearly stellar in the eyepiece, which can make it difficult to discriminate the nebula from the field stars that surround it. One easy way to identify the nebula is to "blink" it by passing your narrowband or O-III filter between your eye and the eyepiece. The filter dims background stars dramatically, but has little dimming effect on the nebula, which makes the nebula "blink" in and out as the filter passes back and forth under your eye.

NGC 2655	★★	⊖⊖	GX	MBUDR
Chart 07-7, 07-8	Figure 07-5	m11.0, 6.6' x 4.8'	08h 55.6m	+78° 13'

NGC 2655 is a moderately small, moderately bright galaxy. At 125X in our 10" scope, NGC 2655 shows a bright 3' x 2' halo elongated east-west with a distinctly brighter extended core.

NGC 2655 can be difficult to locate, because it lies in the middle of a large area that is mostly devoid of bright stars. The key to finding NGC 2655 is first to locate the m4 star SAO 1551, which is highlighted with a finder circle in Chart 07-7. SAO 1551 has a visual magnitude of 4.26, which means it's easily visible to the naked eye from a moderately dark site. SAO 1551 lies just W of the line from Polaris to Dubhe, and about a third of the way along it. Once you locate SAO 1551, center it in your finder cross-hairs and then move it to the NE edge of the finder field. NGC 2655 lies on the S edge of the finder field, about 9' NW of the E star in a widely-spaced E-W pair of m7 stars.

FIGURE 07-5.

NGC 2655 (60' field width)

Image reproduced from Digitized Sky Survey courtesy Palomar Observatory and Space Telescope Science Institute

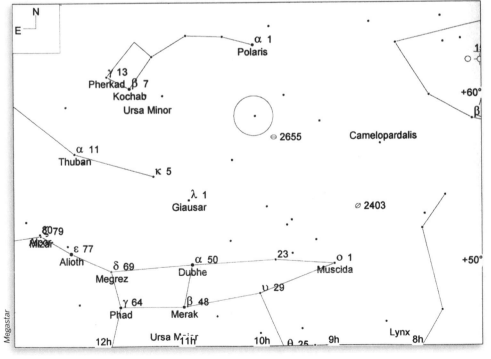

CHART 07-7.

NGC 2655 overview (60° field width, 5.0° finder field, LM 5.0)

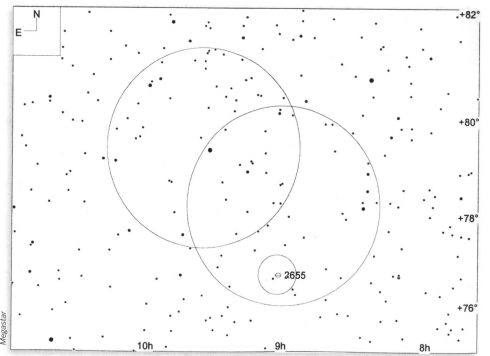

Megastar

Multiple Stars

1 (STF 550)	★★★	☙☙☙	MS	UD
Chart 07-9		m5.8/6.9, 10.3", PA 308°	04h 32.0m	+53° 55'

1-Camelopardalis (STF 550) is a pretty double, easily split in even a small scope. At 100X in our 90mm (3.5") refractor, is white and the companion a pale bluish-white. At 90X in our 10" reflector, both stars appear pure white.

Although STF 550 lies in Camelopardalis, the easiest way to locate it is to begin from the bright stars 47-lambda (λ) Per and 51-mu (π) Per. To do so, place 47-lambda (λ) and 51-mu (π) on the edge of the SW quadrant of your finder field, as shown in Chart 07-9, and move the finder NE until the prominent m5 pair 2- and 3-Camelopardalis come into view. STF 550 is the m7 star that lies about 1.3° WNW (west northwest) of that pair, and is by far the most prominent star in the vicinity.

32 (STF 1694)	★★★	☙☙☙	MS	UD
Chart 07-10		m5.4/5.9, 21.5", PA 329°	12h 49.1m	+83° 25'

32-Camelopardalis (STF 1694) is a beautiful, widely-spaced, evenly-matched pair. Although some observers report that the primary has a yellowish tinge and the companion a bluish tinge, we've always seen these stars as pristine white in instruments ranging from 50mm binoculars to a 17.5" reflector.

This double star has special meaning for us. Although only the International Astronomical Union (IAU) can officially name celestial objects, Robert chose to name this pair LAFTWET in honor of his late parents, Lenore Agnes Fulkerson Thompson and William Ewing Thompson (see http://www.ttgnet.com/daynotes/2003/2003-34.html#Monday).

32-Camelopardalis is somewhat difficult to locate with a traditional star-hop, but easy to locate using geometry. It forms nearly an equilateral triangle with Polaris and 22-epsilon (ε) UMi. Simply place your finder about where the apex of the triangle should be and look for an m5 star that appears prominently near the center of the field. That's 32-Camelopardalis.

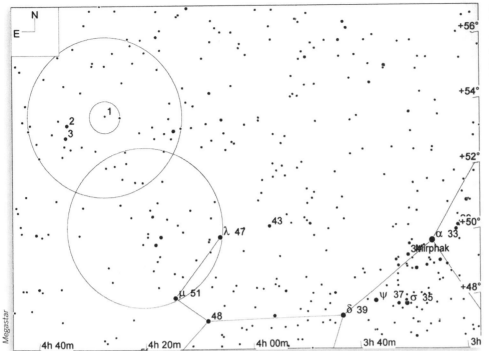

CHART 07-9.

1-Camelopardalis (STF 550) (15° field width, 5.0° finder fields, 1.0° eyepiece field, LM 9.0)

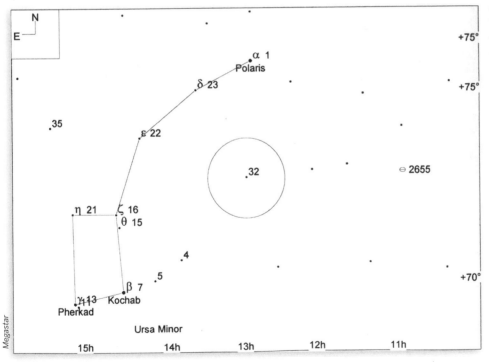

CHART 07-10.

32-Camelopardalis (STF 1694) (30° field width, 5.0° finder field; LM 5.0)

08
Cancer, The Crab

NAME: Cancer (CAN-sur)

SEASON: Winter

CULMINATION: midnight, 30 January

ABBREVIATION: Cnc

GENITIVE: Cancri (CAN-cree)

NEIGHBORS: CMi, Gem, Hya, Leo, Lyn

BINOCULAR OBJECTS: NGC 2632 (M44), NGC 2682 (M67)

URBAN OBJECTS: NGC 2632 (M44), NGC 2682 (M67)

Cancer is a dim winter constellation of medium size, ranking 31st in size among the 88 constellations. It covers 506 square degrees of the celestial sphere, or about 1.2%. Its brightest star is only fourth magnitude, and it has few naked-eye stars. Despite all of this, Cancer is an ancient constellation. In contrast to the 88 modern constellations, which cover the entire celestial sphere, the ancient constellations covered only areas of the sky with bright stars or interesting patterns. Areas with no prominent stars or other notable features were simply ignored.

We suspect that Cancer avoided this fate for two reasons. First, Cancer lies directly on the ecliptic (the apparent path of the sun's orbit across the celestial sphere, and the plane near which the planets lie), so the ancients probably felt compelled to assign that region, sparse though it is, to a Zodiacal constellation. Second, Cancer is home to the extraordinary open cluster M 44, also known as Praesepe or the Beehive. From the dark skies enjoyed by the ancients, M 44 is visible to the naked eye as a prominent nebulous patch three times the diameter of the full Moon.

In Greek mythology, Cancer the crab was the ally of the poisonous, nine-headed monster Hydra. With the help of Pallas Athena, the Hero Herakles located the monster's lair in the marsh at Lerna. Herakles and Hydra fought an evenly-matched battle, with Hydra regrowing a head each time Herakles cut one off. The goddess Hera, wife of Zeus, hated Herakles, and sent Cancer to nip at Herakles' feet to distract him. Herakles slew Cancer and then turned again to Hydra, cutting off its ninth (and immortal) head and finally killing it. Hera, taking pity on Cancer despite its failure to trip up Herakles, gave it a place in the heavens near Hydra.

The heart of Cancer lies between the Scyth of Leo to the east, the bright pair Castor and Pollux in Gemini to the northwest, and brilliant Procyon in Canis Minor to the southwest. Cancer contains only two featured DSOs, but both of them are spectacular open clusters. Cancer is best viewed during evening hours from winter through spring.

TABLE 08-1.

Featured star clusters, nebulae, and galaxies in Cancer

Object	Type	Mv	Size	RA	Dec	M	B	U	D	R	Notes
NGC 2632	OC	3.1	95.0	08 40.4	+19 40	◉	◉	◉			M 44; Cr 189; Class II 3 m
NGC 2682	OC	6.9	29.0	08 51.4	+11 49	◉	◉	◉			M 67; Cr 204; Class II 3 r

TABLE 08-2.

Featured multiple stars in Cancer

Object	Pair	M1	M2	Sep	PA	Year	RA	Dec	UO	DS	Notes
16-zeta*	STF 1196AB-C	5.1	6.2	5.9	70	2006	08 12.2	+17 39		◉	
48-iota	STF 1268	4.1	6.0	30.7	308	2003	08 46.7	+28 45		◉	

CHART 08-1.

The constellation Cancer (field width 40°)

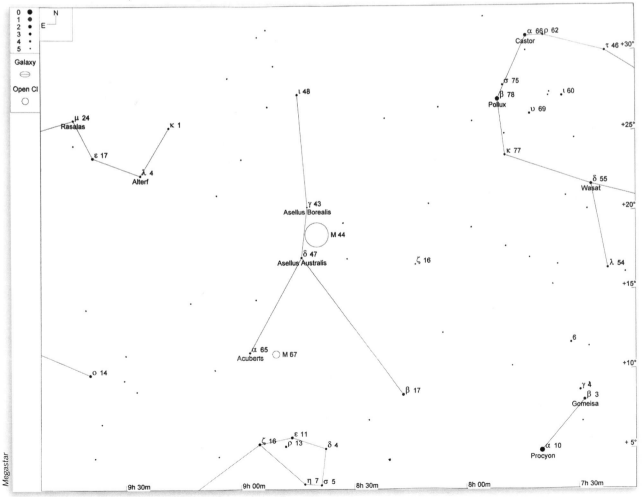

Clusters, Nebulae, and Galaxies

M 44 (NGC 2632)	★★★★	☙☙☙☙	OC	MBUDR
Chart 08-1	Figure 08-1	m3.1, 95.0'	08h 40.4m	+19° 40'

The naked-eye Trumpler Class II 3 m open cluster NGC 2632 (M 44) has been known since ancient times, under various names including the Beehive Cluster and Praesepe (The Manger). M 44 lies only about 525 light years from us. Because it is so close, M 44 ranks third in apparent size and brightness of all the open clusters visible to us, surpassed only by the Pleiades (M 45) and the Hyades, both in Taurus. At about 750 million years old—ten times the age of the Pleiades—M 44 is ancient for an open cluster.

Because it covers more than 1.5° of sky—more than three times the size of the full Moon—M 44 is best viewed with a binocular or at very low power in a short focal length telescope. With a 50mm binocular, about three dozen scattered stars from m6 to m9 are visible, with a prominent group of three m6 stars forming an 8' equilateral triangle near the center of the cluster. Unresolved dimmer stars give the cluster a nebulous appearance

At 31X in our 10" scope (true field 2.2°), there is barely enough margin around the cluster to put it into context as a cluster. At 42X (true field 1.7°), M 44 fills the entire field and appears simply as a large number of bright field stars, with no concentration or structure. The cluster comprises about 80 stars down to m14, with multiple doubles, triples, and chains of stars. A prominent m6/7 triple star lies near the center.

As a naked-eye object, M 44 is easy to locate. Look near the center of the triangle formed by Castor and Pollux in the NW, Procyon in the SW, and Regulus in the SE. Under moderately dark conditions, M 44 is a fairly prominent naked-eye object with direct vision. Even under moderately light-polluted suburban and exurban skies, M 44 is clearly visible with averted vision as a faint, hazy patch. M 44 stands out prominently with the slightest of optical aid, such as a small binocular.

FIGURE 08-1.

NGC 2632 (M 44) (60' field width)

Image reproduced from Digitized Sky Survey courtesy Palomar Observatory and Space Telescope Science Institute

M 67 (NGC 2682)	★★★★	♔♔♔♔	OC	MBUDR
Chart 08-2	Figure 08-2	m6.9, 29.0'	08h 51.4m	+11° 49'

NGC 2682 (M 67) is smaller and dimmer than M 44, but considerably richer, as indicated by its Trumpler Class II 3 r designation. Despite their identical Class II designations, a much stronger central concentration is apparent visually for M 67 than for M 44.

With a 50mm binocular, M 67 appears as a moderately bright nebulous patch, with only a few stars resolved with averted vision. At 42X in our 10" scope, M 67 comprises 100+ stars from m7 down to m14. The cluster is dominated by an m7 star in the NE quadrant and an m8 star SSW of center. A handful of m10 stars are grouped near the center of the cluster, with scores of m11 through m14 stars filling out the cluster. At 125X, the dimmer members seem to consolidate into numerous clumps and chains, with dark voids interspersed.

To locate M 67, simply move your finder about 8° SSE (south southeast) from M 44, as shown in Chart 08-2, until M 67 comes into view, 1.7° dead W of m5 65-alpha (α) Cancri (Acubens).

FIGURE 08-2.

NGC 2682 (M 67) (60' field width)

Image reproduced from Digitized Sky Survey courtesy Palomar Observatory and Space Telescope Science Institute

CHART 08-2.

NGC 2682 (M 67) (12° field width, 5.0° finder fields, 1.0° eyepiece field, LM 9.0)

Multiple Stars

16-zeta (ζ) (STF 1196AB-C)	★★★	❀❀❀		MS	UD
Chart 08-3		m5.1/6.2, 5.9", PA 70° (2006)		08h 12.2m	+17° 39'

16-zeta (ζ) Cancri (STF 1196AB-C) is a pretty visual double—what we call a "fast mover"—that is actually triple. As of early 2006, the AB-C pair is magnitude 5.1/6.2, separated 5.9" at a position angle of 70°. The AB pair appears straw yellow, and the C star warm white. At 90X in our 10" scope, the AB-C pair splits cleanly, but the AB pair appears single. At 180X, the AB pair shows a hint of elongation, which is more evident at 250X, but we were unable to split this pair cleanly.

To locate 16-zeta (ζ) Cnc, put M 44 on the E edge of the finder and move the finder dead W until m5 10-mu (μ) Cancri comes into view on the NW edge of the field and m5 16-zeta (ζ) at the S edge. Alternatively, simply pivot the finder most of a field WSW (west southwest) until 16-zeta (ζ) is centered in the crosshairs. 16-zeta (ζ) is by far the most prominent star in the near vicinity.

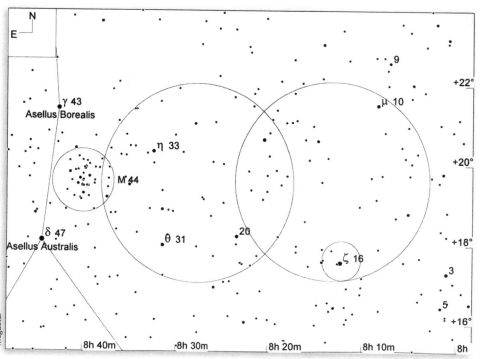

CHART 08-3.

16-Zeta (ζ) Cancri (12° field width, 5.0° finder fields, 1.0° eyepiece field, LM 9.0)

48-iota (ι) (STF 1268)	★★★★	✪✪✪✪	MS	UD
Chart 08-1		m4.1/6.0, 30.7", PA 308° (2003)	08h 46.7m	+28° 45'

48-iota (ι) Cancri (STF 1268) is a magnificent double star that we call the "Winter Albireo." The pair is easily split even with a small scope at low magnification. The m4.1 primary is a bright golden yellow and the m6 companion is distinctly blue with a greenish tinge. 48-iota (ι) Cnc is fourth magnitude, and reasonably prominent to the naked-eye. It lies 9° N of M 44.

09
Canes Venatici, The Hunting Dogs

NAME:	Canes Venatici (CAWN-es ven-AT-ih-see)
SEASON:	Spring
CULMINATION:	midnight, 7 April
ABBREVIATION:	CVn
GENITIVE:	Canum Venaticorum (CAWN-um ven-AT-ih-KOR-um)
NEIGHBORS:	Boo, Com, UMa
BINOCULAR OBJECTS:	NGC 4258 (M106), NGC 4736 (M94), NGC 5055 (M63), NGC 5194 (M51), NGC 5272 (M3)
URBAN OBJECTS:	none

Canes Venatici is a dim spring constellation of medium size, ranking 38th in size among the 88 constellations. It covers 465 square degrees of the celestial sphere, or about 1.1%. Its brightest star, Cor Caroli, is only third magnitude, and it has only a handful of other stars easily visible to the naked eye. Canes Venatici lies nestled beneath the handle of the Big Dipper, bounded on the east by Boötes, on the south by Coma Berenices, and on the north and west by Ursa Major.

TABLE 09-1.

Featured star clusters, nebulae, and galaxies in Canes Venatici

Object	Type	Mv	Size	RA	Dec	M	B	U	D	R	Notes
NGC 5194	Gx	9.0	10.3 x 8.1	13 29.9	+47 12	◉	◉				M 51; Class SA(s)bc pec; SB 12.5
NGC 5005	Gx	10.6	6.5 x 2.7	13 11.0	+37 03					◉	Class SAB(rs)bc; SB 11.9
NGC 5033	Gx	10.8	12.4 x 5.0	13 13.5	+36 36					◉	Class SA(s)c; SB 13.4
NGC 5055	Gx	9.3	13.7 x 7.3	13 15.8	+42 02	◉	◉				M 63; Sunflower Galaxy; Class SA(rs)bc; SB 12.5
NGC 4736	Gx	9.0	14.3 x 12.1	12 50.9	+41 07	◉	◉	◉			M 94; Class (R)SA(r)ab; SB 10.0
NGC 4490	Gx	10.2	6.3 x 2.7	12 30.6	+41 39					◉	Cocoon Galaxy; Class SB(s)d pec; SB 12.0
NGC 4449	Gx	10.0	6.1 x 4.3	12 28.2	+44 06					◉	Class Ibm; SB 11.5
NGC 4111	Gx	11.6	5.2 x 1.2	12 07.0	+43 04					◉	Class SA(r)0+: sp
NGC 4244	Gx	10.3	17.7 x 1.9	12 17.5	+37 48					◉	Class SA(s)cd: sp; SB 13.8
NGC 4214	Gx	10.2	7.4 x 6.5	12 15.7	+36 20					◉	Class IAB(s)m
NGC 4631	Gx	9.8	15.4 x 2.6	12 42.1	+32 32					◉	Class SB(s)d sp; SB 13.3
NGC 4656	Gx	11.0	9.1 x 1.7	12 44.0	+32 10					◉	Class SB(s)m pec; SB 13.9
NGC 5272	GC	6.3	18.0	13 42.2	+28 23	◉	◉	◉			M 3; Class VI
NGC 4258	Gx	9.1	18.8 x 7.3	12 19.0	+47 19	◉	◉				M 106; Class Sb; SB 12.6

With Canis Major and Canis Minor, Canes Venatici is one of three constellations named for dogs. In contrast to those two ancient constellations, Canes Venatici is a modern constellation, first described in 1690 by Johannes Hevelius in his book *Firmamentum Sobiescianum*. As a modern constellation, Canes Venatici has no myths or legends associated with it, although it is considered to represent the two hunting dogs of the herdsman Boötes, Asterion and Chara, as they chase the Great Bear of Ursa Major.

Despite its lack of prominence, Canes Venatici is home to numerous notable DSOs, including four bright Messier Galaxies—among them the spectacular Whirlpool Galaxy, M 51—and the bright Messier globular cluster M 3, which is one of the best globs visible in the night sky. For observers at mid-northern latitudes, Canes Venatici is best placed for evening viewing during the spring and summer months.

CHART 09-1.

The constellation Canes Venatici (field width 45°)

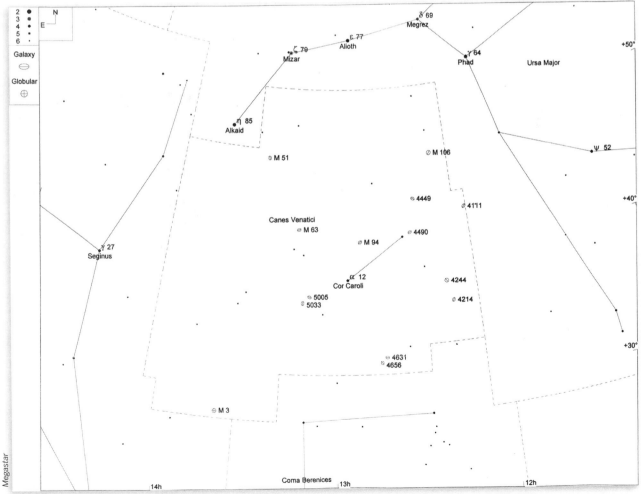

TABLE 09-2.

Featured multiple stars in Canes Venatici

Object	Pair	M1	M2	Sep	PA	Year	RA	Dec	UO	DS	Notes
12-alpha	STF 1692	2.9	5.5	19.3	229	2004	12 56.0	+38 19		◉	Cor Caroli

Clusters, Nebulae, and Galaxies

M 51 (NGC 5194)	★★★★	✿✿✿✿	**GX**	**MBUDR**
Chart 09-2	Figure 09-1	m9.0, 10.3' x 8.1'	13h 29.9m	+47° 12'

NGC 5194 (M 51), also known as the Whirlpool Galaxy, is a magnificent, bright, face-on spiral galaxy. M 51 has about the same size and mass as the Great Andromeda Galaxy (M 31), although, at about 35 million light years, M 51 is a dozen times more distant from us than M 31. Like M 31, M 51 has a major satellite galaxy of its own, NGC 5195. These galaxies are quite close to each other in celestial terms, and are actually interacting. The connector between them is visible with averted vision in a 10" scope under excellent observing conditions, and is quite prominent in 16" and larger scopes.

Visible even with a 50mm binocular, M 51 has surface brightness high enough to make it a very rewarding object in mid-size and larger amateur instruments. With our 50mm binoculars, M 51 is visible as a fuzzy m9 star. At 125X in our 10" scope, M 51 shows a wealth of detail, including subtle mottling, dust lanes, and the connector to NGC 5195. The halo is bright and well concentrated, with significant brightening to the circular core and stellar nucleus. The companion galaxy NGC 5195 is significantly smaller and dimmer, and shows little detail other than some brightening toward the core.

M 51 is easy to find, as it forms nearly a right triangle with Alkaid and Mizar in Ursa Major. To locate M 51, place 85-eta (η) UMa (Alkaid), the last star in the handle of the Big Dipper, on the E edge of your finder field. The m5 star 24-CVn appears prominently near the center of the field. M 51 lies near the SW edge of the finder field, and shows faintly with averted vision in a 50mm finder scope as a fuzzy star.

FIGURE 09-1.

NGC 5194 (M 51) (60' field width)

Image reproduced from Digitized Sky Survey courtesy Palomar Observatory and Space Telescope Science Institute

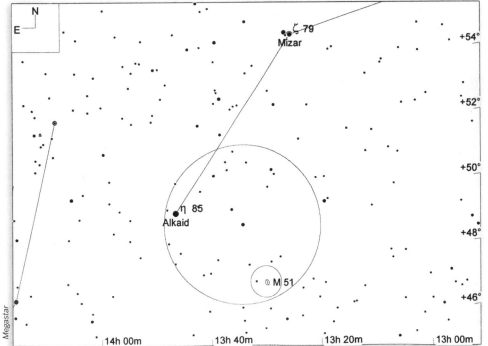

NGC 5005	★★	☉☉☉	GX	MBUDR
Chart 09-3	Figure 09-2	m10.6, 6.5' x 2.7'	13h 11.0m	+37° 03'

NGC 5033	★★	☉☉☉	GX	MBUDR
Chart 09-3	Figure 09-3	m10.8, 12.4' x 5.0'	13h 13.5m	+36° 36'

Located about 3.5° ESE (east southeast) of Cor Caroli, the galaxy pair NGC 5005 and NGC 5033 are visible in the same 1° eyepiece field, but require higher magnification to see much detail. NGC 5005 has much higher surface brightness than NGC 5033. At 125X in our 10" scope, NGC 5005 shows a bright oval central core extending about 2' x 0.5' ENE-WSW (from east northeast to west southwest) with a stellar nucleus. The halo extends about 5' x 2' with gradual dimming toward the edges and narrowing at both ends. Despite its cataloged size, NGC 5033 appears considerably smaller than NGC 5005, with a 2' x 0.5' halo extending N-S, and a small, concentrated, moderately bright central core with a stellar nucleus.

To locate NGC 5005 and NGC 5033, put Cor Caroli on the W edge of your finder field, with m5 14-CVn on the SSW (south southwest) edge of the field and the close m6/7 triple 15-, 16-, 17-CVn near the center of the field. The m6.5 star SAO 63414 shows prominently in the SE quadrant of the finder field. NGC 5033 lies about 17' dead S of that star, with NGC 5005 about 34' WNW (west northwest) of that star and about 26' SE of the m6 star SAO 63372.

FIGURE 09-2.

NGC 5005 (60' field width)

FIGURE 09-3.

NGC 5033 (60' field width)

Image reproduced from Digitized Sky Survey courtesy Palomar Observatory and Space Telescope Science Institute

Image reproduced from Digitized Sky Survey courtesy Palomar Observatory and Space Telescope Science Institute

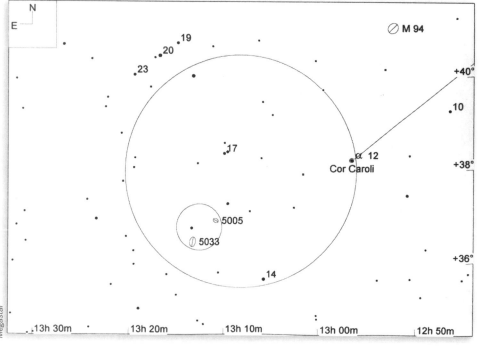

CHART 09-3.

NGC 5005 and NGC 5033 (10° field width, 5.0° finder field, 1.0° eyepiece field, LM 9.0)

M 63 (NGC 5055)	★★★	✧✧✧	GX	MBUDR
Chart 09-4	Figure 09-4	m9.3, 13.7' x 7.3'	13h 15.8m	+42° 02'

NGC 5055 (M 63) is also known as the Sunflower Galaxy because its spiral arms resemble the petals of a flower. In our 10" scope at 125X, M 63 shows a large, bright core elongated E-W (east to west) with a stellar nucleus. The 3.5' x 1.5' halo extends WNW-ESE. Some mottling is visible in the halo, although the spiral arms visible in photographs are below our detection threshold.

To locate M 63, place Cor Caroli on the WSW edge of your finder field and look for the prominent line of stars 23-, 20-, and 19-CVn. (These three stars, along with 18-CVn to the W, form one of the hunting dogs, Asterion, also known as "Starry".) M 63 lies 1.1° N of 19-CVn, where it shows weakly in a 50mm finder.

FIGURE 09-4.

NGC 5055 (M 63) (60' field width)

Image reproduced from Digitized Sky Survey courtesy Palomar Observatory and Space Telescope Science Institute

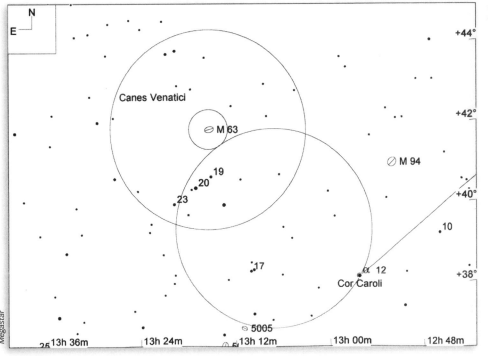

CHART 09-4.

NGC 5055 (M 63) (12° field width, 5.0° finder fields, 1.0° eyepiece field, LM 9.0)

M 94 (NGC 4736)	★★	❖❖❖	GX	MBUDR
Chart 09-5	Figure 09-5	m9.0, 14.3' x 12.1'	12h 50.9m	+41° 07'

NGC 4736 (M 94) is a large, bright galaxy, but one that reveals little detail in typical amateur scopes. In our 10" scope at 90X, M 94 at first glance resembles a large, bright, unresolved globular cluster. The central core is very large, almost circular, of even brightness fading very slightly towards the edges, and reveals no stellar nucleus. An extremely faint oval halo surrounds the core, extending NW-SE.

To locate M 94, place Cor Caroli at the SSE edge of your finder field and look to the center or north-center of the field. M 94 shows weakly with averted vision in a 50mm finder as a fuzzy star with very faint nebulosity.

FIGURE 09-5.

NGC 4736 (M 94) (60' field width)

Image reproduced from Digitized Sky Survey courtesy Palomar Observatory and Space Telescope Science Institute

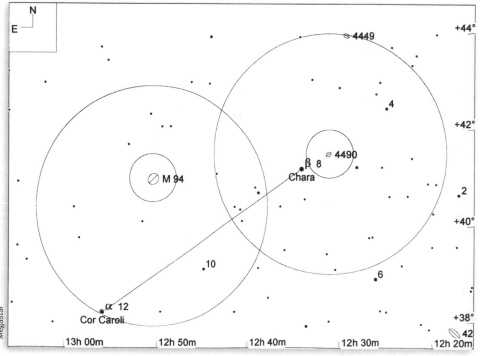

CHART 09-5.

NGC 4736 (M 94) and NGC 4490 (10° field width, 5.0° finder field, 1.0° eyepiece field, LM 9.0)

NGC 4490	★★	✦✦✦✦	GX	MBUDR
Chart 09-5	Figure 09-6	m10.2, 6.3' x 2.7'	12h 30.6m	+41° 39'

NGC 4490, the Cocoon Galaxy, is another galaxy that is bright enough to see easily, but reveals little detail. At 125X in our 10" scope, NGC 4490 is visible as a moderately bright, narrow slash of light extending about 5' x 1.5' NW-SE. Slight central brightening is visible, but no stellar nucleus. NGC 4490 has a companion, NGC 4485, with which it interacts. NGC 4485 is located in the same high-power eyepiece field, about 3' N of NGC 4490. NGC 4485 is much smaller and much dimmer than NGC 4490, and appears to us as a small, moderately dim circular patch of nebulosity without detail.

NGC 4490 is very easy to find. It lies 39' WNW of 8-beta (β) CVn (Chara), and is visible in the same low-power eyepiece field.

FIGURE 09-6.

NGC 4490 (60' field width)

Image reproduced from Digitized Sky Survey courtesy Palomar Observatory and Space Telescope Science Institute

NGC 4449	★★	✦✦✦	GX	MBUDR
Chart 09-6	Figure 09-7	m10.0, 6.1' x 4.3'	12h 28.2m	+44° 06'

NGC 4449 is a small, moderately bright galaxy. At 125X in our 10" scope, NGC 4449 shows a 3' x 1.5' halo extending NE-SW with very slight central brightening, some faint mottling near the edges, and a stellar nucleus.

NGC 4449 is relatively easy to locate. Put 8-beta (β) CVn (Chara) on the SSE edge of your finder field, and look N of center for three m7 stars that form a right triangle. NGC 4449 forms the SW apex of another right triangle with the two southernmost of those stars, lying about 0.75° dead S of the star that forms the N apex and 1° dead W of the star that forms the E apex.

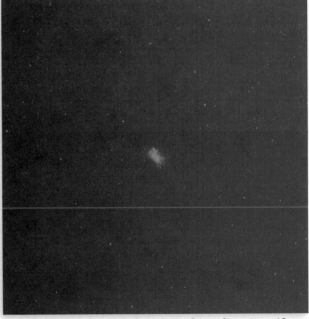

FIGURE 09-7.

NGC 4449 (60' field width)

Image reproduced from Digitized Sky Survey courtesy Palomar Observatory and Space Telescope Science Institute

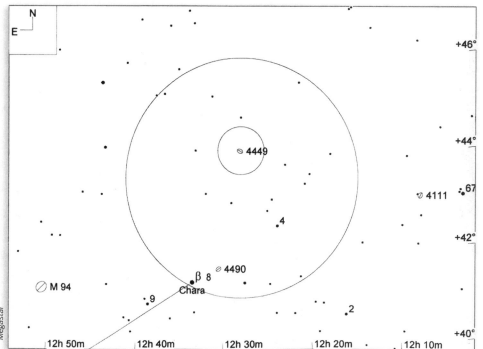

NGC 4111	★★★	✦✦✦	GX	MBUDR
Chart 09-7	Figure 09-8	m11.6, 5.2' x 1.2'	12h 07.0m	+43° 04'

NGC 4111 is a small galaxy with relatively high surface brightness. In our 10" scope at 125X, NGC 4111 shows a moderately faint halo extending 3' x 0.5' NNW-SSE, with strong brightening to an extended central core, with a very bright stellar nucleus.

To locate NGC 4111, put 8-beta (β) CVn (Chara) on the E edge of your finder field, and look for the m6 stars 2- and 4-CVn, which appear prominently in the finder. Move the finder to place 4-CVn on the E edge and 2-CVn on the SE edge, as shown in Chart 09-7. The m5 star 67-UMa appears prominently in the E part of the finder field. NGC 4111 lies 54' dead E of 67-UMa, about 4' WSW of an m8 star and about 25' dead N of another m8 star.

FIGURE 09-8.

NGC 4111 (60' field width)

Image reproduced from Digitized Sky Survey courtesy Palomar Observatory and Space Telescope Science Institute

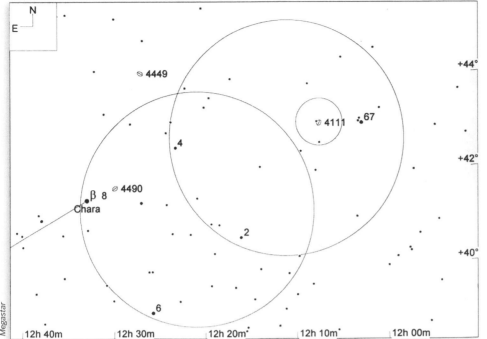

NGC 4244	★★★	◦◦◦	GX	MBUDR
Chart 09-8	Figure 09-9	m10.3, 17.7' x 1.9'	12h 17.5m	+37° 48'

NGC 4244 is an extraordinary spindle galaxy with moderately high surface brightness. At 125X in our 10" scope, the halo of NGC 4244 is visible as a very long, narrow, moderately bright slash of light about 15' x 1', extending NE-SW, with a noticeably brighter extended core. No stellar nucleus is visible.

To locate NGC 4244, place 8-beta (β) CVn (Chara) on the NE edge of your finder field, and follow the line from Chara SW to m5 6-CVn, which appears prominently near the center of the finder. Extend that line SW and look for look for a widely-space m7 double about 1.5° SW of 6-CVn. NGC 4244 lies about 0.5° W of that double, and is visible in the same low-power eyepiece field.

FIGURE 09-9.

NGC 4244 (60' field width)

Image reproduced from Digitized Sky Survey courtesy Palomar Observatory and Space Telescope Science Institute

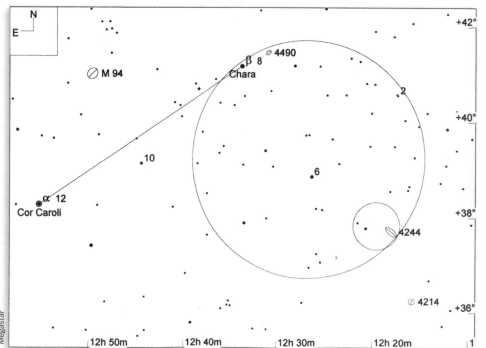

NGC 4214	★★	⊙⊙	GX	MBUDR
Chart 09-8, 09-9	Figure 09-10	m10.2, 7.4' x 6.5'	12h 15.7m	+36° 20'

NGC 4214 is a small galaxy with moderately high surface brightness. At 125X in our 10" scope, the halo of NGC 4214 is visible as a moderately bright, nearly circular patch of light about 3' x 2.5', extending NW-SE, with a strong brightening to the central core.

The easiest way to locate NGC 4214 is by using an eyepiece star hop from NGC 4244, as shown in Chart 09-9. With NGC 4244 on the N edge of a 1° eyepiece field, a pattern of m9/10 stars is visible in the S half of the field. Move the eyepiece field S to place that pattern on the N edge of the field, and look for an arc of three m10 stars in the S half of the field. With that arc placed at the N edge of the eyepiece field, NGC 4214 is visible on the S edge of the field.

FIGURE 09-10.

NGC 4214 (60' field width)

Image reproduced from Digitized Sky Survey courtesy Palomar Observatory and Space Telescope Science Institute

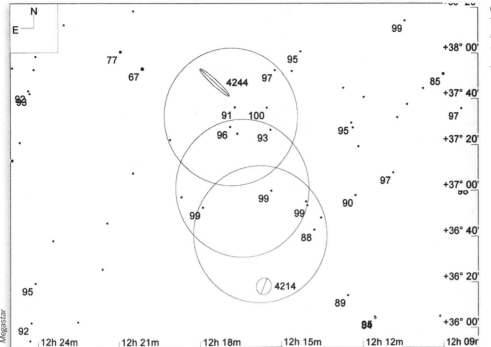

NGC 4631	★★★	◔◔	GX	MBUDR
Chart 09-10	Figure 09-11	m9.8, 15.4' x 2.6'	12h 42.1m	+32° 32'

NGC 4656	★★	◔◔	GX	MBUDR
Chart 09-10	Figure 09-12	m11.0, 9.1' x 1.7'	12h 44.0m	+32° 10'

The galaxy pair NGC 4631 and NGC 4656 provide an interesting contrast. In our 10" scope at 90X, both galaxies are visible in one eyepiece field. NGC 4631 is by far the more impressive of the two. It is a gorgeous edge-on spiral galaxy that to us resembles a dagger. The halo is extremely elongated, extending about 15' x 1.5' E-W. The W side of the halo is longer and narrower, and forms the blade, with the stubbier E side forming the handle. The extended core is very bright, but no stellar nucleus is visible. NGC 4656 is much smaller and considerably dimmer than NGC 4631. It appears as a thin, irregular streak of light extending about 6' x 1' NE-SW. The SW end is far brighter than the NE end, and the very faint central portion is visible only with difficulty with averted vision.

Although NGC 4631 and NGC 4656 lie on the S border of Canes Venatici, the best way to find them is by starting at the N edge of Coma Berenices. To do so, first locate the m5/6 pair 30- and 31-Com, which lie about halfway along the line between the m4 stars 43-beta (β) Com and 15-gamma (γ) Com. Place 30/31 Com on the S edge of your finder field, with m5 37-Com near the NE edge. With our 5° finder, the galaxy pair lie just outside the NW edge of the field. If you move the finder NW until the prominent m5/6 pair 54-CVn comes into view on the WNW edge of the finder field, the galaxy pair will be about centered in the finder, and NGC 4631 will be easily visible in your low-power eyepiece.

FIGURE 09-11.

NGC 4631 (60' field width)

Image reproduced from Digitized Sky Survey courtesy Palomar Observatory and Space Telescope Science Institute

FIGURE 09-12.

NGC 4656 (60' field width)

Image reproduced from Digitized Sky Survey courtesy Palomar Observatory and Space Telescope Science Institute

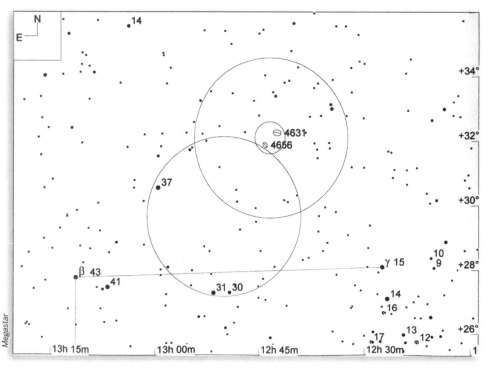

CHART 09-10.

NGC 4631 and NGC 4656 (15° field width, 5.0° finder fields, 1.0° eyepiece field, LM 9.0)

M 3 (NGC 5272)	★★★★	✸✸✸	GX	MBUDR
Chart 09-11	Figure 09-13	m6.3, 18.0'	13h 42.2m	+28° 23'

NGC 5272 (M 3) is, in our opinion and that of many other observers, the second-finest globular cluster visible to northern observers, exceeded only by the magnificent M 13 in Hercules. At 125X in our 10" scope, M 3 is simply stunning. The dense, very bright 6' core is surrounded by an outer halo that covers about 12', with the star density tapering off evenly from core to edge. Although M 3 is cataloged as being Shapley-Sawyer Concentration Class VI, which indicates a moderately tight globular cluster, it's possible to resolve stars almost to the core in a 10" scope.

M 3 is difficult to find by star hopping, but easy to find using celestial geometry. M 3 lies almost exactly halfway on a line from Arcturus in Boötes to Cor Caroli in Canes Venatici, slightly nearer Arcturus. Simply put your finder about where it needs to be, and look for M 3 in the finder. If it's not there, wave the finder around a bit until you see it.

FIGURE 09-13.

NGC 5272 (M 3) (60' field width)

Image reproduced from Digitized Sky Survey courtesy Palomar Observatory and Space Telescope Science Institute

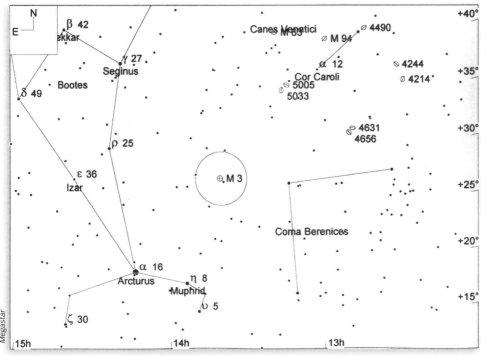

CHART 09-11.

NGC 5272 (M 3) (45° field width, 5.0° finder field, LM 6.0)

M 106 (NGC 4258)	★★★	✧✧✧	GX	MBUDR
Chart 09-12	Figure 09-14	m9.1, 18.8' x 7.3'	12h 19.0m	+47° 19'

NGC 4258 (M 106) is a large galaxy with surface brightness high enough to reveal significant detail in typical amateur scopes. In our 10" scope at 90X, M 106 is an impressive sight. A very bright 1' central core without a stellar nucleus is surrounded by a bright 5' x 3' inner halo that is in turn surrounded by a much dimmer and more diffuse 14' x 4' outer halo that extends NNW-SSE.

Although M 106 lies in CVn, it's easier to locate by starting from the bowl of the Big Dipper in UMa. To do so, place 64-gamma (γ) UMa (Phad) on the WNW edge of your finder field and look for the m5 star 5-CVn, which appears prominently on the ESE edge of the field. Place 5-CVn on the N edge of your finder field, with m5 3-CVn near the center of the field. The m6 star SAO 44141 appears prominently near the S edge of the finder field, with M 106 lying about 0.5° W, where it shows weakly with averted vision in a 50mm finder.

FIGURE 09-14.

NGC 4258 (M 106) (60' field width)

Image reproduced from Digitized Sky Survey courtesy Palomar Observatory and Space Telescope Science Institute

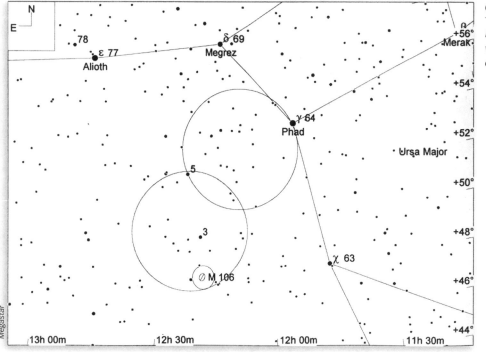

CHART 09-12.

NGC 4258 (M 106) (20° field width, 5.0° finder fields, 1.0° eyepiece field, LM 9.0)

Multiple Stars

Cor Caroli (STF 1692)	★★★	✧✧✧✧	MS	UD
Chart 09-1		m2.9/5.5, 19.3", PA 229° (2004)	12h 56.0m	+38° 19'

Cor Caroli is a fine, wide-spaced double that can be split easily even with a small telescope. At 100X in our 90mm (3.5") refractor, both stars have a pale yellowish cast. At 90X in our 10" reflector, both stars appear pure white, although some observers have reported detecting a bluish or greenish cast in one or both stars.

10
Canis Major, The Larger Dog

NAME: Canis Major (CAWN-is MAY-jur)	
SEASON: Winter	
CULMINATION: midnight, 1 January	
ABBREVIATION: CMa	
GENITIVE: Canis Majoris (CAWN-is muh-JOR-is)	
NEIGHBORS: Col, Lep, Mon, Pup	
BINOCULAR OBJECTS: NGC 2287 (M41), NGC 2360	
URBAN OBJECTS: none	

Canis Major is a mid-size constellation, ranking 43rd in size among the 88 constellations. It covers 380 square degrees of the celestial sphere, or about 0.9%. In mythology, Canis Major is the larger of Orion's two hunting dogs. Canis Major rests at the feet of Orion, waiting to pounce on Lepus, The Hare.

The eye of Canis Major is Sirius, the brightest star in the night sky, also known as the Dog Star. (The name Sirius comes from the Greek for "searing" or "shining.") Sirius shines so brightly both because it is inherently bright and because, in celestial terms, it's our next-door neighbor. The close proximity of Sirius to us means that it also shows very high proper motion (actual movement relative to much more distant background stars.) Two million years ago, Sirius was in what is now the constellation of Lynx, and was barely visible to the naked eye. Only 90,000 years ago—an eye-blink in celestial terms—Sirius appeared near Procyon in Canis Minor, which it barely outshone. Only 80,000 years from now,

Sirius will have brightened still more and moved nearly 40° SSW into the constellation Pictor, where, in close proximity to zeroth-magnitude Canopus, it will dominate the night sky.

Sirius is interesting for other reasons as well. With Procyon in Canis Minor and Betelgeuse in Orion, Sirius makes up the Winter Triangle, the most prominent grouping of bright stars in the night sky. Sirius is believed by some astronomers to be a member of the Ursa Major Moving Group, which includes all of the members of the Big Dipper asterism except Dubhe and Alkaid.

Sirius is also a double star. Its companion, the magnitude 8.5 carbon-core white dwarf star Sirius B, is known colloquially as the Pup. The Pup is very small—about 92% of Earth's diameter—and orbits very near Sirius A. On average, Sirius B is located only 19.8 astronomical units from Sirius A, or about the same distance as the orbit of Uranus from Sol.

TABLE 10-1.

Featured star clusters, nebulae, and galaxies in Canis Major

Object	Type	Mv	Size	RA	Dec	M	B	U	D	R	Notes
NGC 2287	OC	4.5	38.0	06 46.0	-20 45	◉	◉	◉			M 41; Cr 118; Class I 3 r
NGC 2360	OC	7.2	12.0	07 17.7	-15 39				◉		Cr 134; Mel 64; Class I 3 r
NGC 2359	EN	99.9	13.0 x 11.0	07 18.5	-13 14					◉	

TABLE 10-2.

Featured multiple stars in Canis Major

Object	Pair	M1	M2	Sep	PA	Year	RA	Dec	UO	DS	Notes
21-epsilon	CPO 7	1.5	7.5	7.0	161	2000	06 58.6	-28 58		◉	Adhara

As of 2006, the separation of Sirius AB is about 7 arcseconds and growing. The separation reaches its maximum of 11.3 arcseconds in 2022. A 7 arcsecond separation might lead you to believe Sirius AB would be an easy split, but it's not. Far from it. Although the Pup is relatively bright in absolute terms, it's 8,400 times dimmer than Sirius A. This huge difference in brightness makes it very difficult to split the pair, because Sirius B simply disappears in the glare from Sirius A. Very skilled observers using excellent optics on nights of superb seeing have reported splitting Sirius as early as 2001 using a 4" scope. For the rest of us, splitting Sirius is still extraordinarily difficult even with a 10" or larger scope.

Finally, there is one inexplicable oddity about Sirius. The ancient Egyptians knew Sirius well. The rising of Sirius in the early morning sky foretold the coming of the annual Nile Flood, by far the most important recurring event in the life of ancient Egypt. Current cosmology (and our own eyes) tells us that Sirius is a brilliant blue-white star, and that that would also have been true during the time of the ancient Egyptians, three thousand to five thousand years ago. And yet, contemporaneous reports recorded by those ancient Egyptians almost universally report the appearance of Sirius as "red" or even "blood red." It's possible that those reports were based on viewing Sirius when it lay very near the horizon,

CHART 10-1.

The constellation Canis Major (field width 30°)

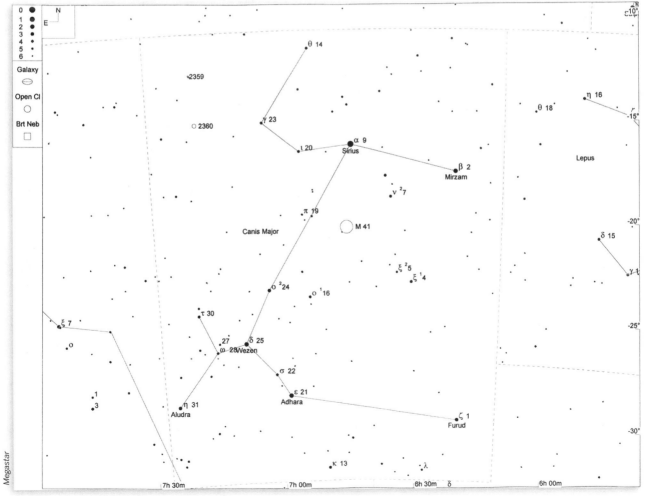

where the dense atmosphere might have given it a reddish cast, but we think that's unlikely to be the whole explanation. The Egyptians must certainly have been intimately familiar with the appearance of Sirius when it lay high in the sky. So why did they see it as red? No one knows.

The location of Canis Major in the Winter Milky Way means that it's rich in open clusters, of which M41 and NGC 2360 feature in this book. There are also several emission/reflection nebulae, of which NGC 2359 is the most impressive. Canis Major is also replete with multiple star systems, although only 21-epsilon (ε) CMa is featured in this book. For observers at mid-northern latitudes, Canis Major is best placed for observing from early winter, when it rises a couple hours after dark, to early spring, when it sets a couple hours after dark.

Clusters, Nebulae, and Galaxies

M 41 (NGC 2287)	★★★★	✦✦✦✦	OC	MBUDR
Chart 10-2	Figure 10-1	m4.5, 38.0'	06h 46.0m	-20° 45'

NGC 2287, better known as Messier 41 or M 41, is a very large, very bright open cluster, cataloged as Trumpler Class I 3 r. M 41 has been known since ancient times, because from even a moderately dark site it is readily visible to the naked eye as a nebulous patch of light about the size of the full Moon. M41 is prominent in a 50mm binocular, with 15+ stars resolvable from m7 down to m9.5. In our 10" scope at 42X, M 41 is a magnificent cluster. 100+ stars are visible, from m7 down to m12, with many forming clumps, arcs, knots, and chains. An m6 star lies on the SE edge of cluster, but is not a member.

M 41 is easy to find. It lies 4° dead S of Sirius, and is visible even in a 30mm finder. Put Sirius at the N edge of your finder, and look for the prominent m4 pair 7-nu^2 (ν^2) CMa and 8-nu^3 (ν^3) CMa at the W and WNW edge of the finder. M 41 is shows distinctly at the S edge of the finder.

FIGURE 10-1.

NGC 2287 (M 41) (60' field width)

Image reproduced from Digitized Sky Survey courtesy Anglo-Australian Observatory and Space Telescope Science Institute

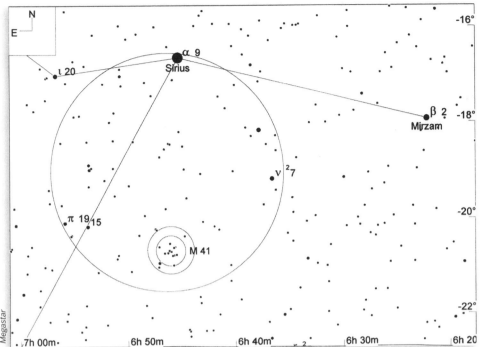

NGC 2360	★★★	◊◊◊◊	OC	MBUDR
Chart 10-3	Figure 10-2	m7.2, 12.0'	07h 17.7m	-15° 39'

NGC 2360 is a very pretty open cluster, cataloged as Trumpler Class I 3 r, and visible even with a small binocular. We're rather surprised that Charles Messier didn't discover NGC 2360 as he was panning the sky around M 41, which lies about 9° SW.

With a 7X50 binocular, NGC 2360 appears as a small, fairly bright nebulous patch about half a field dead E of 23-gamma (γ) CMa. No stars are visible with direct vision. With averted version, an m10 pair is visible near the center of the nebulosity, with an m9 star near the E edge. At 90X in our 10" scope, NGC 2360 is very prominent centered in the same eyepiece field as a bright m5.5 field star lying on the edge of the eyepiece field, about 20' W. The cluster is quite rich and has strong central concentration. One m10 star lies near the center of the cluster, with about 50 m11-m14 members filling out the cluster in irregular chains and knots. The m9 star SAO 152691 lies at the far E edge of the cluster.

NGC 2360 lies 3.4° dead E of 23-gamma (γ) CMa. To locate NGC 2360, center 23-gamma (γ) in the finder crosshairs and move the finder E until 23-gamma (γ) lies near the W edge of the finder. NGC 2360 is visible in a 50mm finder as a moderately bright nebulous patch, just E of an m5 field star.

FIGURE 10-2.

NGC 2360 (DSS image, 60' field width)

Image reproduced from Digitized Sky Survey courtesy Anglo-Australian Observatory and Space Telescope Science Institute

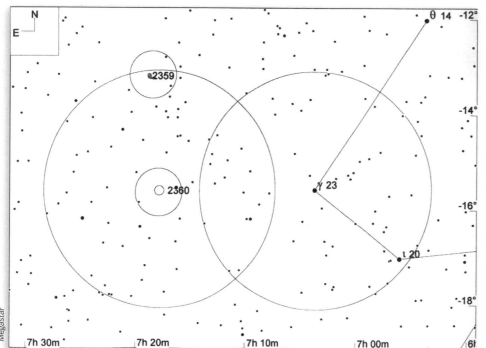

NGC 2359	★★	◦◦◦	EN	MBUDR
Chart 10-3, 10-4	Figure 10-3	m99.9, 13.0' x 11.0'	07h 18.5m	-13° 14'

NGC 2359 is a fairly large, faint emission nebula also known as the Duck Bill Nebula. It looks to us more like a comma. At 90X in our 10" scope with a narrowband filter, NGC 2360 appears as a larger, dimmer, more diffuse northern section that forms a NE-SW (northeast to southwest) oval, and a narrow E-W southern section that extends W from an m10 star on the S end and is noticeably brighter and denser.

To locate NGC 2359, begin by centering the finder on NGC 2360, as described in the preceding section. In a 5° finder, NGC 2359 lies at the extreme N edge of the field. NGC 2359 lies about 30' NE of two m7 field stars that appear prominently in the finder, and about the same distance NNE (north northeast) of another prominent m7 star (which appears as a small cluster at the S edge of the eyepiece field in Chart 10-4).

FIGURE 10-3.

NGC 2359 (60' field width)

Image reproduced from Digitized Sky Survey courtesy Anglo-Australian Observatory and Space Telescope Science Institute

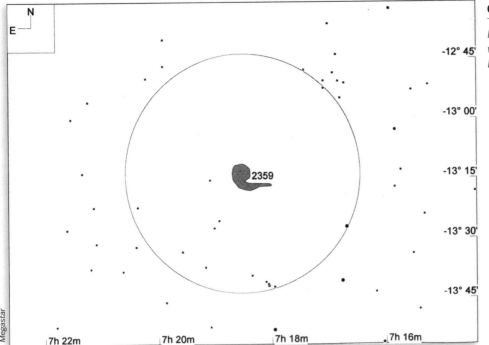

-12° 45'

-13° 00'

-13° 15'

-13° 30'

-13° 45'

7h 22m   7h 20m   7h 18m   7h 16m

Megastar

Multiple Stars

21-epsilon (ε)	★★	◔◔◔◔	MS	UD
Chart 10-1		m1.5/7.5, 7.0", PA 161°	06h 58.6m	-28° 58'

21-epsilon (ε) CMa (Adhara) is a much more difficult split than its relatively wide separation suggests. At 90X in our 10", there is just a hint of elongation or a bump in Adhara. At 180X, it is apparent that the star is binary, but we were able to get only a dirty split because the dazzling glare of Adhara swamps its much dimmer companion. Getting a clean split on this binary requires high magnification on a night of excellent seeing and using pristine optics.

11
Capricornus, The Sea Goat

NAME: Capricornus (CAP-rih-CORN-us)

SEASON: Summer

CULMINATION: midnight, 5 August

ABBREVIATION: Cap

GENITIVE: Capricorni (CAP-rih-CORN-ee)

NEIGHBORS: Aql, Aqr, Mic, PsA, Sgr

BINOCULAR OBJECTS: NGC 7099 (M30)

URBAN OBJECTS: none

Capricornus is a dim summer constellation of medium size, ranking 40th among the 88 constellations. It covers 414 square degrees of sky, or about 1%. Its brightest stars are of only third and fourth magnitude, making it the second dimmest of the zodiacal constellations after Cancer. Despite its lack of prominent stars, Capricornus is an ancient constellation, recognized by the ancient Sumerians, Babylonians, and Egyptians, all of whom saw it as a chimera that melded the front half of a goat with the back half of a fish.

Despite the fact that it lies just east of Sagittarius, with its amazing wealth of deep-sky objects, Capricornus is nearly devoid of DSOs of interest to beginning and intermediate amateur astronomers. The only DSO readily visible in amateur telescopes is the Messier Globular M 30 (NGC 7099), which is noted for being among the last objects that must be logged in the early morning sky during a Messier Marathon. Capricornus is somewhat richer in double and multiple stars. The region around Algedi and Dabih in the northwestern corner of the constellation contains numerous multiple stars that are easily visible with a binocular or small telescope. In fact, at low power in our rich-field 4.5" reflector, this area resembles a very loose open cluster. For observers at mid-northern latitudes, Capricornus is best placed for evening viewing during the autumn months.

TABLE 11-1.

Featured star clusters, nebulae, and galaxies in Capricornus

Object	Type	Mv	Size	RA	Dec	M	B	U	D	R	Notes
NGC 7099	GC	6.9	12.0	21 40.4	-23 11	◉	◉				M 30; Class V

TABLE 11-2.

Featured multiple stars in Capricornus

Object	Pair	M1	M2	Sep	PA	Year	RA	Dec	UO	DS	Notes
5-alpha1, 6-alpha2	STFA 51AE	3.7	4.3	381.2	292	2002	20 18.1	-12 33		◉	Algedi
9-beta1, beta2	STFA 52Aa-Ba	3.2	6.1	207.0	267	2001	20 21.0	-14 47		◉	Dabih

CHART 11-1.

The constellation Capricornus (field width 30°)

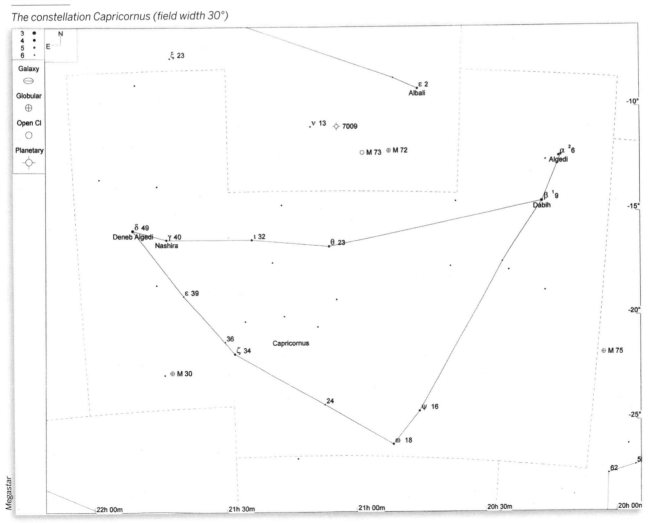

Clusters, Nebulae, and Galaxies

M 30 (NGC 7099)	★★★	◊◊◊	GC	MBUDR
Chart 11-2		m6.9, 12.0'	21h 40.4m	-23° 11'

Capricornus is nearly bereft of interesting deep-sky objects for beginning and intermediate observers. The sole exception is the excellent Messier globular cluster M 30 (NGC 7099). To locate M 30, begin at the prominent m3 star Deneb Algedi (Chart 11-1) at the eastern edge of Capricornus. Look about 8° SW for the m4 pair 34-ζ (zeta) and 36, which stand prominently alone to the naked eye from a dark site. Place those two stars on the NW edge of your finder field, as shown in Chart 11-2. Look in the NE quadrant of the field for the m5 star 41-Cap, which appears bright in the finder. M 30 lies about 23' W of that star, and appears in the finder as a fuzzy m7 star.

With our 7X50 binoculars, M 30 is visible only as a fuzzy m7 star. In any telescope at medium or higher magnification, M 30 reveals itself as a moderately concentrated globular cluster. At 90X in our 10" Dob, M 30 shows a bright unresolvable core of about 3' extent, with a loose scattering of outliers extending to about 5' diameter. The m9 star SAO 190531 appears prominent at the W edge of the cluster, lying just beyond the extent of the halo visible with a 10" scope under dark conditions. Increasing magnification to 125X, 180X, and 240X did not allow us to resolve this cluster to the core in our 10" Dob. More aperture helps. With a 17.5" Dob, we can easily resolve M 30 at 158X.

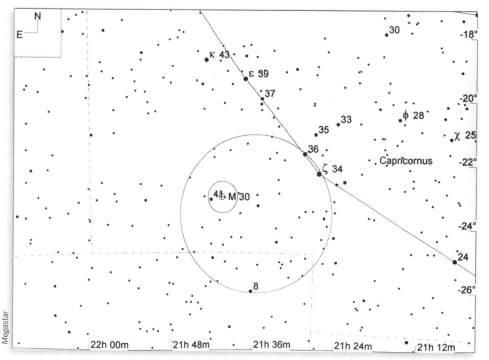

CHART 11-2.

M 30 (15° field width; 5° finder circle; 1° eyepiece circle; LM 9.0)

Multiple Stars

5-α¹, 6-α² (STFA 51AE)	★★★	✴✴✴✴	MS	UD
Chart 11-1		m3.7/4.3, 381.2", PA 292°	20h 18.1m	-12° 33'

The naked-eye pair 5-α¹ and 6-α² (Algedi), located at the NW corner of Capricornus, are separated by 381.2", or about a fifth the diameter of the full moon. This is an optical double, with m3.6 Algedi lying about 100 LY distant and m4.3 α¹ at about five times that distance. Both stars appear warm white in our 7X50 binocular, and show a deep golden color in our 10" scope at 90X.

9-β¹, β² (STFA 52Aa-Ba)	★★★	✴✴✴✴	MS	UD
Chart 11-1		m3.2/6.1, 207.0", PA 267°	20h 21.0m	-14° 47'

Located about 2.4° SSE of the alpha pair, the beta pair offers a striking color contrast in our 10" scope at 90X. The blue-white m6 companion β² lies about 3.5' dead W of the brilliant yellow primary 9-β¹ (Dabih). This is actually a trinary system, with a tertiary m9 star lying about 3.8' SE of Dabih.

12
Cassiopeia, The Queen

NAME: Cassiopeia (cass-ee-oh-PEE-uh)

SEASON: Autumn

CULMINATION: midnight, 9 October

ABBREVIATION: Cas

GENITIVE: Cassiopeiae (cass-ee-oh-PEE-eye)

NEIGHBORS: And, Cam, Cep, Lac, Per

BINOCULAR OBJECTS: NGC 129, NGC 457, NGC 581 (M 103), NGC 663, Cr 463, St 2, Mrk 6, Mel 15, Tr 3, NGC 7654 (M 52), NGC 7789

URBAN OBJECTS: NGC 129, NGC 457, NGC 663, Cr 463, St 2, Tr 3, NGC 7789, 24-eta (STF 60)

Cassiopeia is a bright circumpolar autumn constellation of medium size, ranking 25th among the 88 constellations. It covers 598 square degrees of sky, or about 1.4%. The prominent shape of Cassiopeia (like the letter M or W, depending on the time of year) makes it among the most easily recognized constellations, familiar even to casual observers of the night sky. Cassiopeia lies directly opposite Ursa Major from Polaris, which means that when Cassiopeia is high in the sky, Ursa Major (and the Big Dipper) are low, and vice versa.

Cassiopeia has been recognized as a constellation since remote antiquity. In Greek mythology, Cassiopeia was the Queen of Æthiopia, the wife of King Cepheus, and the mother of Princess Andromeda. Cassiopeia's vanity and boastfulness offended the Nereids, who called down the wrath of Poseidon upon Æthiopia in the person of Cetus the sea monster. Cepheus and Cassiopeia consulted an oracle, who told them that the only way to appease the sea gods was to sacrifice their daughter Andromeda to Cetus.

They chained Andromeda to a rock on shore to await her fate. As fierce Cetus approached Andromeda, the Hero Perseus happened to be passing. Perseus slew Cetus, rescued Andromeda, and married her. The frustrated Nereids punished Cassiopeia by having Poseidon cast her in her chair into the heavens, where she is doomed to spend half of eternity upside down.

Cassiopeia lies nestled in the autumn Milky Way, surrounded by Cepheus, Andromeda, and Perseus. Even for a Milky Way constellation, Cassiopeia is unusually rich in open clusters. Many of these open clusters are small and hard to differentiate from the rich Milky Way starfields, but several are among the finest open clusters in the night sky. Cassiopeia is also home to several planetary and emission nebulae, an interesting galaxy, and numerous multiple stars. For observers at mid-northern latitudes, Cassiopeia is best placed for evening viewing during the late autumn and early winter months.

TABLE 12-1.

Featured star clusters, nebulae, and galaxies in Cassiopeia

Object	Type	Mv	Size	RA	Dec	M	B	U	D	R	Notes
NGC 7654	OC	6.9	12.0	23 24.8	+61 36	◉	◉				M 52; Cr 455; Class II 2 r
NGC 7635	EN	99.9	16.0 x 6.0	23 20.7	+61 12					◉	Bubble Nebula
NGC 7789	OC	6.7	15.0	23 57.4	+56 43			◉	◉	◉	Cr 460; Mel 245; Class II 2 r
NGC 129	OC	6.5	21.0	00 29.9	+60 13			◉	◉		Cr 2; Class III 2 m
NGC 281	OC/EN	99.9	28.0 x 21.0	00 53.0	+56 38					◉	
NGC 457	OC	6.4	13.0	01 19.6	+58 17			◉	◉	◉	Cr 12; Mel 7; Class II 3 r
NGC 581	OC	7.4	6.0	01 33.4	+60 40	◉	◉				M 103; Cr 14; Class II 2 m
NGC 663	OC	7.1	16.0	01 46.3	+61 13			◉	◉	◉	Cr 20; Mel 12; Class III 3 r
St 2	OC	4.4	60.0	02 15.6	+59 32			◉	◉		Class I 2 m
Mrk 6	OC	7.1	4.5	02 29.6	+60 39				◉		Class IV 2 p
Cr 26	OC	6.5	21.0	02 32.7	+61 27				◉		Mel 15; Class II 3 m n
Cr 36	OC	7.0	23.0	03 12.0	+63 12			◉	◉		Tr 3; Harvard 1; Class III 2 m
IC 0289	PN	12.3	0.6	03 10.3	+61 19					◉	Hubble 1; PK 138+2.1; PNG 138.8+2.8
NGC 185	Gx	10.1	12.9 x 10.1	00 39.0	+48 20					◉	Class E3 pec
Cr 463	OC	5.7	36.0	01 47.6	+71 46			◉	◉		Class III 2 m

TABLE 12-2.

Featured multiple stars in Cassiopeia

Object	Pair	M1	M2	Sep	PA	Year	RA	Dec	UO	DS	Notes
24-eta*	STF 60AB	3.5	7.4	13.0	320	2006	00 49.1	+57 49	◉	◉	
8-sigma	STF 3049AB	5.0	7.2	3.2	327	2004	23 59.0	+55 45		◉	

CHART 12-1.

The constellation Cassiopeia (field width 50°)

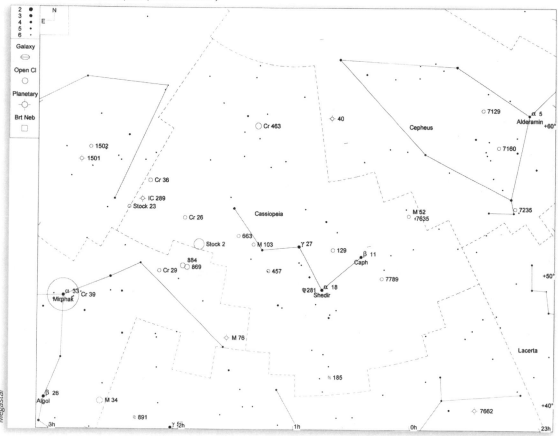

Clusters, Nebulae, and Galaxies

M 52 (NGC 7654)	★★★	⦿⦿⦿⦿	OC	MBUDR
Chart 12-2	Figure 12-1	m6.9, 12'	23h 24.8m	+61° 36'

NGC 7654 (M 52) lies on the extreme western edge of Cassiopeia, near the border with Cepheus. Charles Messier discovered M 52 by chance on the evening of 7 September 1774 as he scanned that area of the sky to locate a comet that had been reported by other observers.

M 52 is one of the easier Messier objects to find. To locate M 52, move your binocular or finder along the line from Shedir to Caph past Caph about one field (6.0°), as shown in Chart 12-2, and look for the m5 star 4-Cas, which is by far the most prominent in the vicinity. M 52 lies about 40' dead S of 4-Cas.

In a 50mm binocular or finder, M 52 is visible with direct vision as a moderately bright patch of nebulosity, with an m8 star prominent near the W edge. Averted vision reveals a hint of a few dim stars being on the edge of being resolved, but the object remains nebulous. With our 10" scope at 42X and 90X, M 52

reveals itself as a bright, rich, highly condensed open cluster that stands out well from the surrounding star field. About 75 stars are visible, most of which are white or blue-white. A few contrasting yellowish stars are visible, including the yellowish m8 star on the W edge of the cluster. The presence of many dim stars that are not individually resolved contributes to the appearance of patchy nebulosity throughout the cluster.

The distance to M 52 is not well known. Various sources quote it as between 3,000 and 7,000 ly, with most estimates clustering around 5,000 ly. At that distance, the 13' apparent size of this object translates to an actual extent of about 19 ly. As you might expect from the preponderance of bluish stars, M 52 is a relatively young open cluster, with estimates of its age ranging from 35 million to 50 million years, or about half the age of M 45 (the Pleiades).

FIGURE 12-1.

NGC 7654 (M 52) at upper left and NGC 7635 at lower right (60' field width)

Image reproduced from Digitized Sky Survey courtesy Palomar Observatory and Space Telescope Science Institute

NGC 7635	★	⊖⊖⊖⊖	EN	MBUDR
Chart 12-2	Figure 12-1	m99.9, 16.0' x 6.0'	23h 20.7m	+61° 12'

NGC 7635, also known as the Bubble Nebula, is easy to find but notoriously difficult to see. In fact, we consider the Bubble Nebula to be the most difficult of the objects covered in this book. Although very skilled observers under ideal conditions have reported seeing the Bubble Nebula in scopes as small as 4", most experienced observers agree that this object is extremely difficult to see even under good conditions in 10" and 12" scopes, and challenging even in large scopes.

We tried in many sessions at dark-sky sites over several years to log the Bubble Nebula, with scopes ranging from 10" to 17.5", at various magnifications, and with or without filters. We never saw the faintest hint of nebulosity, despite knowing exactly where to look. We finally succeeded one very clear night in autumn 2006, using a friend's 17.5" telescope from, oddly enough, one of our suburban observing sites. The combination of no moon or local lights, full dark adaptation, extremely transparent skies, a narrowband filter, and a large telescope with clean optics finally allowed us to see the Bubble Nebula. Even at that, we were able to see only the faintest hints of nebulosity using averted vision. But, hey, that counts. So we logged NGC 7635, greatly relieved to have finally bagged it.

To find NGC 7635, begin by locating the m7 star SAO 20562 in your finder or a wide-field eyepiece. SAO 20562 lies about 45' SW of M 52 and is by far the most prominent star in the vicinity. The m9 star SAO 20575 lies about 6' NE of SAO 20562, and is centered in the nebulosity of NGC 7635. Unfortunately, SAO 20562 is bright enough that its glare may subsume the dim tendrils of NGC 7635 unless your optics are pristine and the transparency is excellent. To avoid this, we suggest you use a high power eyepiece and place SAO 20562 just outside the field of view.

Don't be surprised if you see little or no nebulosity. Although NGC 7635 is listed as 16.0' x 6.0', that is a photographic extent. Visually, NGC 7635 appears much smaller to us, with faint nebulosity extending perhaps 3' x 1' NNW (north northwest) to SSE (south southeast) around SAO 20575.

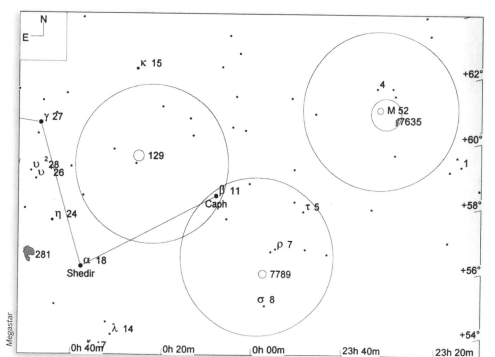

NGC 7789	★★★	⬤⬤⬤⬤		OC		MBUDR
Chart 12-2	Figure 12-2	m6.7, 15.0'		23h 57.4m		+56° 43'

We're surprised Messier missed the open cluster NGC 7789. Although we have never seen it without optical aid, various observers have reported seeing it naked-eye from very dark locations as a tiny patch of nebulosity. NGC 7789 was discovered by Caroline Herschel in 1783, one of her several original discoveries, and logged by her brother William as object H VI.30. At about two billion years old, NGC 7789 is one of the oldest known open clusters. Its distance is not well known, but is estimated at about 6,000 ly, which puts the actual size of the cluster between 45 and 50 ly in diameter.

To locate NGC 7789, place Caph at the NE edge of your binocular or finder field, and look for the prominent NNW-SSE (from north northwest to south southeast) chain of three m5 stars, 5-τ (tau), 7-ρ (rho), and 8-σ (sigma), shown in Chart 12-2. NGC 7789 lies about halfway between 7-ρ (rho) and 8-σ (sigma), and is visible with direct vision in a 50mm instrument as a bright nebulous patch with no stars resolved. At 90X in our 10" scope, NGC 7789 presents a striking appearance. It is extremely rich and very condensed, with 100+ stars from m11 to m13 visible, embedded in the nebulosity of hundreds of dimmer, unresolved stars. The stars that comprise the 5' core are uniformly distributed, and are surrounded by a more sparsely populated band with knots and gaps. For an interesting visual experience, compare this object to M 37 in Auriga. (As long as you have NGC 7789 in view, you might just as well observe the nearby multiple star 8-σ, which lies less than 1° S of NGC 7789 and is described at the end of this chapter.)

FIGURE 12-2.

NGC 7789 (60' field width)

Image reproduced from Digitized Sky Survey courtesy Palomar Observatory and Space Telescope Science Institute

NGC 129	★★	✧✧✧✧	OC	MBUDR
Chart 12-2	Figure 12-3	m6.5, 21.0'	00h 29.9m	+60° 13'

NGC 129 is a large, bright, open cluster that is easy to locate, about halfway on the line between m2.3 Caph and m2.2 27-γ (gamma). The center of the cluster lies about 15' N of a prominent m6 field star. NGC 129 is extremely loose and scattered, with little noticeable concentration, which makes it very difficult to discriminate its members from the background star field.

With a 7X50 binocular using averted vision, about half a dozen stars of magnitude 9 through 12 are visible, with the light from many unresolved stars providing a noticeable surrounding nebulosity. A pair of m9 stars separated by about 1.5' lie on the E edge of the cluster, but are not members. At 42X in our 10" reflector, about 50 stars are visible from m9 through m13 in the form of a wedge spanning about 15' with a dark central gap. A prominent triangle of three m9 stars lies at the center of the cluster, with a fourth m9 star near the NE edge of the cluster.

FIGURE 12-3.

NGC 129 (60' field width)

Image reproduced from Digitized Sky Survey courtesy Palomar Observatory and Space Telescope Science Institute

NGC 281	★★	✧✧✧✧	OC/EN	MBUDR
Chart 12-3	Figure 12-4	m99.9, 28.0' x 21.0'	00h 53.0m	+56° 38'

NGC 281 is a very dim emission nebula embedded in a rather ordinary open cluster. To locate NGC 281, place m2.3 Shedir near the W edge of your finder field and locate m3.5 24-η (eta) about 1.7° to the NE. With Shedir as the apex of an isosceles triangle, 24-η forms one corner of the base and NGC 281 the other. No hint of a cluster is visible in our 9X50 finder, but merely a random field of dim stars.

At 42X in our 10" reflector, the cluster becomes apparent, although it is very poor, loose, and sparse. Eight stars are visible from m7 through m10, with another 15 or so visible down to m13. The cluster is not concentrated, and is difficult to discriminate from background field stars. Without filtration, no nebulosity is visible with direct or averted vision. With a narrowband or O-III filter, patches of faint nebulosity are visible near the NW edge of the cluster. (As long as you have NGC 281 in view, you might just as well observe the nearby multiple star 24-η, which lies about 1.3° NNW of NGC 281 and is described at the end of this chapter.)

FIGURE 12-4.

NGC 281 (60' field width)

Image reproduced from Digitized Sky Survey courtesy Palomar Observatory and Space Telescope Science Institute

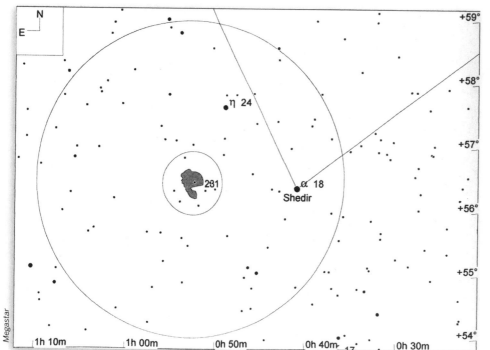

Megastar

CHART 12-3.

NGC 281 (7.5° field width; 5° finder circle; 1° eyepiece circle; LM 9.0)

NGC 457	★★★	✸✸✸✸	OC	MBUDR
Chart 12-4	Figure 12-5	m6.4, 13.0'	01h 19.6m	+58° 17'

NGC 457 is one of our favorite open clusters. It lies 2.1° SSW of m2.7 Ruchbah, where it is easy to locate by placing Ruchbah in the center of the finder field and looking for the nebulous glow around the lovely yellow/blue m5.0 double star 34-ϕ (phi) near the SSW edge of the finder field.

With a 7X50 binocular using averted vision, a bright scattering of a dozen or so stars of magnitude 8 through 12 are visible, extending generally NW from 34-ϕ (phi) and its companion. At 90X in our 10" reflector, the cluster is gorgeous. About three dozen stars are visible from m9 through m13. Most are members of the chain that extends NW from 34-ϕ (phi), curves N, and finally E, where it terminates in an m9 star. Barbara describes this cluster as a brilliant large diamond set in a scattering of smaller diamonds.

FIGURE 12-5.

NGC 457 (60' field width)

Image reproduced from Digitized Sky Survey courtesy Palomar Observatory and Space Telescope Science Institute

CHART 12-4.

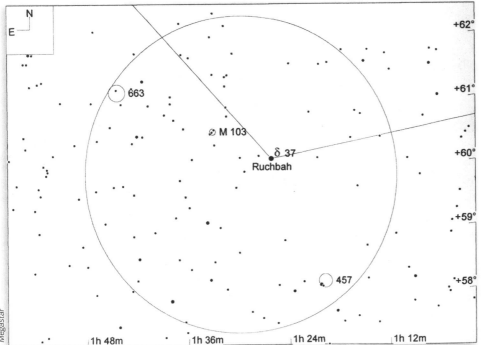

*NGC 457, NGC 581 (M 103), and
NGC 663 (7.5° field width; 5°
finder circle; LM 9.0)*

M 103 (NGC 581)	★★★	✦✦✦✦	OC	MBUDR
Chart 12-4	Figure 12-6	m7.4, 6.0'	01h 33.4m	+60° 40'

The open cluster NGC 581 (M 103) was first logged by Pierre Méchain in early 1781, and later added by Charles Messier as the final object in his list. (Messier objects M 104 through M 110 were added to Messier's list retroactively after his death.) M 103 has a spotted history. Until the 1930s, astronomers debated whether M 103 was a true open cluster or merely a random collection of field stars. The status of M 103 was finally resolved when measurements of the proper motion of its stars confirmed that they were gravitationally bound and therefore members of the same cluster. Even today, the stellar population of M 103 is in question, with various sources giving values that range from as few as 25 stars to as many as 172.

M 103 is easy to locate. It lies 1° ENE (east northeast) of m2.7 Ruchbah, where it is visible as a distinct hazy patch in a 50mm binocular or finder. Beginners sometimes confuse M 103 with NGC 663, which is more prominent in a binocular or finder, but lies 2.7° ENE of Ruchbah.

Our 7X50 binoculars reveal M 103 as a moderately bright triangular or fan-shaped patch of nebulosity with no individual stars visible that stands out prominently from the background field stars. At 90X in our 10" reflector, the cluster is hard to isolate from the extremely rich surrounding star field. Talk about not being able to see the forest for the trees! To identify the cluster, look for a prominent m7.3 star that is the NW end of a 5' line to the SE formed by two more m8 stars. The m7.3 star lies near the NW edge of M 103. The third (SE) star in that line lies just outside the SE edge of the cluster.

FIGURE 12-6.

NGC 581 (M 103) (60' field width)

Image reproduced from Digitized Sky Survey courtesy Palomar Observatory and Space Telescope Science Institute

NGC 663	★★★	⚭⚭⚭⚭	OC	MBUDR
Chart 12-4	Figure 12-7	m7.1, 16.0'	01h 46.3m	+61° 13'

NGC 663 is another of the objects that we're surprised Messier missed. It is larger and brighter than M 103, richer, and somewhat better detached from the surrounding star field. NGC 663 lies just 1.7° ENE of M 103, and is visible in the same binocular or finder field. (In fact, using a 40mm Pentax XL eyepiece in our 10" reflector, we're just able to fit both M 103 and NGC 663 in the same eyepiece field.)

With Ruchbah centered in the binocular or finder field, NGC 663 is visible as a rather large nebulous patch at or just beyond the ENE edge of the field. With a 7X50 binocular, we count 3 m8 stars and 8 m9/m10 stars embedded in the nebulosity. At 90X in our 10" reflector, about 50 stars are visible down to m13. Several fine double stars are embedded in the cluster, including the pretty ninth magnitude pairs STF 151 and STF 152 just W of the cluster center, and STF 153 on the NE edge of the core.

FIGURE 12-7.

NGC 663 (DSS image, 60' field width)

Stock 2	★★	⚭⚭⚭	OC	MBUDR
Chart 12-5	Figure 12-8	m4.4, 60.0'	02h 15.6m	+59° 32'

Stock 2 is a large, bright, moderately rich, scattered open cluster that is best viewed with a binocular or at very low power in a telescope. Although its visual magnitude is relatively bright at 4.4, that light is spread across a full degree of sky, so the surface brightness is lower than you might expect.

Stock 2 is a bit harder to locate than the preceding objects we've covered in this chapter, because it's located farther from bright stars. It lies 6.2° dead E of Ruchbah, so you can center Stock 2 in your binocular or finder by placing Ruchbah on the W edge of the field and sweeping E about half a field. Alternatively, if you locate the Double Cluster (see the Perseus chapter), you can center the Double Cluster in your finder or binocular field and look for Stock 2 at the NNW edge of the field.

With a 7X50 binocular, about a dozen m8/m9 stars are visible, most densely scattered across the middle and S parts of the cluster, with another 20 or so stars down to m10 filling out the cluster. At 42X in our 10" reflector, about 50 stars are visible down to m12, with numerous chains, clumps, and dark gaps.

FIGURE 12-8.

Stock 2 (60' field width)

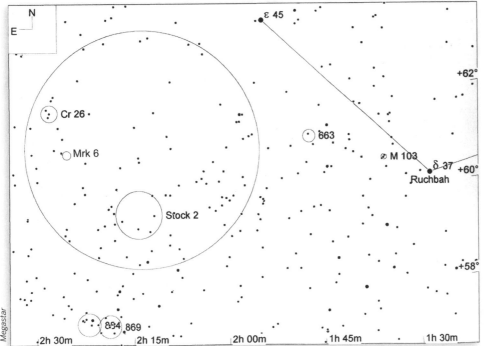

Megastar

Markarian 6	★	✦✦✦	OC	MBUDR
Chart 12-5	Figure 12-9	m7.1, 4.5'	02h 29.6m	+60° 39'

Markarian 6 is a 4.5' N-S line of five stars, the S two of which form a hook to the W and are both double. Markarian 6 lies about 2° NE of Stock 2 and just under 1° SSW of Collinder 26. The easiest way to locate Markarian 6 is to place Stock 2 at the bottom of your finder or binocular field and look for Collinder 26 on the E edge of the field. Markarian 6 is visible as a short, tight, N-S line of m8/m9 stars 1° SSW of Collinder 26, with no nebulosity visible. We were surprised that this object had been assigned a label. It appears simply as a small group of stars that happens to form a line. In fact, this object is so obscure that it isn't cataloged by our MegaStar software, which is otherwise comprehensive.

FIGURE 12-9.

Markarian 6 (60' field width)

Image reproduced from Digitized Sky Survey courtesy Palomar Observatory and Space Telescope Science Institute

Collinder 26	★★	✺✺✺	OC	MBUDR
Chart 12-5	Figure 12-10	m6.5, 21.0'	02h 32.7m	+61° 27'

Collinder 26 is a moderately large, moderately bright open cluster that is one of the required objects for the Astronomical League Deep-Sky Binocular club. With our 7X50 binoculars, we see Cr 26 as a fairly prominent hazy patch of moderate size with 3 or 4 embedded m8/m9 stars. It's very unlikely that any of the nebulosity we see in our binoculars is the actual nebulosity associated with this cluster and visible in Figure 12-10. Instead, the visible haze is probably just the 40 or so unresolved stars of m10 and dimmer that make up the open cluster.

To locate Collinder 26, put Stock 2 on the S edge of your binocular field and look for Collinder 26 as a hazy patch on the E edge of the field. We have never observed Cr 26 telescopically. The night we logged it, we'd taken only our binoculars along to work on the Deep-Sky Binocular list.

FIGURE 12-10.

Collinder 26 (60' field width)

Image reproduced from Digitized Sky Survey courtesy Palomar Observatory and Space Telescope Science Institute

Collinder 36	★★	✺✺✺	OC	MBUDR
Chart 12-6	Figure 12-11	m7.0, 23.0'	03h 12.0m	+63° 12'

The open cluster Collinder 36 is a bit larger and a bit dimmer than Cr 26, but otherwise similar. In a 50mm finder or binocular, we see Cr 36 as a moderately large, relatively faint hazy patch with no stars resolved. At 90X in our 10" reflector, about two dozen m10/m11 stars are visible, scattered loosely over the 23.0' extent of the cluster.

We located Collinder 36 by placing the m3.4 star 45-ε (epsilon) on the W edge of our binocular or finder field and sweeping just under 9° (a bit less than two full fields) dead E. Alternatively, if you have already located Cr 26, you can locate Cr 36 by placing Cr 26 on the WSW edge of your finder or binocular field and looking for the nebulous patch of Cr 36 on the ENE edge of the field, about 4.9° distant.

FIGURE 12-11.

Collinder 36 (60' field width)

Image reproduced from Digitized Sky Survey courtesy Palomar Observatory and Space Telescope Science Institute

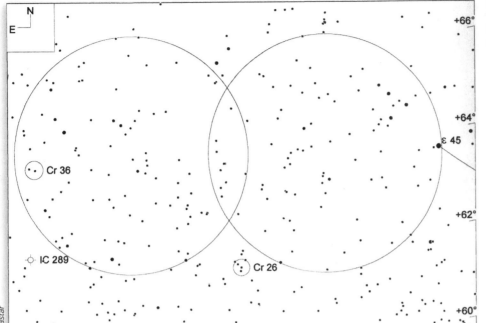

CHART 12-6.

Cr 36 (10° field width; 5° finder circles; LM 9.0)

Megastar

IC 289	★	◔◔	PN	MBUDR
Chart 12-7	Figure 12-12	m12.3, 0.6'	03h 10.3m	+61° 19'

IC 289 is an unremarkable planetary nebula that lies 1.9° S of Collinder 36. To locate IC 289, center Cr 36 in your finder and look near the S edge of the finder for the arc of three m8 stars shown in Chart 12-7. IC 289 lies about 23' N of the W star in that arc. Center that area in your finder, and use your low-power eyepiece to locate IC 289. At 42X in our 10" Newtonian, IC 289 appears stellar, with perhaps a hint of fuzziness. (You can identify IC 289 by "blinking" it, or passing a narrowband or O-III filter between your eye and eyepiece.) At 125X with a narrowband filter, the object is obviously non-stellar, but even averted vision reveals no detail. When we boosted magnification to 240X, the object disappeared entirely.

FIGURE 12-12.

IC 289 (60' field width)

Image reproduced from Digitized Sky Survey courtesy Palomar Observatory and Space Telescope Science Institute

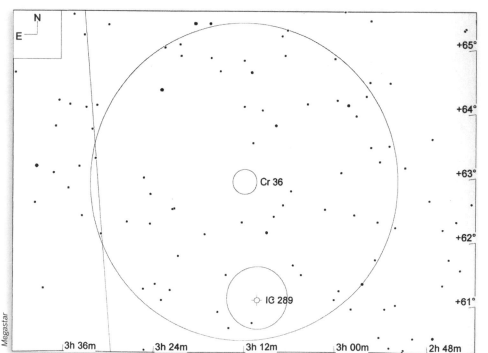

Megastar

NGC 185	★★	❂❂❂	GX	MBUDR
Chart 12-8, 12-9	Figure 12-13	m10.1, 12.9' x 10.1'	00h 39.0m	+48° 20'

NGC 185 is a small, faint galaxy located in Cassiopeia near the border with Andromeda. Although its listed magnitude of 10.1 implies that it is moderately bright, its surface brightness of 13.7 makes it a challenging object in small scopes. The cataloged extent of 12.9' x 10.1' is photographic; visually, NGC 185 appears much smaller. At 90X in our 10" reflector, NGC 185 appears as a faint oval of about 2.5' x 1.0', oriented NE-SW, with some brightening toward the nearly stellar core, and lying about midway on a 20' line between two m8/9 stars. No mottling or other surface detail is visible. Boosting magnification to 125X and 180X reveals no additional detail.

To locate NGC 185, put Shedir on the N edge of your finder, as shown in Chart 12-8, and look for m3.7 17-ζ (zeta), which appears prominently near the center of the finder field. Shift the finder S to put 17-ζ (zeta) at the N edge of the field, and look for the fifth magnitude stars 25-ν (nu) and 19-ξ (xi), which appear prominently in the SE quadrant of the finder. Continue shifting the finder S until 25-ν (nu) and 19-ξ (xi) are near the N edge of the field, and look for the prominent m5 stars 22-o (omicron) and 20-π (pi) to come into view. NGC 185 is invisible in the finder, but lies about 1° dead W of 22-o (omicron), about 12' E of an m8 star that is visible in the finder.

Alternatively, you can find NGC 185 easily if you can locate the m4.3 star 42-φ (phi) Andromedae. Place that star on the E edge of your finder and look for the two Cassiopeia stars 22-o (omicron) and 20-π (pi) on the W edge of the finder field. Locate NGC 185 from 22-o as described in the preceding paragraph.

FIGURE 12-13.

NGC 185 (60' field width)

Image reproduced from Digitized Sky Survey courtesy Palomar Observatory and Space Telescope Science Institute

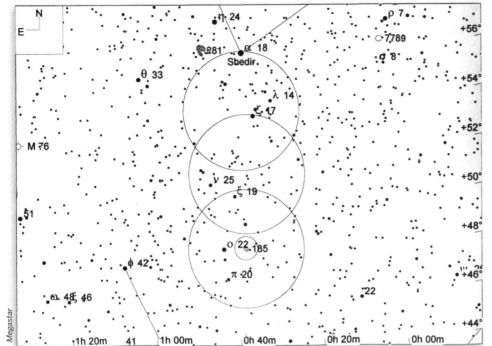

CHART 12-8.

NGC 185 from Shedir (20° field width; 5° finder circles; 1° eyepiece circle; LM 9.0)

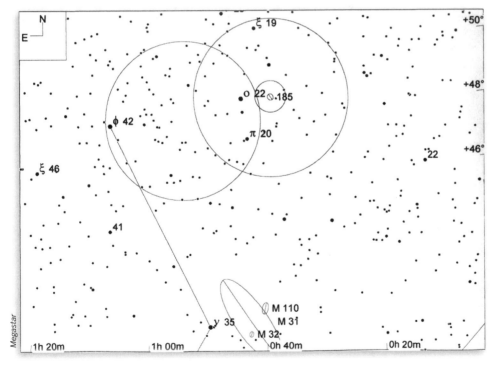

CHART 12-9.

NGC 185 from 42 Andromedae (15° field width; 5° finder circles; 1° eyepiece circle; LM 9.0)

Collinder 463	★★	✦✦✦	OC	MBUDR
Chart 12-10	Figure 12-14	m5.7, 36.0'	01h 47.6m	+71° 46'

Collinder 463 is a large, bright, loose open cluster located near the N border of Cassiopeia with Cepheus. Like many open clusters, Collinder 463 is best viewed with a binocular or at low magnification in a telescope. Cr 463 lies 8.1° N of m 3.4 45-ε (epsilon) Cas, the nearest bright star, and nestled at the N end of the arc of the four m4 and m5 stars 36-ψ (psi), 46-ω (omega), 48-, and 50-Cas. To locate Cr 463, place 45-ε (epsilon) at the S edge of your binocular or finder, and move it about one full field N until the N-S (north to south) chain 46-ω (omega), 48-, and 50-Cas comes into view. Cr 463 forms an equilateral triangle with the two N stars (48 and 50), and is visible in a 50mm binocular or finder.

With a 50mm binocular, we see two m8 and a dozen or so m9/m10 stars scattered across a field about the size of the full moon, with unresolved dimmer stars providing a faint nebulous haze. At 90X in our 10" reflector, about three dozen stars are visible from m8 down to m13, with the magnitudes distributed fairly evenly over that range. A chain of m9/m10 stars arcs through the SW quadrant, and several prominent clumps exist in the W quadrants. The cluster is quite loose and scattered, and it's difficult to tell at the eyepiece where the cluster ends and the background field stars begin. One of the m8 stars lies just W of the cluster center, with another m8 star just inside the NW boundary of the cluster and a prominent m9 star just inside the SE boundary.

FIGURE 12-14.

Collinder 463 (60' field width)

Image reproduced from Digitized Sky Survey courtesy Palomar Observatory and Space Telescope Science Institute

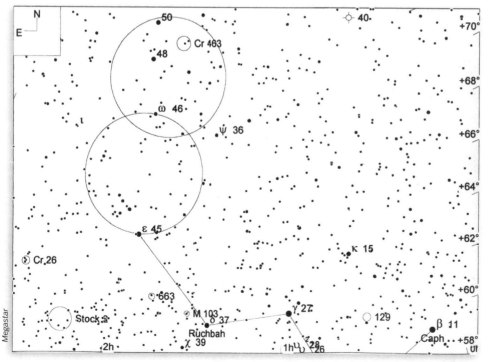

CHART 12-10.

Cr 463 (20° field width; 5° finder circles; LM 9.0

Multiple Stars

24-eta (STF 60AB)	★★★	✹✹✹✹	MS	UD
Chart 12-3		m3.5/7.4, 13.0", PA 320° (2006)	00h 49.1m	+57° 49'

The magnificent m3.5 double star 24-η (eta) lies 1.7° NE of Shedir, where it was first logged by William Herschel in 1779. At 125X in our 10" reflector, 24-η splits easily and provides a beautiful color contrast, with a golden-yellow primary and a reddish-purple secondary. 24-η is a visual double, what we call a "fast mover." Its orbital period is about 500 years, during which the apparent separation of the primary and companion range from 5" to 16". Other than Albireo in Cygnus and 145-CMa (often called the Winter Albireo), this is one of our favorite double stars.

8-sigma (STF 3049AB)	★★★	✹✹✹✹	MS	UD
Chart 12-2		m5.0/7.2, 3.2", PA 327° (2004)	23h 59.0m	+55° 45'

The m5 double star 8-σ (sigma) lies 3.7° SSW of Caph and about 1° S of the prominent open cluster NGC 7789, so it is easy to locate. At 42X in our 10" reflector, 8-σ appears elongated, but does not split. At 125X, the pair splits cleanly and shows a striking color contrast, with the blue-white m5 primary and a lemon-yellow m7 companion lying in a star field so rich it almost appears to be an open cluster.

13

Cepheus, The King

NAME: Cepheus (SEE-fee-us)

SEASON: Autumn

CULMINATION: midnight, 29 September

ABBREVIATION: Cep

GENITIVE: Cephei (SEE-fee-ee)

NEIGHBORS: Cam, Cas, Cyg, Dra, Lac, UMi

BINOCULAR OBJECTS: NGC 7160, NGC 7235

URBAN OBJECTS: NGC 7160, 27-delta (STFA 58)

Cepheus is a relatively dim circumpolar constellation of medium size, ranking 27th among the 88 constellations. It covers 588 square degrees of sky, or about 1.4%. Despite the fact that its brightest star is only magnitude 2.5, Cepheus is easy to recognize because its five brightest stars form a shape that resembles an ice cream cone or a child's drawing of a house.

Cepheus is an ancient constellation that represents King Cepheus of Æthiopia, the husband of Cassiopeia and the father of Andromeda. To learn more about the myths surrounding Cepheus, read the introduction of the Cassiopeia chapter.

Although Cepheus is not particularly rich in DSOs or interesting multiple stars, it is notable for the presence of the star 27-δ (delta) Cephei, which was the first of the Cepheid variables (although they take their name from Cepheus, the name describes a class of variables that are found all over the heavens) to be discovered and measured. The importance of Cepheid variables to cosmology cannot be overstated. Cepheid variables have extremely regular periods that correlate directly with their absolute magnitude.

That means that if the period of a Cepheid variable is known, so too is its actual luminosity. Measuring the apparent luminosity tells astronomers the distance of the Cepheid variable with high accuracy. Because Cepheid variables are inherently bright, it's possible to observe Cepheid variables embedded in distant clusters and galaxies, which allows the distance of those objects to be calculated very accurately.

Cepheus culminates at midnight on 29 September, and, for observers at mid-northern latitudes, is best placed for evening viewing during the late autumn and early winter months.

TABLE 13-1.

Featured star clusters, nebulae, and galaxies in Cepheus

Object	Type	Mv	Size	RA	Dec	M	B	U	D	R	Notes
NGC 0040	PN	10.7	1.1 x 1.0	00 13.0	+72 31					◉	Class 3b+2; central star m11.5
NGC 6939	OC	7.8	7.0	20 31.5	+60 39					◉	Cr 423; Mel 231; Class II 1 r
NGC 6946	Gx	9.6	11.6 x 9.8	20 34.9	+60 09					◉	Class SAB(rs)cd; SB 14.0
NGC 7129	OC	11.5	8.0	21 42.0	+66 05					◉	Cr 441; Class IV 2 p n; listed as EN
NGC 7160	OC	6.1	7.0	21 53.8	+62 36			◉	◉		Cr 443; Class I 3 p
NGC 7235	OC	7.7	4.0	22 12.6	+57 17				◉		Cr 447; Class II 3 m

TABLE 13-2.

Featured multiple stars in Cepheus

Object	Pair	M1	M2	Sep	PA	Year	RA	Dec	UO	DS	Notes
8-beta	STF 2806Aa-B	3.2	8.6	13.2	248	1999	21 28.7	+70 34		◉	Alfirk
HD 206267	STF 2816AD	5.7	8.1	19.7	339	2003	21 39.0	+57 29		◉	
HD 206267	STF 2816AC	5.7	8.1	11.7	120	2003	21 39.0	+57 29		◉	
17-xi*	STF 2863Aa-B	4.4	6.5	8.3	274	2006	22 03.8	+64 38		◉	
27-delta	STFA 58AC	4.1	6.1	40.6	191	2004	22 29.2	+58 25	◉	◉	

CHART 13-1.

The southern (main) portion of the constellation Cepheus (field width 35°)

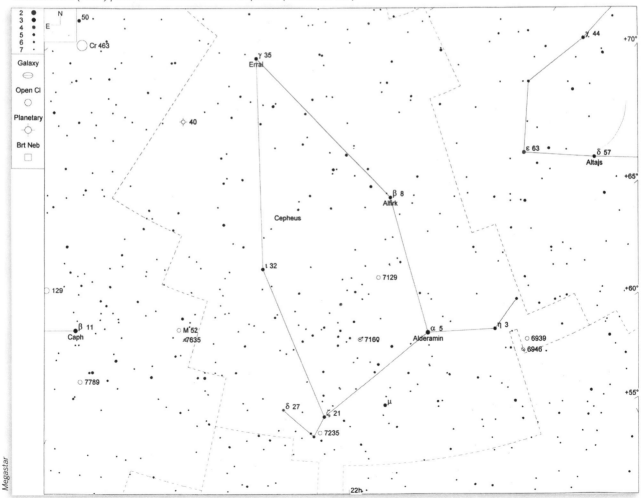

Megastar

Clusters, Nebulae, and Galaxies

NGC 40	★★	◉◉	PN	MBUDR
Chart 13-2	Figure 13-1	m10.7, 1.1' x 1.0'	00h 13.0m	+72° 31'

NGC 40 is a large, bright planetary nebula located near the m3.2 star Errai (35-γ), the point of the "ice cream cone" of Cepheus. To locate NGC 40, put 35-γ (gamma) on the NW edge of your finder field and look for the m5/m6 pair 21- and 23-Cas near the SE edge of the field. Once you locate that pair, move the finder field S until the m5.9 star SAO4229/HD4440 appears prominently near the E edge of the field. Although it is invisible in a 50mm finder, NGC 40 lies 2.6° (about half a finder field) W of that star, at the S end of a long N-S (north to south) chain of m7/m8 stars that are visible in the finder. Center the finder in that area and use a narrowband or O-III filter with your lowest power eyepiece to identify the planetary by "blinking" it (moving the filter between your eye and the eyepiece).

At 42X in our 10" reflector, NGC 40 appears as a slightly fuzzy m10/11 star that forms a flat triangle with an m9 star about 4' to the ENE (east northeast) and a second m9 star about 4' to the SW. After confirming its identity by blinking it with our Ultrablock narrowband filter, we increased magnification. At 125X, NGC 40 appears as a flat, featureless disk, slightly elongated NE-SW. Bumping the magnification to 240X showed no additional detail. With the Ultrablock filter at 125X the disc appeared evenly illuminated, but at 240X with the Ultrablock we were able to see some brightening on the E and W edges of the disc. Although the central star is listed at m11.5, we were unable to see any hint of it in our 10" scope, although that may have been due to mediocre seeing. Others have reported seeing the central star in scopes as small as 8", and we have seen that star clearly in an observing buddy's 17.5" reflector.

FIGURE 13-1.

NGC 40 (60' field width)

Image reproduced from Digitized Sky Survey courtesy Palomar Observatory and Space Telescope Science Institute

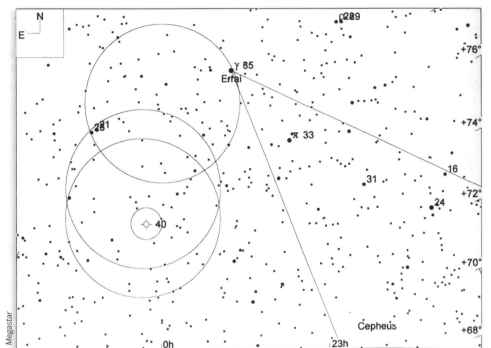

CHART 13-2.

NGC 40 (15° field width; 5° finder circles; 1° eyepiece circle; LM 9.0

NGC 6939	★★	ᕷᕷᕷᕷ	OC	MBUDR
Chart 13-3, 13-4	Figure 13-2	m7.8, 7.0'	20h 31.5m	+60° 39'

NGC 6939 is a moderately large, fairly bright, condensed open cluster of stars of relatively uniform brightness that is well detached from the surrounding star fields and moderately well concentrated. To find NGC 6939, locate the m3.4 star 3-η (eta) and the m4.2 star 2-θ (theta), both of which are naked-eye stars and lie a finder field or so W of m2.5 Alderamin. Place 2-θ at the N edge of your finder field and 3-η at the NE edge, and NGC 6939 should be about centered in the finder. Use your lowest power eyepiece to locate NGC 6939. If that eyepiece has a true field of 40' or greater, you'll be able to put both NGC 6939 and NGC 6946 (see the following section) in the same eyepiece field.

At 42X in our 10" reflector, NGC 6939 appears as a grainy patch of nebulosity with a dozen or so m11 stars resolved. Boosting magnification to 125X reveals 50+ stars, mostly of m12 and dimmer that are grouped in lumps and chains with darker gaps separating them. NGC 6939 is a very pretty, albeit subtle, open cluster for those with dark skies and 8" or larger instruments.

FIGURE 13-2.

NGC 6939 at upper right and NGC 6946 at lower left (60' field width)

Image reproduced from Digitized Sky Survey courtesy Palomar Observatory and Space Telescope Science Institute

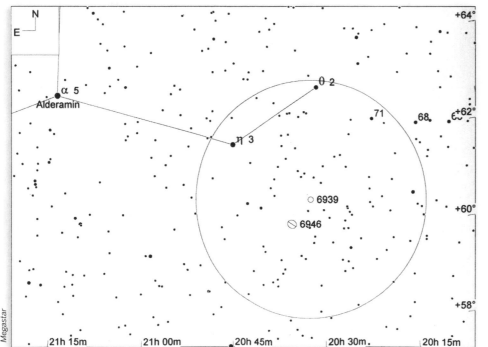

CHART 13-3.

NGC 6939 (10° field width; 5° finder circle; LM 9.0)

NGC 6946	★★	✧✧✧✧	GX	MBUDR
Chart 13-4, 13-3	Figure 13-2	m9.6, 11.6' x 9.8'	20h 34.9m	+60° 09'

NGC 6946 is a large, faint galaxy. Its cataloged magnitude of 9.6 is deceptive, as is the cataloged photographic extent of 11.6' x 9.8'. Its 14.0 surface brightness better indicates the difficulty of observing this object. In fact, the first time we observed NGC 6939, we had NGC 6946 in the same low-power eyepiece field, but didn't realize it was there until we referred to our charts.

NGC 6946 is easy to locate, lying about 40' SE of NGC 6939 (see the preceding section). At 42X in our 10" reflector, NGC 6946 is visible with averted vision as a small, very faint hazy patch with some central brightening. Boosting magnification to 125X makes the galaxy stand out better. Visually, the object appears as a 2.5' x 1.0' diffuse oval halo oriented NE-SW with noticeable brightening toward a nearly stellar core. No mottling or other surface detail is visible, and no hint of the spiral arms.

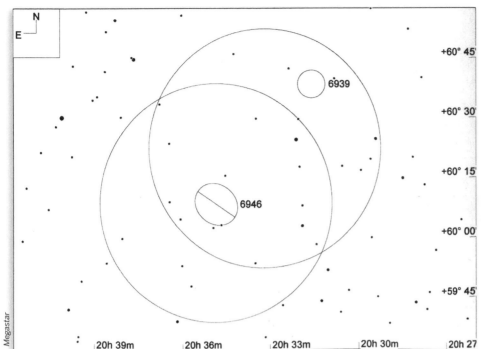

CHART 13-4.

NGC 6946 (2° field width; 1° eyepiece circles; LM 11.0)

Megastar

NGC 7129	★	◌	OC	MBUDR
Chart 13-5	Figure 13-3	m11.5, 8.0'	21h 42.0m	+66° 05'

NGC 7129 is a singularly unimpressive open cluster that is difficult to find because it is merely a small group of half a dozen dim stars embedded in a surrounding star field of stars of similar brightness. NGC 7129 lies in a region devoid of bright stars, 4.3° NE of m2.5 Alderamin, 2.7° NW of m 4.4 17-ξ (xi), and 4.6° SSE of m 3.2 Alfirk. The cluster contains 10 stars, of which the brightest six (at m9 to m11) form a pattern reminiscent of the Little Dipper or Delphinus. Unfortunately, this pattern is too dim to locate in a 50mm finder.

The easiest way we found to locate NGC 7129 was to place m4.4 17-ξ (xi) on the SE edge of our finder field and look for m5.4 7-Cep, which appears prominently near the NW edge. NGC 7129 lies on a line between those two stars, about two-thirds of the way from 17-ξ to 7-Cep. By placing the finder crosshairs at about that location and using our low-power eyepiece, we were able to locate the pattern described and identify the object unambiguously. Having done so, we wondered why we'd bothered, other than to log it as one of the required objects for completing the RASC list.

FIGURE 13-3.

NGC 7129 (60' field width)

Image reproduced from Digitized Sky Survey courtesy Palomar Observatory and Space Telescope Science Institute

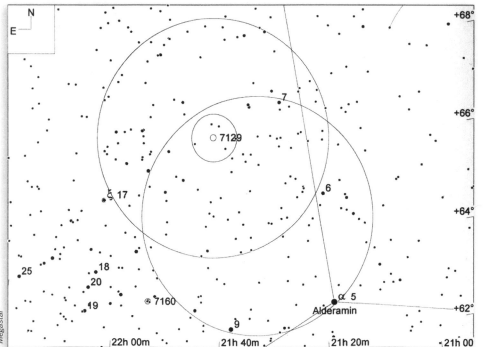

NGC 7160	★★	❀❀❀	OC	MBUDR
Chart 13-6	Figure 13-4	m6.1, 7.0'	21h 53.8m	+62° 36'

NGC 7160 is a small, bright, condensed open cluster of about a dozen stars of widely varying brightness that is well detached from the surrounding star fields. To find NGC 7160, place the m4.3 star 10-ν (nu) and the m4.8 star 9-Cep near the SW edge of your binocular finder field, and locate the m4.4 star 17-ξ (xi) near the NNW edge of the field. NGC 7160 is visible in a 50mm binocular or finder as a faint hazy patch about halfway on a line from 10-ν to 17-ξ.

With a 50mm binocular, we see NGC 7160 as a nebulous patch with five stars resolved, the brightest of which are an m7/m8 pair located near the center of the cluster. (The m6 star that appears prominently about 10' NW of the cluster is not a member.) An arc of three m10 stars is also faintly visible with averted vision on the W edge of the cluster. At 90X in our 10" reflector, the apparent nebulosity disappears, and another dozen or so stars of magnitude 11 and dimmer become visible with averted vision.

FIGURE 13-4.

NGC 7160 (60' field width)

Image reproduced from Digitized Sky Survey courtesy Palomar Observatory and Space Telescope Science Institute

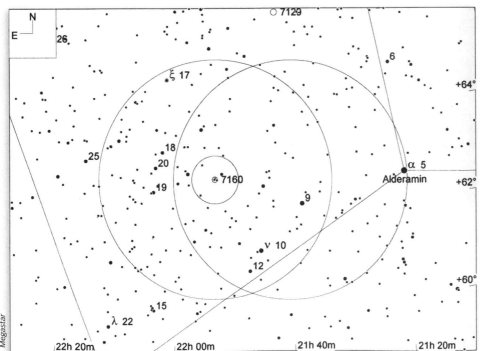

NGC 7235	★★	✦✦✦✦	OC	MBUDR
Chart 13-7	Figure 13-5	m7.7, 4.0'	22h 12.6m	+57° 17'

NGC 7235 is a relatively small, relatively bright, scattered open cluster with little concentration. NGC 7235 is very easy to locate because it lies just 25' NW of the naked-eye star 23-ε (epsilon), and is distinctly visible in a 50mm finder or binocular.

With a 50mm binocular, we see NGC 7235 as a relatively bright nebulous patch with only one or two m9/m10 stars resolved with averted vision. At 90X in our 10" reflector, another dozen or so stars from m10 through m12 are visible with averted vision.

FIGURE 13-5.

NGC 7235, with the planetary nebula Minkowski 2-51 visible at the upper left edge (60' field width)

Image reproduced from Digitized Sky Survey courtesy Palomar Observatory and Space Telescope Science Institute

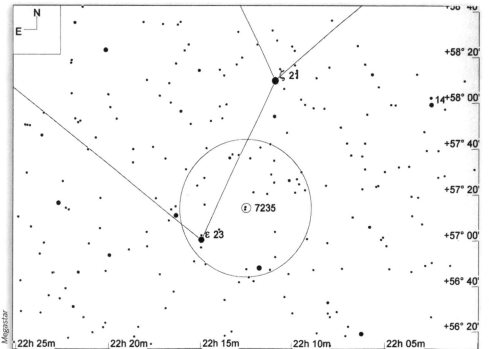

CHART 13-7.

NGC 7235 (3.5° field width; 1° eyepiece circle; LM 11.0)

Megastar

Multiple Stars

8-beta (STF 2806Aa-B)	★★	✧✧✧✧		MS		UD
Chart 13-1		m3.2/8.6, 13.2", PA 248° (1999)		21h 28.7m		+70° 34'

Beta Cephei, also called Alfirk, is the middle bright star on the W side of the Cepheus ice cream cone. At 90X in our 10" reflector, the primary is a pure white diamond, with the much dimmer companion a pretty aqua shade.

HD 206267 (STF 2816AD)	★★	✧✧✧		MS		UD
Chart 13-8		m5.7/8.1, 19.7", PA 339° (2003)		21h 39.0m		+57° 29'

HD 206267 (STF 2816AC)	★★	✧✧✧		MS		UD
Chart 13-8		m5.7/8.1, 11.7", PA 120° (2003)		21h 39.0m		+57° 29'

To locate HD 206267, place m4.2 23-ε (epsilon) on the E edge of your finder field and look for m4.0 μ (mu) Cephei near the NW edge of the field. Move the finder about half a field W and look for a m5.7 star that appears prominent in the finder field about 1.4° SSW of mu Cephei. That star is HD 206267, which is actually a triple star system. At 90X in our 10" reflector, the primary is a warm white, with the two much dimmer companions showing a cool white color tending toward bluish.

Mu Cephei, incidentally, is a more rewarding target than this multiple star. Mu Cephei is a long-period variable star with a striking color. At 90X in our 10" reflector, the color is a saturated yellow-orange. We thought that was odd, because Mu Cephei is also known as Herschel's Garnet Star, and it certainly didn't look garnet-colored to us. Later, we found out that in smaller instruments such as our 90mm refractor, Mu Cephei has a striking crimson red coloration. This apparently is a rare case of larger aperture being undesirable; our 10" reflector simply gathers too much light to allow us to see the star as the saturated red reported by Herschel and many later observers.

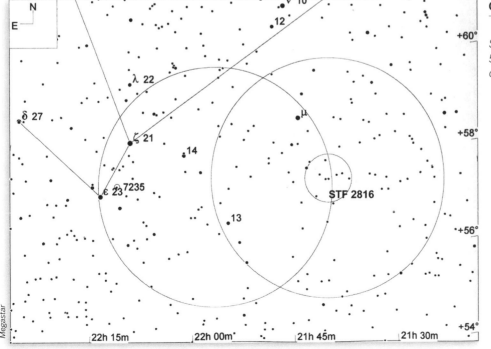

CHART 13-8.

STF 2816 (10° field width; 5° finder circles; 1° eyepiece circle; LM 9.0)

17-xi (STF 2863Aa-B)	★★	✺✺✺✺	MS	UD
Chart 13-9		m4.4/6.5, 8.3", PA 274° (2006)	22h 03.8m	+64° 38'

17-xi is a pretty double star that's easy to locate (it's a naked-eye star) and easy to split at 90X in our 10" reflector. The primary is yellowish, and the companion a distinct reddish-orange.

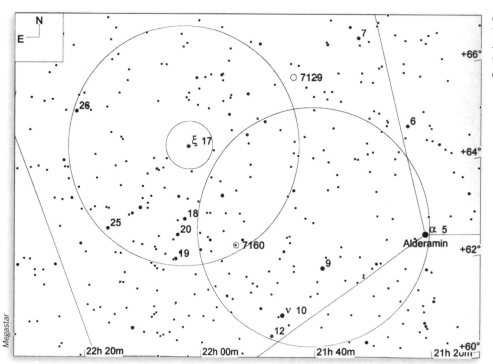

CHART 13-9.

17-xi Cephei (10° field width; 5° finder circles; 1° eyepiece circle; LM 9.0)

27-delta (STFA 58AC)	★★★	✺✺✺✺	MS	UD
Chart 13-1		m4.1/6.1, 40.6", PA 191° (2004)	22h 29.2m	+58° 25'

27-delta Cephei is a naked-eye star just off the SE corner of the Cepheus ice cream cone. Although we list the magnitude of the primary as 4.1, in fact it varies from m3.48 to m4.37 with a period of 5.4 days. The primary is warm white, with a bluish-white secondary.

14
Cetus, The Sea Monster

NAME: Cetus (SEE-tus)

SEASON: Autumn

CULMINATION: midnight, 15 October

ABBREVIATION: Cet

GENITIVE: Ceti (SEE-tee)

NEIGHBORS: Aqr, Ari, Eri, For, Psc, Scl, Tau

BINOCULAR OBJECTS: NGC 1068 (M 77)

URBAN OBJECTS: NGC 1068 (M 77)

Cetus is a large but relatively dim equatorial constellation that ranks 4th in size among the 88 constellations. It covers 1,231 square degrees of sky, or about 3.0%. Despite its relative paucity of bright stars, Cetus is easy to recognize because several of its brighter stars form a naked-eye pentagon pattern in an area of the sky that is nearly devoid of bright stars.

Cetus is an ancient constellation that represents the Sea Monster in the tale of King Cepheus of Æthiopia, his wife Cassiopeia, and his daughter Andromeda. On the advice of an oracle, Cepheus chained Andromeda to a rock near the shore as a sacrifice to Cetus. As Cetus approached a screaming Andromeda, the Hero Perseus happened to be sailing past and heard her cries. Perseus lept between Andromeda and Cetus, whipped out the head of Medusa, and showed it to Cetus, who promptly turned to stone. Perseus freed Andromeda from her chains, claimed his Hero's rights, and sailed off with Andromeda to live happily ever after.

Although Cetus has few interesting DSOs or multiple stars, it is notable for the presence of the star 68-o (omicron) Ceti, better known as Mira. Mira is the archetype for the nearly 6,000 known Mira variables. These are all long-period variable red giant stars with periods ranging from about 80 days to more than 1,000 days. (Mira has a period of 332 days, which may be slightly variable.) In addition to being the first discovered Mira variable, Mira is extraordinary for the range of magnitudes it covers during its period. The maximum and minimum magnitudes of Mira vary from cycle to cycle. At minima, Mira ranges from magnitude 8.6 to 10.1, the latter making it challenging to see in a 50mm finder or binocular. At maxima, Mira ranges from magnitude 2.0 (about as bright as Polaris) to 4.9.

Cetus culminates at midnight on 15 October, and, for observers at mid-northern latitudes, is best placed for evening viewing during the late autumn and early winter months.

TABLE 14-1.

Featured star clusters, nebulae, and galaxies in Cetus

Object	Type	Mv	Size	RA	Dec	M	B	U	D	R	Notes
NGC 246	PN	8.0	4.1	00 47.1	-11 52					◉	Class 3b
NGC 936	Gx	11.1	4.7 x 4.0	02 27.6	-01 09					◉	Class SB(rs)0+; SB 12.4
NGC 1068	Gx	9.6	7.1 x 6.0	02 42.7	-00 01	◉	◉	◉			M 77; Class (R)SA(rs)b; SB 10.6

TABLE 14-2.

Featured multiple stars in Cetus

Object	Pair	M1	M2	Sep	PA	Year	RA	Dec	UO	DS	Notes
86-gamma	STF 299AB	3.6	6.2	2.3	299	2002	02 43.3	+03 14		◉	

CHART 14-1.

The constellation Cetus (field width 60°)

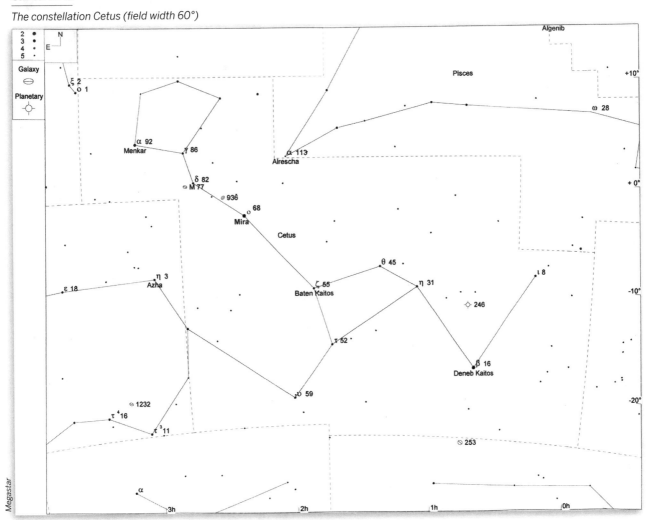

Megastar

Clusters, Nebulae, and Galaxies

NGC 246	★★★	◇◇◇	PN	MBUDR
Chart 14-2	Figure 14-1	m8.0, 4.1'	00h 47.1m	-11° 52'

NGC 246 is a large, bright planetary nebula. Although it appears round in photographs, with our 10" reflector at 125X it appears more like an apple with a large bite taken out of the SE edge. With our Ultrablock narrowband filter, the annulus edge is visible all the way around the circle, although extremely faint in the SE quadrant, but the disc itself in the SE quadrant is too dim to be visible. Two m13 stars are visible within the disc itself. An m12 star lies very near the NW edge of the disc, with two m13 stars just outside the disc on the SW and S edges.

NGC 246 is relatively easy to find. Put m3.5 31-η (eta) on the E edge of your finder field and follow the E-W chain of m5 stars 23-ϕ^4 (phi^4), 22-ϕ^3, 19-ϕ^2, and 17-ϕ^1. Although we see no hint of it in our 50mm finder, NGC 246 forms the S apex of an equilateral triangle with 19-ϕ^2 and 17-ϕ^1.

FIGURE 14-1.

NGC 246 with NGC 255 visible at upper left (60' field width)

Image reproduced from Digitized Sky Survey courtesy Anglo-Australian Observatory and Space Telescope Science Institute

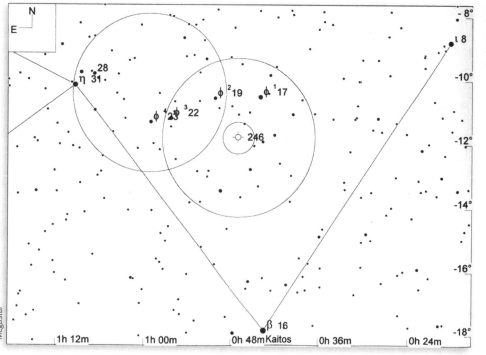

CHART 14-2.

NGC 246 (15° field width; 5° finder circles; 1° eyepiece circle; LM 9.0)

NGC 936	★★	☉☉☉	GX	MBUDR
Chart 14-3	Figure 14-2	m11.1, 4.7' x 4.0'	02h 27.6m	-01° 09'

NGC 936 is a moderately large, bright galaxy with cataloged surface brightness of 12.4. Despite this relatively high surface brightness, we find NGC 936 to be relatively elusive in the eyepiece. At 125X in our 10" reflector, a faint E-W (from east to west) oval halo about 2.5'x1.5' is visible, with a much brighter non-stellar core. No mottling or other surface detail is visible.

To locate NGC 936, begin by placing m4.1 82-δ (delta) at the NE edge of your finder field. Look for m5.4 75-Cet, which appears prominently near the center of the finder field. Move the finder W until the m5 pair 69- and 70-Cet come into view. NGC 936 lies 1.1 dead W of 75-Cet, almost exactly halfway between 75- and 70-Cet, where it should be visible with averted vision in your low power eyepiece.

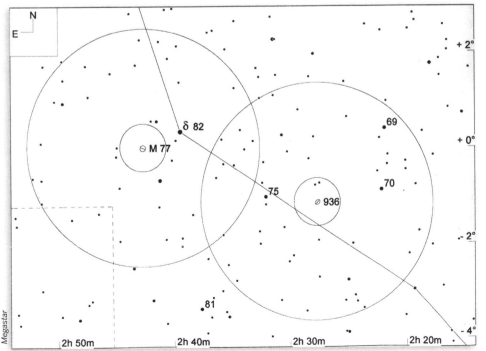

FIGURE 14-2.

NGC 936 with NGC 941 to its left and UGC 1945 below and left (60' field width)

Image reproduced from Digitized Sky Survey courtesy Anglo-Australian Observatory and Space Telescope Science Institute

CHART 14-3.

NGC 936 and NGC 1068 (M 77) (10° field width; 5° finder circles; 1° eyepiece circles; LM 9.0)

M 77 (NGC 1068)	★★★	✦✦✦✦	GX	MBUDR
Chart 14-3	Figure 14-3	m9.6, 7.1' x 6.0'	02h 42.7m	-00° 01'

NGC 1068, better known as Messier 77 or M 77, is a large, bright galaxy with a very high surface brightness of 10.6. Pierre Méchain discovered this galaxy on 29 October, 1780, and described its location to Charles Messier, who added it to his catalog as object 77 in December of that year. M 77 is very easy to find. It lies 52' ESE of the m4.1 star 82-δ (delta) Ceti, where it is visible in the same low-power eyepiece field. At 125X in our 10" reflector, M 77 shows a relatively bright 3' round halo surrounding a much brighter central core that covers more than half the visible disc. The spiral arms and dust lanes that are visible in photographs are too dim to resolve in our 10" scope.

Although the Astronomical League lists M 77 as a binocular object, they classify it as a "tougher" object with an 11X80 binocular and don't list it at all for 35mm and 50mm binoculars. We have tried several times to bag M 77 with our 50mm binoculars, but haven't succeeded even from relatively dark sites.

FIGURE 14-3.

NGC 1068 (M 77) (60' field width)

Image reproduced from Digitized Sky Survey courtesy Anglo-Australian Observatory and Space Telescope Science Institute

Multiple Stars

86-gamma (STF 299AB)	★★	◑◑◑◑		MS	UD
Chart 14-1		m3.6/6.2, 2.3", PA 299° (2002)		02h 43.3m	+03° 14'

Gamma Ceti is a pretty physical binary, with a pure white primary and a yellowish companion. Gamma Ceti is actually trinary, with the third member an m10.1 red dwarf lying 13.9' at position angle 306°. Gamma Ceti is easy to find. It's the southernmost star of the pentagon that forms the head of the sea monster.

15

Coma Berenices, Berenice's Hair

NAME: Coma Berenices (KOE-muh BAIR-uh-NEES-us)

SEASON: Spring

CULMINATION: midnight, 2 April

ABBREVIATION: Com

GENITIVE: Comae Berenices (KOE-meye BAIR-uh-NEES-us)

NEIGHBORS: Boo, CVn, Leo, UMa, Vir

BINOCULAR OBJECTS: NGC 4254 (M 99), NGC 4321 (M 100), Mel 111, NGC 4382 (M 85), NGC 4501 (M 88), NGC 4826 (M 64), NGC 5024 (M 53)

URBAN OBJECTS: Mel 111, NGC 4826 (M 64)

Coma Berenices is an inconspicuous, mid-size constellation, ranking 42nd in size among the 88 constellations. It covers 386 square degrees of the celestial sphere, or about 0.9%. Coma—most astronomers drop the "Berenices" part in casual conversation—occupies the center of the box defined by 0th magnitude Arcturus in Boötes, 2nd magnitude Denebola in Leo, 3rd magnitude Vindemiatrix in Virgo, and 3rd magnitude Cor Caroli in Canes Venatici. The brightest stars in Coma are only 4th magnitude, so with direct vision Coma appears to be a very sparsely populated area of the night sky. Nothing could be further from the truth. Coma is home to a multitude of objects.

Despite its lack of prominence, Coma is an ancient constellation, and is unique in at least one respect. While many constellations are named for mythological figures, Coma Berenices has the distinction of being named for an historical personage, Queen Berenice of Egypt. When her husband, Ptolemy III, departed on a military campaign, Berenice swore an oath to the goddess Venus to ensure Ptolemy's safe return. When he returned safely, Berenice made good on that promise by cutting off her beautiful hair. In admiration of her, the god Jupiter scattered her tresses amongst the stars, where they are still visible on a dark night as Coma Berenices, or Berenice's Hair.

With the exception of the huge, bright open cluster for which the constellation was named, and a few globular clusters, Coma is the realm of galaxies. Coma is home to scores of galaxies that are visible in amateur telescopes, including seven bright Messier galaxies, as well as the bright Messier globular cluster M 53.

In Coma, we go from star-hopping with a finder to galaxy-hopping with an eyepiece. In many constellations, the challenge is often simply to find an object that lurks dimly in an otherwise empty field of stars. In Coma, finding objects isn't difficult; scores of bright galaxies are visible even with a small telescope. The challenge is to identify which object (or objects) are in the field of view.

A low-power "finder" eyepiece is essential for locating objects in Coma, few of which are visible in a 50mm finder. The trick is to locate and identify a particular object in the eyepiece, and then eyepiece-hop from that object to nearby objects. Using an eyepiece with the widest possible field of view makes that job much easier.

Coma culminates at midnight on 2 April, and, for observers at mid-northern latitudes, is best placed for evening viewing during the late winter through mid-summer months.

TABLE 15-1.

Featured star clusters, nebulae, and galaxies in Coma Berenices

Object	Type	Mv	Size	RA	Dec	M	B	U	D	R	Notes
Mel 111	OC	1.8	275	12 25.1	+26 07			◉	◉		Class III 3 r
NGC 4559	Gx	10.5	10.8 x 4.3	12 36.0	+27 58					◉	Class SAB(rs)cd; SB 13.3
NGC 4494	Gx	9.8	4.8 x 3.6	12 31.4	+25 47					◉	Class E1-2; SB 12.3
NGC 4565	Gx	10.4	15.9 x 1.8	12 36.3	+25 59					◉	Class SA(s)b? sp; SB 13.1
NGC 4274	Gx	11.3	6.8 x 2.5	12 19.8	+29 37					◉	Class (R)SB(r)ab; SB 12.7
NGC 4414	Gx	11.0	4.3 x 3.1	12 26.4	+31 13					◉	Class SA(rs)c?; SB 11.3
NGC 4725	Gx	10.1	10.7 x 8.0	12 50.4	+25 30					◉	Class SAB(r)ab pec; SB 13.2
NGC 4826	Gx	9.4	10.1 x 5.4	12 56.7	+21 41	◉	◉	◉			M 64; Class (R)SA(rs)ab; SB 11.8
NGC 5024	GC	7.7	13.0	13 12.9	+18 10	◉	◉				M 53
NGC 4192	Gx	11.0	9.8 x 2.7	12 13.8	+14 54	◉					M 98; Class SAB(s)ab; SB 13.8
NGC 4254	Gx	10.4	5.4 x 4.7	12 18.8	+14 25	◉	◉				M 99; Class SA(s)c; SB 12.6
NGC 4321	Gx	10.1	7.5 x 6.3	12 22.9	+15 49	◉	◉				M 100; Class SAB(s)bc; SB 13.3
NGC 4382	Gx	9.1	7.1 x 5.5	12 25.4	+18 12	◉	◉				M 85; Class SA(s)0+ pec; SB 11.9
NGC 4501	Gx	10.4	7.0 x 3.7	12 32.0	+14 25	◉	◉				M 88; Class SA(rs)b; SB 12.7
NGC 4548	Gx	11.0	5.4 x 4.2	12 35.4	+14 30	◉					M 91; Class SB(rs)b; SB 13.4

TABLE 15-2.

Featured multiple stars in Coma Berenices

Object	Pair	M1	M2	Sep	PA	Year	RA	Dec	UO	DS	Notes
24	STF 1657	5.1	6.3	20.1	270	2004	12 35.1	+18 22		◉	

CHART 15-1.

The constellation Coma Berenices (field width 35°)

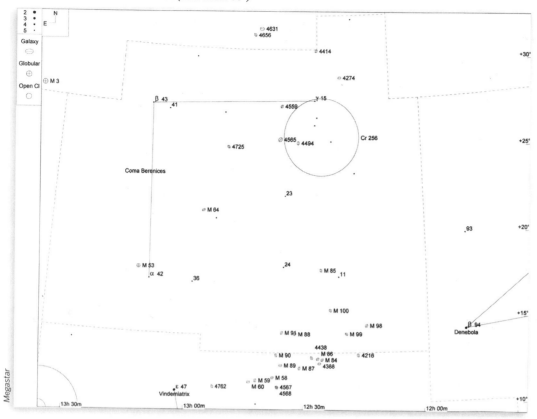

Clusters, Nebulae, and Galaxies

Cr 256 (Melotte 111)	★★★	✧✧✧✧	OC	MBUDR
Chart 15-1		m1.8, 275'	12h 25.1m	+26° 07'

Melotte 111, also designated Collinder 256, is the brightest object in Coma, and the one for which it was named. Scattered like diamonds on black velvet, Melotte 111 was seen by the ancients as Berenice's Hair. Mel 111 is a gigantic, bright Trumpler Class III 3 r open cluster that spans 275'—more than 4.5°—and shines at magnitude 1.8. That magnitude is deceptive, though. Because the light from Mel 111 is distributed more or less evenly across a huge area of the night sky, the object itself is not prominent, although it is visible to the naked eye from a dark site.

To locate Mel 111, draw an imaginary line between the prominent stars Cor Caroli in Canes Venatici and Denebola in Leo. Mel 111 is located at the halfway point, 14° from either star. From a dark site, you can glimpse Mel 111 with direct vision, and with averted vision it's clearly visible as a dim, nebulous patch of haze with many tiny embedded points of light. A binocular offers the best view of Mel 111. Even at very low power, a telescope simply has too narrow a field to take in the entire object.

Mel 111 is a beautiful sight in our 7X50 binoculars. A dozen or so m5 and m6 stars blaze prominently through the field, with a chain twisting from the N boundary to the S boundary and branching at the midpoint of the cluster into an E-W chain. About two dozen dimmer stars, down to m9, fill out the NW and SW quadrants of the cluster. Because of its large extent, the cluster appears scattered and sparse, but beautiful nonetheless. In a telescope, even at low magnification, the cluster loses its coherency as a cluster and appears as just random field stars.

Three of the other featured objects in Coma lie near the edges of Melotte 111, one of them, NGC 4494, actually within its boundaries. The m4 naked-eye star 15-γ (gamma) Com is the bright star near the N edge of Mel 111, and provides us with a jumping-off point to locate these other objects in the northern part of Coma.

NGC 4559	★★★	✹✹✹✹	GX	MBUDR
Chart 15-2	Figure 15-1	m10.5, 10.8' x 4.3'	12h 36.0m	+27° 58'

NGC 4559 is an interesting galaxy, although fainter than its magnitude of 10.5 and surface brightness of 13.3 implies. To locate NGC 4559, place m4 15-γ (gamma) Com near the W edge of the finder field, with the chain of m5 stars 14-, 16-, and 17-Com lying near the edge of the finder field in its SW quadrant, as shown in Chart 15-2. (Note that in Chart 15-2 the large circle at the lower right is not a finder circle, but delimits the extent of the open cluster Cr 256/Mel 111.) NGC 4559 is near the center of the finder field, where it should be visible in your low-power eyepiece. At 42X in our 10" Dob, NGC 4559 is faintly visible as a NW-SE (from northwest to southeast) elongated oval without surface detail. At 90X and 125X with averted vision, a 6' x 2' halo is visible, with considerably mottling, gradually brightening to a non-stellar central core. The halo appears to rest in a cup at the SE end formed by an arc of three stars, the two end stars m12 and the center star m13.

FIGURE 15-1.

NGC 4559 (60' field width)

Image reproduced from Digitized Sky Survey courtesy Palomar Observatory and Space Telescope Science Institute

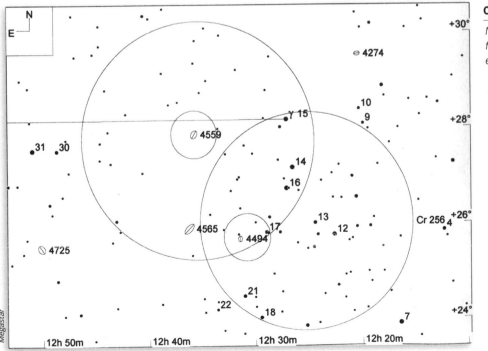

CHART 15-2.

NGC 4559, NGC 4494 (10° field width, 5.0° finder field, 1.0° eyepiece fields, LM 9.0)

NGC 4494	★★	✧✧✧	GX	MBUDR
Chart 15-2, 15-3	Figure 15-2	m9.8, 4.8' x 3.6'	12h 31.4m	+25° 47'

NGC 4494 is a bright galaxy (SB 12.3) that lies at the SE end of the arc of m5 stars 14-, 16-, and 17-Com shown in Chart 15-2. To locate NGC 4494, center the star 17-Com in your finder. (17-Com is a pretty m5.3/6.6 close double in the finder.) NGC 4494 lies 35' E of that star and 6' SSW of the m7.9 star SAO 82354, which is clearly visible in a 50mm finder.

At 42X in our 10" reflector, NGC 4494 is visible with direct vision as a bright nearly circular N-S oval core with a very faint surrounding halo, but shows no surface detail even with averted vision. At first glance at low magnification, it's easy to mistake NGC 4494 for a small, unresolved globular cluster. Higher magnification reveals no further detail.

FIGURE 15-2.

NGC 4494 (60' field width)

Image reproduced from Digitized Sky Survey courtesy Palomar Observatory and Space Telescope Science Institute

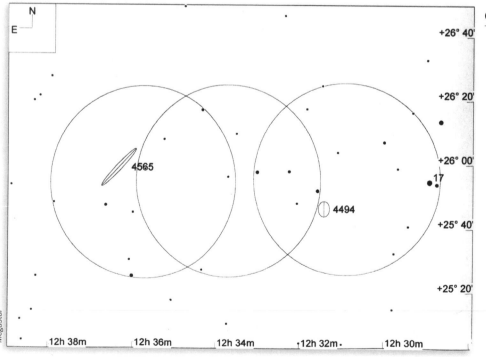

CHART 15-3.

NGC 4494 and NGC 4565 detail (2.5° field width, 1.0° eyepiece fields, LM 11.0)

NGC 4565	★★★	⊖⊖⊖	GX	MBUDR
Chart 15-2, 15-3	Figure 15-3	m10.4, 15.9' x 1.8'	12h 36.3m	+25° 59'

NGC 4565 is one of the most beautiful edge-on galaxies in the night sky. It's one of the best non-Messier galaxies, and many would vote it best, period. Barbara calls it the Little Sombrero, after the better-known M 104 in Virgo. It's as bright as its 10.4 magnitude suggests, and the cataloged surface brightness of 13.1 is, if anything, conservative.

NGC 4565 is relatively easy to locate. It lies 1.7° dead E of 17-Com, and can be reached simply by placing 17-Com on the W edge of the eyepiece field and drifting the scope about one full eyepiece field due E until NGC 4565 comes into view.

At 42X in our 10" reflector, NGC 4565 shows with direct vision a bulging bright 3' x 2' core. With averted vision, a fainter 10' x 1.5' halo is visible extending NW-SE. At 125X, significant additional detail is visible. The dust lane is conspicuous, even with direct vision, and some mottling and knotting is visible, particularly SE of the core. This is a magnificent galaxy, as good as many Messier galaxies.

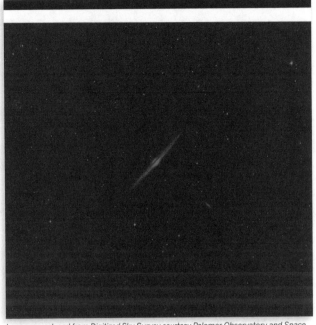

FIGURE 15-3.

NGC 4565 (60' field width)

NGC 4274	★★	⊖⊖⊖⊖	GX	MBUDR
Chart 15-4	Figure 15-4	m11.3, 6.8' x 2.5'	12h 19.8m	+29° 37'

NGC 4574 is a bright, pretty galaxy that, with its surface brightness of 12.7, is as prominent as most of the Messier galaxies in the Coma-Virgo cluster. NGC 4574 is easy to locate. It lies 2.0° NW of the naked-eye star 15-γ (gamma), and can be centered in the finder field simply by placing 15-γ near the SE edge of the field. The m6/7 pair 9-/10-Com lies just over 1.0° S of NGC 4574 and appear prominently in the finder, as does the m5.7 star SAO 82219 about 50' SW of NGC 4274.

At 90X in our 10" Newtonian, NGC 4274 is visible with direct vision as a moderately bright 1' x 4' E-W streak of light. With averted vision, significant additional detail is visible. A fainter 2' x 5' halo extends outward E and W from an oval core.

FIGURE 15-4.

NGC 4274, with IC 779 above it, NGC 4253 to the upper right, NGC 4245 at the right edge, and a line of three galaxies (NGC 4286, 4283, and 4278, left to right) below it (60' field width)

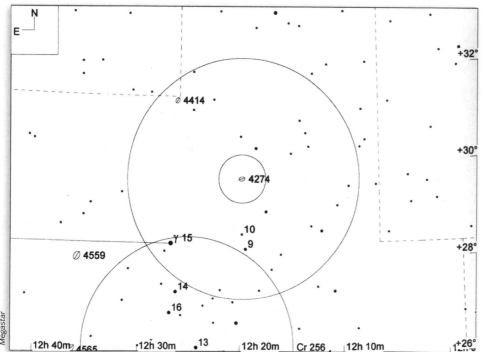

CHART 15-4.

*NGC 4274 (10° field width, 5.0°
finder field, 1.0° eyepiece field,
LM 9.0)*

Megastar

NGC 4414	★★	✺✺✺✺	GX	MBUDR
Chart 15-5	Figure 15-5	m11.0, 4.3' x 3.1'	12h 26.4m	+31° 13'

NGC 4414 is a small, very bright (SB 11.3) galaxy that's very easy
to locate. NGC 4414 lies 3.0° dead N of the naked-eye star 15-γ
(gamma). To center it in your finder field, place 15-γ at the S edge
of the field, and look for the m7 star SAO 62988 near the center
of the field. That star is noticeably brighter than the other stars
visible in the vicinity, and lies about 24' SSW (south southwest) of
NGC 4414.

At 42X in our 10" Dob, NGC 4414 is visible with direct vision as
a small bright patch of nebulosity without surface detail. At 125X
with averted vision, a bright NW-SE 0.75' x 1.5' NW-SE oval core is
visible with a surrounding much fainter halo extending to 1' x 3'.

FIGURE 15-5.

NGC 4414 (60' field width)

*Image reproduced from Digitized Sky Survey courtesy Palomar Observatory and Space
Telescope Science Institute*

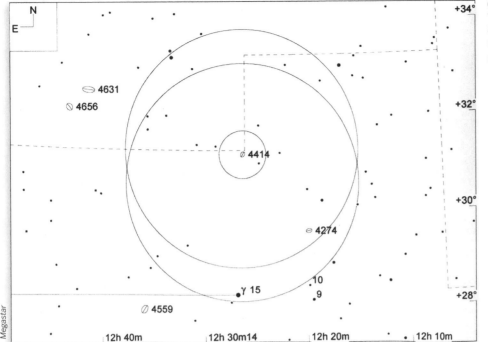

CHART 15-5.

NGC 4414 (10° field width, 5.0° finder fields, 1.0° eyepiece field, LM 9.0)

NGC 4725	★★	✺✺✺	GX	MBUDR
Chart 15-6	Figure 15-6	m10.1, 10.7' x 8.0'	12h 50.4m	+25° 30'

NGC 4725 is a medium size, moderately bright galaxy. To locate NGC 4725, put the naked-eye star 43-β (beta) on the E edge of your finder field and look for the m5/6 pair 30-/31-Com on the W edge of the finder field. With that pair on the N edge of the finder field, NGC 4725 should be nearly centered in the finder field, about 45' NNE (north northeast) of the m6.3 star SAO 82511, which appears prominently in the finder.

At 90X in our 10" Dob, NGC 4414 is visible with averted vision as a bright 5' x 3' NE-SW (from northeast to southwest) oval halo with gradual brightening to a nearly stellar core. Some hints of the spiral arms are very faintly visible to the NE and SW of the core.

FIGURE 15-6.

NGC 4725 (60' field width)

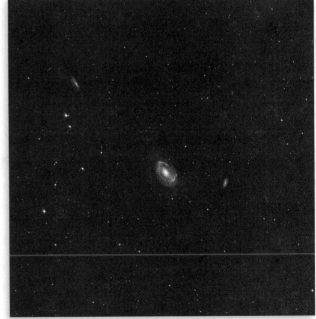

Image reproduced from Digitized Sky Survey courtesy Palomar Observatory and Space Telescope Science Institute

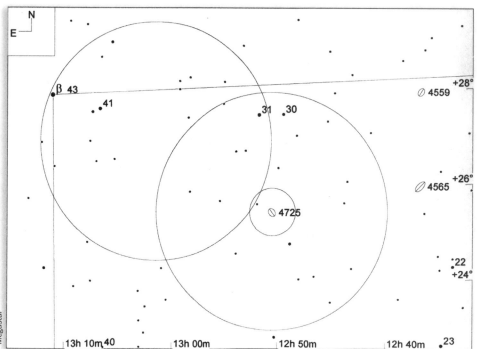

CHART 15-6.

NGC 4725 (10° field width, 5.0°
finder fields, 1.0° eyepiece field,
LM 9.0)

M 64 (NGC 4826)	★★★	✿✿✿	GX	MBUDR
Chart 15-7	Figure 15-7	m9.4, 10.1' x 5.4'	12h 56.7m	+21° 41'

NGC 4826—better known as Messier 64, M 64, or the Black Eye Galaxy—is a big, bright, beautiful galaxy. Although it is a large galaxy, its very bright integrated magnitude of 9.4 translates to a relatively high surface brightness of 11.8. For many years, M 64 was listed as having been discovered on 4 April 1779 by Johann Elert Bode. In fact, M 64 had been observed and logged by Edward Pigott on March 23, 1779, a fact that was uncovered only in 2002. Charles Messier, unaware of either of these earlier observations, re-discovered M 64 independently on the night of 1 March 1780.

M 64 is relatively easy to find, lying 5.2° NW of the m4.3 naked-eye star 42-α (alpha), and about 55' ENE of the m4.9 star 35-Com. To center M 64 in your finder, put 42-α (alpha) on the SE edge of the field and move the finder field NW until the m4.9 star 35-Com comes into view. On a night with excellent transparency, M 64 itself may be just visible in a 50mm finder as a very faint hazy patch. If it is not, place the finder field as shown in Chart 15-7, with m6 39-Com and m5.5 40-Com on the E edge of the finder field to center M 64 in the field.

The Astronomical League Binocular Messier Club lists M 64 as Tougher (their middle category) with both 35/50mm binoculars and 80mm binoculars, and in our judgment that assessment is accurate. Using averted vision with our 50mm binoculars, we see M 64 as a small, very faint patch of fuzziness lying just ENE of m4.9

FIGURE 15-7.

NGC 4826 (60' field width)

Image reproduced from Digitized Sky Survey courtesy Palomar Observatory and Space Telescope Science Institute

35-Com, and nestled W and S of an arc of m7 stars that extends SE-NW-W. At 42X in our 10" Dob, M 64 is prominently visible with direct vision, but higher magnification and averted vision reveals much additional detail. At 125X, the bright oval halo extends 6' x 3' WNW-ESE, with a brighter core noticeably off-center to the SW. The dark patch that gives this galaxy its nickname is just visible on a night with excellent transparency as a noticeable darkening on the NE edge of the core.

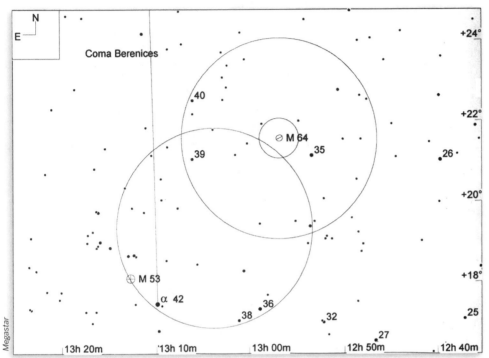

CHART 15-7.

NGC 4826 (M 64) (12° field width, 5.0° finder fields, 1.0° eyepiece field, LM 9.0)

M 53 (NGC 5024)	★★★	◊◊◊◊	GC	MBUDR
Chart 15-8	Figure 15-8	m7.7, 13.0'	13h 12.9m	+18° 10'

NGC 5024, better known as Messier 53 or M 53, isn't the most impressive of the Messier globular clusters, but it's a beautiful object nonetheless. M 53 was discovered on 3 February 1775 by Johann Elert Bode and independently re-discovered two years later by Charles Messier on 26 February 1777. Bode and Messier both described the object as a prominent, round nebula, and no individual stars were resolved in this object until William Herschel later viewed the object in a larger scope. Messier himself thought that M 53 resembled his later discovery M 79, and Herschel compared M 53 to M 10. They were both right, as M 53 bears a strong resemblance to both of these two other Messier globulars.

M 53 is very easy to locate. It lies just under 1° NE of the naked eye star 42-α (alpha), and is visible in the same low-power eyepiece field in many telescopes. The Astronomical League Binocular Messier Club lists M 53 as Tougher (their middle category) with 35/50mm binoculars and Easy with 80mm binoculars, but we find it relatively easy with our 50mm binoculars. Using averted vision, we see M 53 as a relatively faint but clearly visible hazy patch lying just NE of 42-α (alpha).

M 53 is just as easy in a 50mm finder, so you can center it in the finder simply by putting 42-α (alpha) in the cross hairs and moving the finder 1° NE to center the hazy patch that's visible in the finder. At 42X in our 10" Dob, M 53 is prominently visible with direct vision, but only one star can be resolved in the core (and this may be a foreground star rather than an actual member of the cluster). At 90X with averted vision, the core takes on a granular appearance, with many individual stars resolved in the 5' surrounding halo.

FIGURE 15-8.

NGC 5024 (M 53) (60' field width)

Image reproduced from Digitized Sky Survey courtesy Palomar Observatory and Space Telescope Science Institute

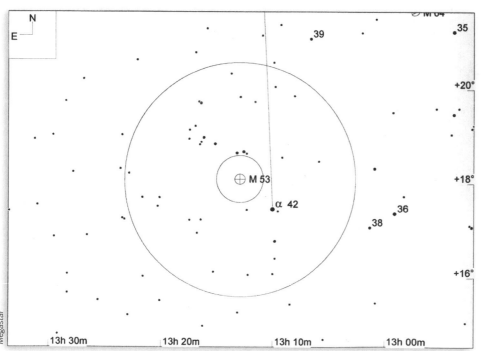

CHART 15-8.

NGC 5024 (M 53) (10° field width, 5.0° finder field, 1.0° eyepiece field, LM 9.0)

The Coma-Virgo Cluster

Chart 15-9 shows the intense concentration of Messier galaxies and other objects in the small area that is not-so-fondly known to amateur astronomers as the "Coma-Virgo Clutter." Use Chart 15-9 to orient yourself to the locations of the objects shown in the smaller-scale finder charts.

CHART 15-9.

Coma-Virgo Cluster (25° field width, 5.0° finder field for scale, LM 9.0)

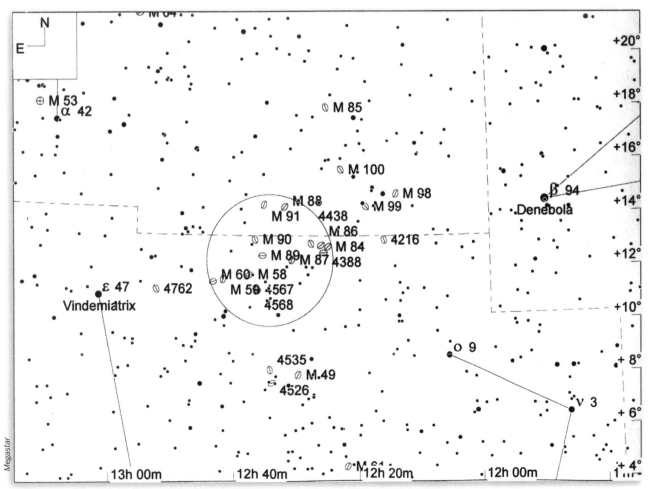

M 98 (NGC 4192)	★★★	✧✧✧	GX	MBUDR
Chart 15-10, 15-11	Figure 15-9	m11.0, 9.8' x 2.7'	12h 13.8m	+14° 54'

NGC 4192, better known as Messier 98 or M 98, is one of the dimmer galaxies in the Messier catalog, but is still an interesting object. Pierre Méchain discovered M 98 (as well as the nearby galaxies M 99 and M 100) on the night of 15 March 1781. Charles Messier confirmed these observations on the night of 13 April 1781, just in time to add them to the final, published edition of his catalog.

M 98 is relatively easy to locate, lying 6° dead E of Denebola in Leo. (In fact, if you're patient, you can center Denebola in a low-power eyepiece and wait about 25 minutes for M 98 to drift into view.) We locate M 98 by working E from Denebola. Place Denebola on the W edge of your finder and move the finder field about half a field E until the m5.1 star 6-Com comes into view. That star is far brighter and more prominent than any other star in the vicinity, so it's difficult to mistake for another star. M 98 lies almost exactly half a degree W of 6-Com, and is visible in the same low-power eyepiece field.

At 42X in our 10" Dob, M 98 is visible with direct vision as a moderately bright slash of light, elongated about 5' x 1.5' and oriented NNW-SSE. Higher magnification and averted vision reveal considerably more detail. At 125X, the oval halo extends 6' x 2', with significant brightening to a mottled, irregularly-shaped bright central core with a stellar nucleus.

FIGURE 15-9.

NGC 4192 (M 98) (60' field width)

Image reproduced from Digitized Sky Survey courtesy Palomar Observatory and Space Telescope Science Institute

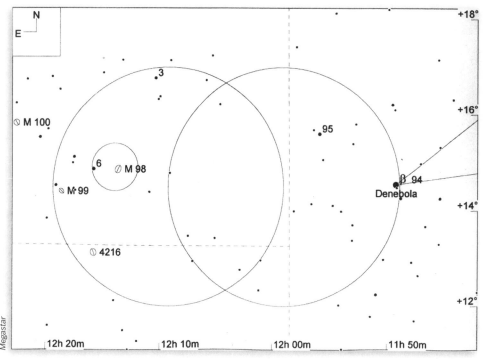

CHART 15-10.

NGC 4192 (M 98) (10° field width, 5.0° finder fields, 1.0° eyepiece field, LM 9.0)

M 99 (NGC 4254)	★★★	◦◦	GX	MBUDR
Chart 15-11	Figure 15-10	m10.4, 5.4' x 4.7'	12h 18.8m	+14° 25'

NGC 4254, better known as Messier 99 or M 99, is considerably brighter than M 98. Like M 98, M 99 was discovered by Pierre Méchain on the night of 15 March 1781 and added by Charles Messier to his catalog on 13 April 1781, just in time to meet the deadline for the final published edition.

The easiest way to locate M 99 is to hop to it immediately after you view M 98. M 98 forms one apex of a flat scalene triangle, as shown in Chart 15-11. With M 98 near the W edge of a low-power eyepiece field, the m5.1 star 6-Com is prominently visible about 32' dead E. M 99 lies 50' SE of that star. If your low-power eyepiece has a true field of 1° or more, 6-Com and M 99 are visible in the same field. If your low-power eyepiece has a smaller true field, it's relatively easy to do an eyepiece starhop using the prominent stars shown in Chart 15-11.

At 42X in our 10" reflector, M 99 is visible with direct vision as a bright oval patch with some brightening toward the center. At 125X, the oval halo extends 4' x 3' E-W. With averted vision, some hint of the spiral arms to the SW and N are faintly visible.

FIGURE 15-10.

NGC 4254 (M 99) (60' field width)

Image reproduced from Digitized Sky Survey courtesy Palomar Observatory and Space Telescope Science Institute

CHART 15-11.

NGC 4254 (M 99) detail (2.5° field width, 1.0° eyepiece fields, LM 11.0)

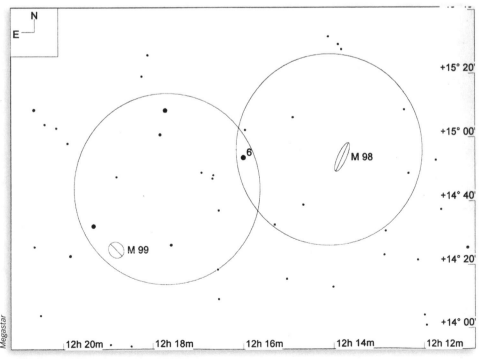

M 100 (NGC 4321)	★★★	◊◊	GX	MBUDR
Chart 15-12	Figure 15-11	m10.1, 7.5' x 6.3'	12h 22.9m	+15° 49'

NGC 4321, better known as Messier 100 or M 100, is fainter than M 99 but brighter than M 98. Like M 98 and M 99, M 100 was discovered by Pierre Méchain on the night of 15 March 1781 and added by Charles Messier to his catalog on 13 April 1781.

The easiest way to locate M 100 is by visualizing the isosceles triangle it forms with the bright stars 6-Com (m5.1) and 11-Com (m4.7), both of which are prominent in the finder. The line 6-Com to 11-Com forms a 3° baseline with M 100 at the SE apex of the triangle, about 2° from either star. Just using this geometric relationship to point your finder should put M 100 in the field of view of your low-power eyepiece.

The Astronomical League Binocular Messier Club does not list M 100 as possible with 35/50mm binoculars, and lists it as a Challenge (their most difficult category) with 80mm binoculars. We have never been able to see M 100 in our 50mm binoculars, so this assessment seems reasonable. At 42X in our 10" reflector, M 100 is visible with direct vision as a moderately bright nearly circular patch with considerable brightening toward the center. At 125X, the relatively uniform halo extends 4' x 3' ESE-WNW, with a very bright central core. With averted vision, no hint of the spiral arms is visible, but some hint of mottling in the halo near the core is faintly visible.

FIGURE 15-11.

NGC 4321 (M 100) (60' field width)

Image reproduced from Digitized Sky Survey courtesy Palomar Observatory and Space Telescope Science Institute

CHART 15-12.

NGC 4321 (M 100) (10° field width, 5.0° finder fields, 1.0° eyepiece field, LM 9.0)

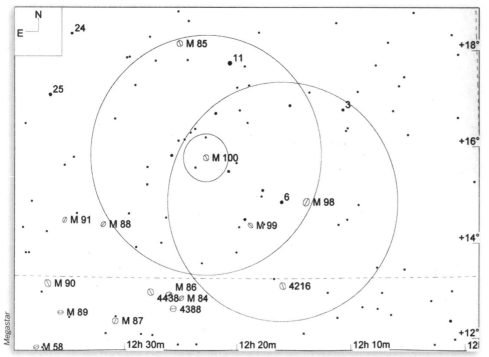

ILLUSTRATED GUIDE TO ASTRONOMICAL WONDERS

M 85 (NGC 4382)	★★★	⊙⊙	GX	MBUDR
Chart 15-12, 15-13	Figure 15-12	m9.1, 7.1' x 5.5'	12h 25.4m	+18° 12'

NGC 4382, better known as Messier 85 or M 85, is a bright galaxy that was discovered on 4 March 1781 by Pierre Méchain, and reported to Charles Messier. Messier immediately began exploring this area of the sky, and logged this galaxy as his 85th object on 18 March 1781. While sweeping this area that night, Messier discovered seven additional objects, all of which he added to his list.

The easiest way to locate M 85 is to start from M 100. With M 100 centered in the eyepiece, the m5 stars 6- and 11-Com are prominent in the finder, as shown in Chart 15-12. Drift the finder NE until the m5 star 24-Com comes into view on the E edge of the finder field, as shown in Chart 15-13. M 85 lies about one third of the way along the line from 11-Com to 24-Com and just slightly N, where it is prominently visible in a low-power eyepiece. (24-Com, described at the end of this chapter, is the one featured multiple star in Coma, so you might just as well observe it while you're in the vicinity.)

The Astronomical League Binocular Messier Club does not list M 85 as possible with 35/50mm binoculars, and lists it as a Challenge (their most difficult category) with 80mm binoculars, which our own attempted observations confirm. At 42X in our 10" Newtonian, M 85 is visible with direct vision as a bright oval core surrounded by a much fainter halo that gradually dims below the threshold of vision. At 125X with averted vision, the halo extends 6' x 4' N-S, but no hint of structure is visible in the halo. An m10 star lies SE of the core, just outside the visible halo. The companion galaxy NGC 4394 (to the left of M 85 in Figure 15-12) is visible in the same eyepiece field, about 7' E of M 85, albeit much fainter than M 85.

Image reproduced from Digitized Sky Survey courtesy Palomar Observatory and Space Telescope Science Institute

FIGURE 15-12.

NGC 4382 (M 85) with NGC 4394 (left) and IC 3292 (the fuzzy star to its right) (60' field width)

CHART 15-13.

NGC 4382 (M 85) (10° field width, 5.0° finder fields, 1.0° eyepiece field, LM 9.0)

Megastar

M 88 (NGC 4501)	★★★	⊙⊙	GX	MBUDR
Chart 15-14, 15-15	Figure 15-13	m10.4, 7.0' x 3.7'	12h 32.0m	+14° 25'

NGC 4501, better known as Messier 88 or M 88, is one of the seven galaxies that Charles Messier discovered on his "big night" of 18 March 1781. Messier described this object as a "nebula without stars" and thought it resembled his earlier discovery, M 58.

M 88 is relatively hard to find, because it lies in an area devoid of nearby bright stars. We locate it by putting the m5 stars 6- and 11-Com on the edge of our finder field, as shown in Chart 15-14. That puts M 88 near the SE edge of the finder field, where it lies about 35' ESE (east southeast) of the m7 star SAO 100127. That star, while not particularly bright, is by far the most prominent star in that quadrant of the finder field. Centering that star in the finder and then bumping the finder half a degree ESE puts M 88 near the center of the field of a low power eyepiece.

The Astronomical League Binocular Messier Club does not list M 88 as possible with 35/50mm binoculars, and lists it as a Challenge (their most difficult category) with 80mm binoculars. It is certainly not visible in our 50mm finder. At 42X in our 10" reflector, M 88 is visible with direct vision as a bright oval core surrounded by a fainter, generally oval halo. At 125X with averted vision, the halo extends 5' x 2' NW-SE and its irregular structure becomes more evident as a teardrop shape, with the blunt end to the NW and the point to the SE. Very faint mottling within the halo is visible with difficulty. A pretty m11 double star lies about 3' S of the S edge of the halo.

FIGURE 15-13.

NGC 4501 (M 88, right) and NGC 4548 (M 91) (60' field width)

Image reproduced from Digitized Sky Survey courtesy Palomar Observatory and Space Telescope Science Institute

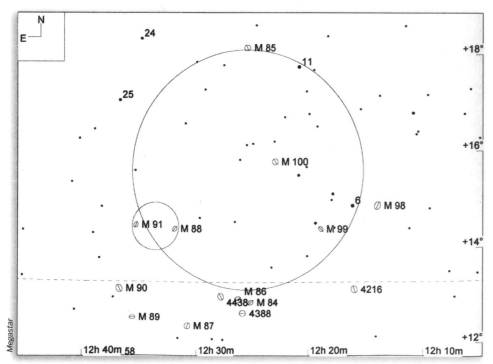

CHART 15-14.

NGC 4501 (M 88) and NGC 4548 (M 91) (10° field width, 5.0° finder field, 1.0° eyepiece field, LM 9.0)

M 91 (NGC 4548)	★★	⊖⊖	GX	MBUDR
Chart 15-14, 15-15	Figure 15-13	m11.0, 5.4' x 4.2'	12h 35.4m	+14° 30'

NGC 4548, better known as Messier 91 or M 91, is the last of the seven galaxies that Charles Messier discovered on 18 March 1781. Messier described this object as a "nebula without stars, fainter than M 90." For many years, M 91 was one of the "missing Messier objects," those that could not be found at the positions Messier listed. There is still some debate as to whether NGC 4548 is the object that Messier logged as M 91. Various other possibilities have been proposed, the most likely of which are that M 91 was a duplicate observation of M 58 or that Messier actually observed a comet that he did not recognize as such.

Whatever the truth, NGC 4548 is now generally accepted as M 91. M 91 is the faintest and least impressive of the Messier galaxies, and is considered by most observers (including us) to be among the most difficult objects on Messier's list, if not the most difficult. In addition to being hard to see, M 91 is relatively hard to find because, like M 88, it lies in a region devoid of nearby bright stars. The best way to locate M 91 is to start from M 88, as described in the preceding section. M 91 lies about 50' dead E of M 88, so both are visible in a 1° eyepiece field.

At 42X in our 10" reflector, M 91 is visible with direct vision as a moderately bright, nearly stellar core surrounded by a small, much fainter, slightly elongated halo. At 125X with averted vision, the halo extends 2' x 1.5' NNW-SSE with no mottling or other detail visible.

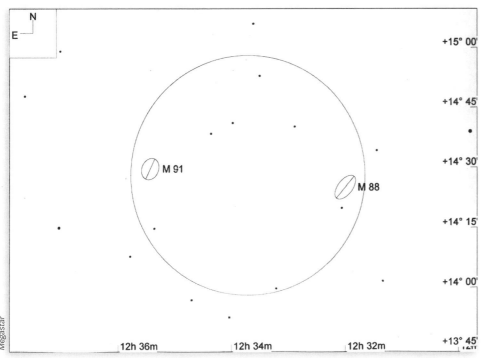

CHART 15-15.

NGC 4501 (M 88) and NGC 4548 (M 91) detail (2° field width, 1.0° eyepiece field, LM 11.0)

Multiple Stars

24-Com (STF 1657)	★★★	✦✦✦		MS	UD
Chart 15-16		m5.1/6.3, 20.1", PA 270° (2004)		12h 35.1m	+18° 22'

We observed 24-Com just after observing M 85, which is nearby. You can also locate 24-Com by star hopping from the naked-eye (m4.2) star 42-α (alpha), as shown in Chart 15-16. To do so, place 42-α (alpha) on the E edge of your finder field and look for the m6/7 double star 32/33-Com near the W edge of the finder field. Move the finder field W to put 32/33-Com on the E edge of the field, and look for 24-Com, which appears prominently near the NW edge of the finder field.

At 125X in our 10" reflector, 24-Com is a beautiful double star, with a yellowish-white primary and a blue-white companion. As pretty as 24-Com is in the 10" scope, it's prettier still in our 90mm refractor, which shows the primary as yellow-orange and the secondary as a more saturated blue. As is often true of double stars, using too much aperture brightens them enough to subdue the colors, while using less aperture allows the more saturated colors to show.

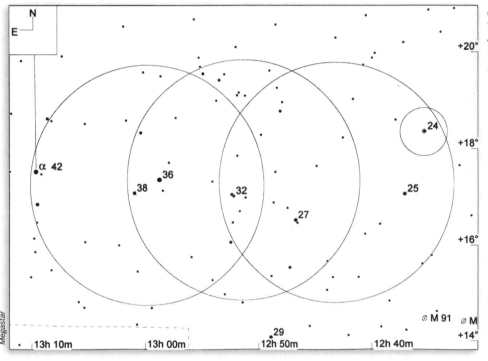

CHART 15-16.

24-Com (10° field width, 5.0° finder fields, 1.0° eyepiece field, LM 9.0)

16

Corona Borealis, The Northern Crown

NAME: Corona Borealis (kuh-ROE-nuh BOR-ee-AL-is)

SEASON: Spring

CULMINATION: midnight, 19 May

ABBREVIATION: CrB

GENITIVE: Coronae Borealis (kuh-ROE-neye BOR-ee-AL-is)

NEIGHBORS: Boo, Her, Ser

BINOCULAR OBJECTS: none

URBAN OBJECTS: none

Corona Borealis (CrB) is a small, relatively inconspicuous mid-northerly constellation, ranking 73rd in size among the 88 constellations. It covers 179 square degrees of the sky, or about 0.4%. CrB lies midway between the square of Hercules and the kite of Boötes. Its brightest star is magnitude 2.2 5-α (alpha), also known as Alphekka. All other stars in CrB are fourth-magnitude or dimmer. Despite its paucity of bright stars, CrB stands out in a relatively empty area of the sky. On a dark night, it's easily visible as a semi-circle of six fourth-magnitude stars, anchored by Alphekka on the southwestern side of the arc.

Corona Borealis is an ancient constellation, although Ptolemy named it simply Corona (or Crown). The Borealis (Northern) part of the name was added later, after Corona Australis (the Southern Crown) was charted by European astronomers. The crown represents not the bejeweled golden crown worn by kings, but the Greek and Roman corona, a wreath of laurel worn by the winners of athletic competitions and victorious Roman generals. In constellation mythology, Corona represents the laurel crown given by the Hero Theseus to Ariadne on their wedding day.

Like most other constellations that lie outside the Milky Way, CrB has many galaxies and few other DSOs. Unfortunately, the galaxies in CrB are so dim that even the brightest of them are impossible to see in any but the largest amateur instruments. We feature only two objects in this constellation, both of which are double stars that are required to complete the Astronomical League Double Star Club requirements.

CrB culminates at midnight on 19 May, and, for observers at mid-northern latitudes, is best placed for evening viewing during the early spring through late autumn months.

TABLE 16-1.

Featured multiple stars in Corona Borealis

Object	Pair	M1	M2	Sep	PA	Year	RA	Dec	UO	DS	Notes
7-zeta	STF 1965	5.0	5.9	6.3	306	2003	15 39.4	+36 38		◉	
17-sigma*	STF 2032AB	5.6	6.5	7.1	237	2006	16 14.7	+33 52		◉	

CHART 16-1.

The constellation Corona Borealis (field width 25°)

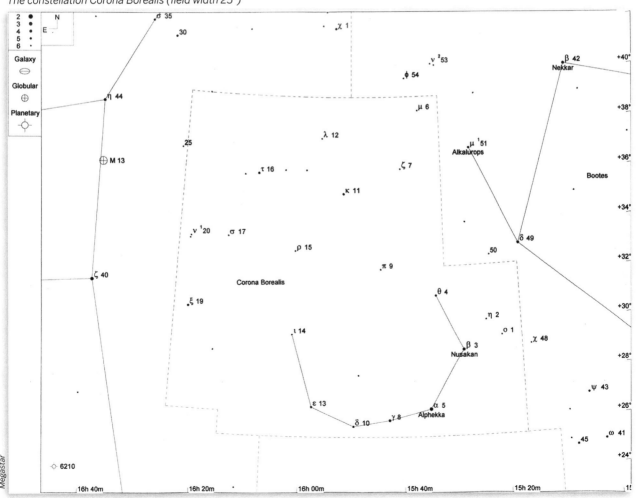

Clusters, Nebulae, and Galaxies

There are no featured clusters, nebulae, or galaxies in Corona Borealis.

Multiple Stars

7-zeta (STF 1965)	★★	✦✦✦		MS	UD
Chart 16-2		m5.0/5.9, 6.3", PA 306° (2003)		15h 39.4m	+36° 38'

7-zeta CrB is a rather ordinary double star. To locate it, put m4.3 Alkalurops (in Bootes) on the W edge of your finder field, as shown in Chart 16-2, with m 5.1 6-μ (mu) CrB near the N edge of the finder field. The m5 star 7-ζ (zeta) is the SE apex of the isosceles triangle formed with Alkalurops and 6-μ (mu). At 125X in our 10" reflector, the primary and companion are easily split. Both appear a cool white color.

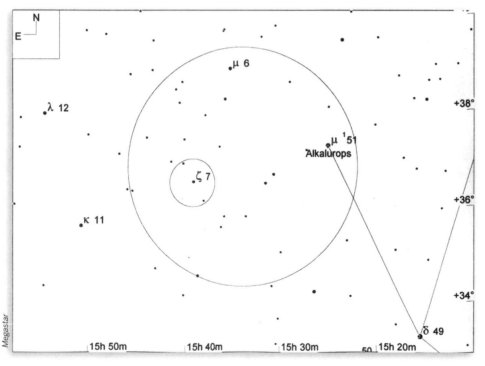

CHART 16-2.

7-zeta CrB (STF 1965) (10° field width, 5.0° finder field, 1.0° eyepiece field, LM 9.0)

17-sigma (STF 2032AB)	★★	✪✪✪		MS	UD
Chart 16-3		m5.6/6.5, 7.1", PA 237° (2006)		16h 14.7m	+33° 52'

17-sigma CrB is another ordinary double star. To locate it, put m5.0 14-ι (iota)—the star at the NE end of the arc of CrB—at the SW edge of your finder field and look for m5.6 17-σ (sigma) near the NE edge of the finder field. You can confirm the identity of 17-σ by moving the finder field slightly NW until the prominent m5 double star 20-ν (nu) comes into the field; 17-σ lies 1.6° dead W of 20-ν. At 125X in our 10" reflector, the primary and companion are easily split. Both appear a yellowish white color.

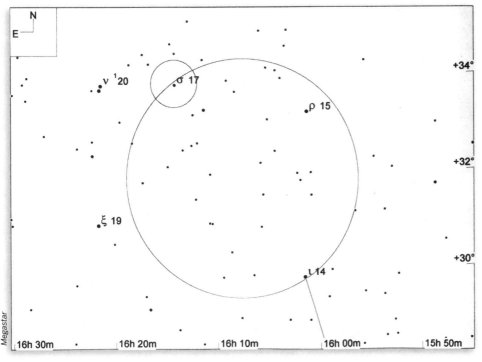

CHART 16-3.

17-sigma CrB (STF 2032AB) (10° field width, 5.0° finder field, 1.0° eyepiece field, LM 9.0)

17

Corvus, The Crow

NAME: Corvus (KOR-vus)

SEASON: Early spring

CULMINATION: midnight, 28 March

ABBREVIATION: Crv

GENITIVE: Corvi (KOR-vee)

NEIGHBORS: Crt, Hya, Vir

BINOCULAR OBJECTS: none

URBAN OBJECTS: none

Corvus is a small, mid-southerly constellation, ranking 70th in size among the 88 constellations. It covers 184 square degrees of sky, or about 0.5%. Because it lies in a star-poor region south and east of Virgo, Corvus is more prominent than its size and the magnitude of its brightest stars might lead you to expect. On a dark night, the distinctive trapezoid shape formed by its four brightest stars—all third magnitude—stands out clearly to the WSW (west-southwest) of first-magnitude Spica in adjacent Virgo.

Corvus the Crow is an ancient constellation, recognized as such from at least the time of the Babylonians and Assyrians. The ancient Greeks knew it as Corax the Raven, and saw it as pecking the tail of Hydra the Serpent, a tale they inherited from the early Mesopotamian civilizations.

Corvus is surprisingly rich in DSOs for a small constellation that lies outside the Milky Way, but most of those objects are too dim to be rewarding in typical amateur telescopes. The exceptions are the bright planetary nebula NGC 4361, and pair of interacting galaxies, NGC 4038/4039, which are collectively known as the Ring Tail Galaxy.

Corvus culminates at midnight on 28 March, and, for observers at mid-northern latitudes, is best placed for evening viewing during the late winter through early summer months.

TABLE 17-1.

Featured star clusters, nebulae, and galaxies in Corvus

Object	Type	Mv	Size	RA	Dec	M	B	U	D	R	Notes
NGC 4361	PN	10.3	118"	12 24.5	-18 47					◉	Class 3a+
NGC 4038	Gx	10.9	3.7' x 1.7'	12 01.9	-18 52					◉	Class SB(s)m pec

TABLE 17-2.

Featured multiple stars in Corvus

Object	Pair	M1	M2	Sep	PA	Year	RA	Dec	UO	DS	Notes
7-delta	SHJ 145	3.0	8.5	24.9	217	2003	12 29.9	-16 31		◉	Algorab

CHART 17-1.

The constellation Corvus (field width 25°)

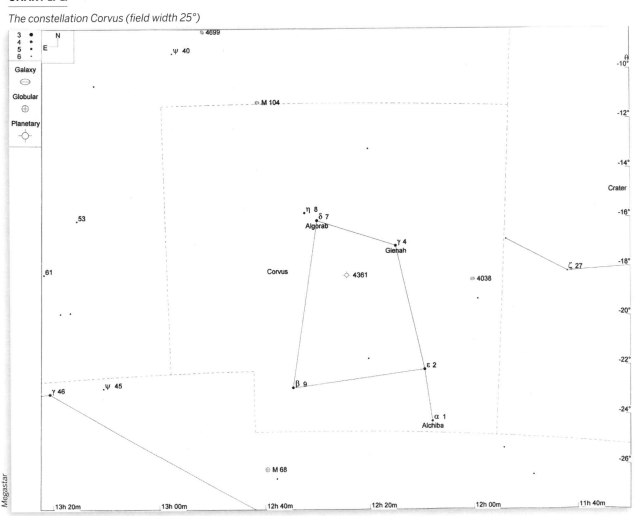

Clusters, Nebulae, and Galaxies

NGC 4361	★★	◊◊◊◊	PN	MBUDR
Chart 17-2	Figure 17-1	m10.3, 118"	12h 24.5m	-18° 47'

NGC 4361 is a relatively large, bright planetary nebula located inside the trapezoid of Corvus. NGC 4361 is easy to locate by noting the geometric pattern it forms with the m3 stars Algorab and Gienah that form the northernmost points of the Corvus trapezoid. If you're using a standard 50mm finder with a 5° or 5.5° field of view, placing Algorab and Gienah as shown in Chart 17-2 puts NGC 4361 almost exactly at the cross hairs, where it is visible in a low-power eyepiece.

At 42X in our 10" reflector, NGC 4361 is visible with direct vision as a small, bright, circular hazy patch that looks like a fuzzy star. At 125X, the m13 central star is visible, surrounded by a bright, circular 30" core of relatively uniform brightness. Our Ultrablock narrowband filter makes the central star invisible, but reveals a very faint halo surrounding the bright central core and extending to a total size of about 60".

FIGURE 17-1.

NGC 4361 (60' field width)

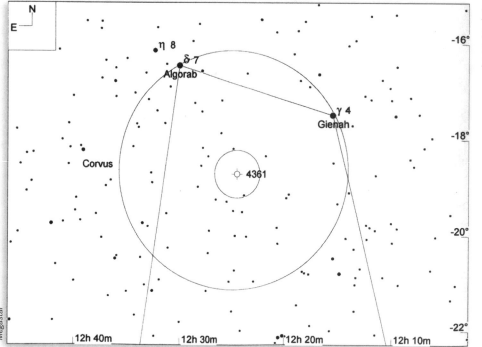

CHART 17-2.

NGC 4361 (10° field width, 5.0° finder field, 1.0° eyepiece field, LM 9.0)

NGC 4038	★	✹✹✹	GX	MBUDR
Chart 17-3	Figure 17-2	m10.9, 3.7' x 1.7'	12h 01.9m	-18° 52'

NGC 4038 is half of the interacting galaxy pair NGC 4038/4039, also known as the Ring Tail Galaxy. The RASC list includes only NGC 4038, although it is actually the fainter of the pair.

To locate NGC 4038, place m2.6 Gienah and 30-η (eta) Crt at opposite edges of the finder field, as shown in Chart 17-3. Look for the m5.3 star TY-Corvi, which appears prominently near the SW edge of the finder field. Center TY-Corvi in your finder and locate it in your low-power eyepiece. Place TY-Corvi at the SW edge of the eyepiece field and scan to the NNE (north northeast) to locate NGC 4038, which lies just 5' SSE (south southeast) of an m9 star that appears prominent in the eyepiece.

At 125X in our 10" reflector, NGC 4038/4039 is visible with averted vision as a relatively bright, irregularly circular 2' patch of haze with a dark notch in the W flank. We see no hint of the "tail" structures, faintly visible in Figure 17-2, that give this object its name.

FIGURE 17-2.

NGC 4038 (60' field width)

Image reproduced from Digitized Sky Survey courtesy Anglo-Australian Observatory and Space Telescope Science Institute

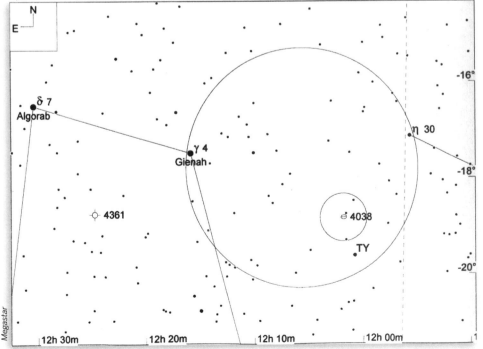

CHART 17-3.

NGC 4038 (10° field width, 5.0° finder field, 1.0° eyepiece field, LM 9.0)

Multiple Stars

7-delta (SHJ 145)	★★	◔◔◔◔		MS		UD
Chart 17-1		m3.0/8.5, 24.9", PA 217° (2003)		12h 29.9m		-16° 31'

7-delta Corvi (Algorab) is a rather ordinary double star that forms the NE apex of the Corvus trapezoid. At 125X in our 10" reflector, the primary is a brilliant white, and the much dimmer companion is yellowish-white.

18
Cygnus, The Swan

NAME: Cygnus (SIG-nus)

SEASON: Summer

CULMINATION: midnight, 29 June

ABBREVIATION: Cyg

GENITIVE: Cygni (SIG-nee)

NEIGHBORS: Cep, Dra, Lac, Lyr, Peg, Vul

BINOCULAR OBJECTS: NGC 6819, NGC 6910, NGC 6913 (M 29), NGC 7063, NGC 7092 (M 39)

URBAN OBJECTS: NGC 6826, NGC 6910, NGC 7092 (M 39), 6-beta (STFA 43)

Cygnus is a large, prominent mid-northerly constellation, ranking 16th in size among the 88 constellations. It covers 804 square degrees of the celestial sphere, or about 1.9%.

Cygnus is an ancient constellation. In Greek mythology, Cygnus represents the swan form that Zeus assumed when he set out to seduce Queen Leda, the wife of King Tyndareus of Sparta. As a result of that union, Leda produced an egg, from which emerged Castor, the mortal son of Tyndareus, Polydeuces (the Roman

Pollux), the immortal son of Zeus, and his immortal daughter Helen, who later as Helen of Troy became the *casus belli* of the Trojan War.

Cygnus, also known as the Northern Cross, dominates the summer Milky Way. Cygnus is incredibly rich in DSOs, lacking only a globular cluster. It is home to scores of bright open clusters, planetary nebulae, reflection and emission nebulae, and supernova remnants. In fact, if you scan Cygnus with a binocular

TABLE 18-1.

Featured star clusters, nebulae, and galaxies in Cygnus

Object	Type	Mv	Size	RA	Dec	M	B	U	D	R	Notes
NGC 6910	OC	7.4	7.0	20 23.1	+40 47		◉	◉			Cr 420; Class I 3 m n
NGC 6913	OC	6.6	6.0	20 24.0	+38 30	◉	◉				M 29; Cr 422; Class II 3 m n
NGC 6888	BN	99.9	18.0 x 8.0	20 12.0	+38 23					◉	Crescent Nebula; Class E
NGC 7000	BN	99.9	120.0 x 100.0	20 58.0	+44 20					◉	North America Nebula; Class E
NGC 7027	PN	10.4	1.0	21 07.0	+42 14					◉	Class 3a
NGC 7092	OC	4.6	31.0	21 32.2	+48 27	◉	◉	◉			M 39; Cr 438; Class III 2 m
NGC 6992/6995	SR/EN	99.9	80.0 x 26.0	20 57.0	+31 30					◉	Veil Nebula (eastern half); Class E
NGC 6960	SR/EN	99.9	60.0 x 9.0	20 45.9	+30 43					◉	Veil Nebula (western half); Class E
NGC 7063	OC	7.0	7.0	21 24.5	+36 30				◉		Cr 435; Class III 1 p
NGC 6819	OC	7.3	5.0	19 41.3	+40 12				◉	◉	Cr 403; Mel 223; Class I 1 r
NGC 6826	PN	9.8	0.6	19 44.8	+50 32			◉		◉	Blinking Planetary Nebula; Class 3a+2

TABLE 18-2.

Featured multiple stars in Cygnus

Object	Pair	M1	M2	Sep	PA	Year	RA	Dec	UO	DS	Notes
6-beta	STFA 43Aa-B	3.4	4.7	34.7	55	2003	19 30.7	+27 57	◉	◉	Albireo
31-omicron1, 30	STFA 50Aa-C	3.9	7.0	107.1	173	2000	20 13.6	+46 44		◉	
31-omicron1, 30	STFA 50Aa-D	3.9	4.8	330.7	324	1998	20 13.6	+46 44	◉		
61*	STF 2758AB	5.4	6.1	31.1	151	2006	21 06.9	+38 45		◉	Piazzi's Flying Star

CHART 18-1.

The heart of the constellation Cygnus (field width 40°)

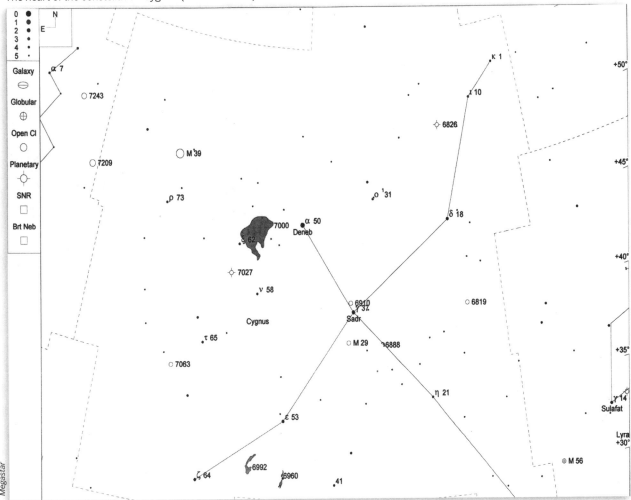

or wide-field telescope, it's almost impossible to find even one field of view that doesn't contain at least one interesting object. On the downside, the incredibly rich Milky Way star field in which Cygnus is embedded often makes it difficult to isolate open clusters from the surrounding field stars. In Cygnus, it really is difficult to tell the forest from the trees.

Cygnus culminates at midnight on 29 June, and, for observers at mid-northern latitudes, is best placed for evening viewing during the early summer through mid-autumn months.

Clusters, Nebulae, and Galaxies

NGC 6910	★★	✦✦✦✦	OC	MBUDR
Chart 18-2	Figure 18-1	m7.4, 7.0'	20h 23.1m	+40° 47'

NGC 6910 is a bright, mid-size open cluster that lies half a degree NNE (north northeast) of m2.2 Sadr, the heart of the swan. Although the cluster is bright in absolute terms, it's diminished in relative terms by the glare from Sadr when that bright star is in the same field of view.

Using averted vision with our 50mm binoculars or 50mm finder, NGC 6910 is visible as a distinct hazy patch. Two embedded m6/7 stars are visible within the patch, with most of the nebulosity lying to their W and SW. Two m8 stars lie just outside the main nebulosity, one to the N and the other to the SW.

At 90X in our 10" reflector, about two dozen stars are visible down to m13. The most prominent are an m6 star that lies just N of center and an m7 star on the SE edge of the cluster. An m8 star lies near the center of the cluster, and an uneven chain of four m9/10 stars extends about 3.5' W from the m7 star at the SE edge. Unresolved dim stars (and possibly actual nebulosity) form a faint nebulosity extending S and W of the two brightest stars.

FIGURE 18-1.

NGC 6910 (Sadr visible at bottom edge) (60' field width)

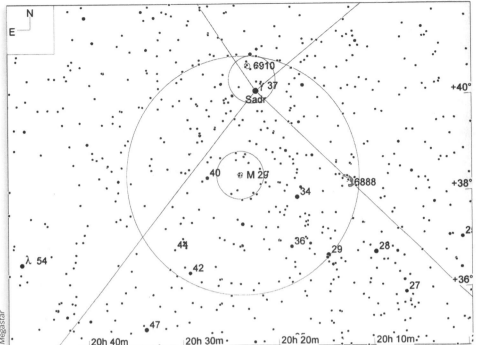

M 29 (NGC 6913)	★★★	✿✿✿✿	OC	MBUDR
Chart 18-2	Figure 18-2	m6.6, 6.0'	20h 24.0m	+38° 30'

NGC 6913, better known as Messier 29 or M 29, is a very bright, smallish, coarse, moderately concentrated open cluster that lies 1.8° almost directly S of Sadr, where it is visible with direct vision in a 50mm binocular or finder as bright hazy patch. Charles Messier discovered and logged M 29 on the night of 29 July 1764. Messier observed this object with his 90mm f/12 refractor, and described it as a "cluster of 7 or 8 very small [dim] stars, which are below Gamma Cygni, which one sees with an ordinary telescope of 3.5-foot in the form of a nebula."

Using averted vision with our 50mm binoculars, NGC 6910 is visible as a prominent pattern of seven m9 stars embedded in a distinct patch of nebulosity. The four southerly bright stars form a nearly square box, with two more bright stars to the N and another to the W. At 90X in our 10" reflector, we can resolve another 15+ stars down to m13, with dimmer stars just on the edge of visibility forming a grainy nebulosity.

FIGURE 18-2.

NGC 6913 (M 29) (60' field width)

Image reproduced from Digitized Sky Survey courtesy Palomar Observatory and Space Telescope Science Institute

NGC 6888	★	๑๑	BN	MBUDR
Chart 18-3	Figure 18-3	m99.9, 18.0' x 8.0'	20h 12.0m	+38° 23'

NGC 6888, also known as the Crescent Nebula, is an extremely faint emission nebula. This object is extremely difficult to see, probably more difficult than any object in this book other than NGC 7635, the Bubble Nebula in Cassiopeia. Despite repeated attempts on nights of excellent transparency, we have never been able to see this object in our 10" reflector.

We finally viewed and logged the Crescent one autumn night in 2006 by begging some eyepiece time on a 17.5" Dob that belongs to an observing buddy. Even though that big Dob goes more than a full magnitude deeper than our 10", we were unable to make out NGC 6888 without filtration at any of the several magnifications we tried. We finally saw this object as a very faint but distinct crescent by using a narrowband filter with a 27mm Panoptic eyepiece at 83X.

NGC 6888 is nearly as hard to find as it is to see. The only bright star anywhere nearby is m4.8 34-Cygni, which lies 1.2° ESE (east southeast). To locate NGC 6888, we started by putting m2.2 37-γ (gamma) Cygni (Sadr) on the NE edge of our finder field. That puts an arc of three m5 stars, 28-, 29-, and 36-Cygni along the S edge of the finder field, with m4.8 34-Cygni lying 1.0° N of 36-Cygni. We centered 34-Cygni in our finder and eyepiece, and then moved 34-Cygni to the ESE edge of the eyepiece field. We panned the eyepiece field about one full field to the WNW (west northwest), looking for a prominent parallelogram pattern of four m7/8 stars,

FIGURE 18-3.

NGC 6888 (60' field width)

Image reproduced from Digitized Sky Survey courtesy Palomar Observatory and Space Telescope Science Institute

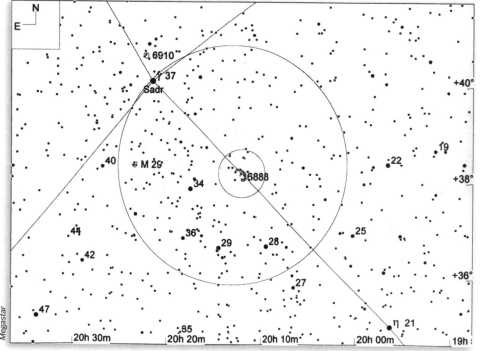

CHART 18-3.

NGC 6888 (10° field width, 5.0° finder field, 1.0° eyepiece field, LM 9.0)

visible in Figure 18-3 surrounding the top section of the nebula. From there, it was merely a matter of keeping that parallelogram pattern centered in the eyepiece and using averted vision with a narrowband filter to reveal the object.

Having viewed it, we can honestly say that it's not worth the time to find it unless you need to log this object to complete the RASC list.

NGC 7000	★★	✵✵✵✵	BN	MBUDR
Chart 18-4	Figure 18-4	m99.9, 120.0' x 100.0'	20h 58.0m	+44° 20'

NGC 7000, better known as the North America Nebula, is a huge, faint emission nebula. The northern portion of the nebula lies from 2.0° to 4.2° E of Deneb, with the body of the nebula extending southward.

This is one of those objects for which proper filtration is a magic bullet. Without filtration, the North America Nebula is at best faintly visible even in large scopes. With a narrowband (or, ideally, an O-III) filter, this nebula is clearly visible even in smaller scopes. On one extraordinarily transparent night, our observing buddy Paul Jones decided to try his new O-III filter on his new Orion 80mm short-tube refractor with his new 32mm Tele Vue Plössl, just for the heck of it. He pointed his short-tube at Cygnus, called Robert over, and asked, "Do you see anything?" Robert's incredulous reply was, "Holy Cow! The North America Nebula!!!!" (Well, Robert didn't actually use the word cow…) So much for conventional wisdom about O-III filters being unusable on small-aperture scopes.

Finding and viewing NGC 7000 is easy enough. Use your lowest-power, widest-field eyepiece with a narrowband or O-III filter. Center your finder on Deneb, and sweep your eyepiece field E until you see the tendrils of NGC 7000 come into view. We used a 2" 40mm Pentax eyepiece with a 2.1° true field of view on our 10" Dob to view this object. We were able to see not just the brighter portions of the nebula around Florida and the Gulf of Mexico (shown in Figure 18-4), but most of Central America and the US east cost.

FIGURE 18-4.

NGC 7000 (60' field width)

Image reproduced from Digitized Sky Survey courtesy Palomar Observatory and Space Telescope Science Institute

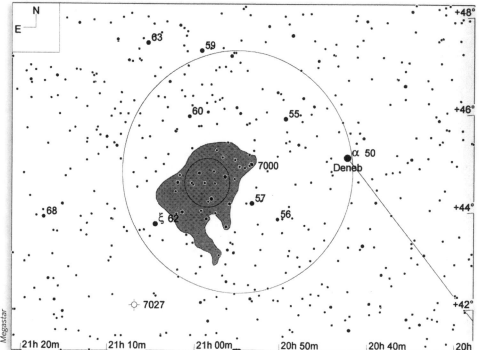

CHART 18-4.

NGC 7000 (North America Nebula) (10° field width, 5.0° finder field, 1.0° eyepiece field, LM 9.0)

Megastar

NGC 7027	★★	◦◦◦	PN	MBUDR
Chart 18-5	Figure 18-5	m10.4, 1.0'	21h 07.0m	+42° 14'

NGC 7027 is a rather ordinary planetary nebula, quite small but with relatively high surface brightness. To locate this object, place Deneb on the NW edge of your finder field and look for the very prominent m4 stars 62-ξ (xi) and 58-ν (nu) which lie near or just past the E and SSE (south southeast) edge of the field. Move the finder field to place those stars as shown in Chart 18-5, with m5.1 68-Cygni appearing prominently at the NE edge of the finder field. NGC 7027 lies near the center of the finder field, where it should be within the field of your low-power eyepiece.

At 42X in our 10" reflector, NGC 7027 appears stellar, without any hint of fuzziness. We used our Ultrablock narrowband filter to "blink" the nebula (pass the filter between your eye and the eyepiece), which dims the surrounding stars without dimming the nebula itself. Having identified the planetary, we centered it in our low-power eyepiece and changed to a 10mm eyepiece for 125X, with the Ultrablock installed. With this combination, the nebula is obviously non-stellar. With averted vision, it appears as a relatively faint 8" x 4" oval patch with a slight blue-green tinge, but no surface detail visible.

FIGURE 18-5.

NGC 7027 (60' field width)

Image reproduced from Digitized Sky Survey courtesy Palomar Observatory and Space Telescope Science Institute

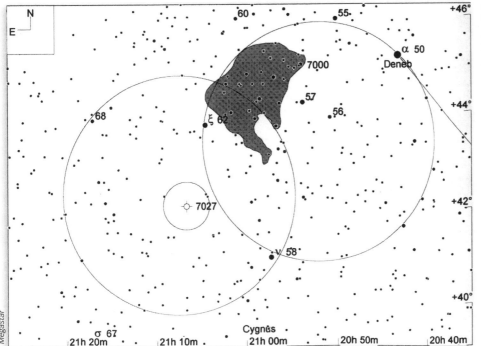

CHART 18-5.

*NGC 7027 (10° field width, 5.0°
finder fields, 1.0° eyepiece field,
LM 9.0)*

M 39 (NGC 7092)	★★★	☾☾☾	OC	MBUDR
Chart 18-6	Figure 18-6	m4.6, 31.0'	21h 32.2m	+48° 27'

NGC 7092, better known as Messier 39 or M 39, is a large, bright, loose, moderately rich open cluster. Under very dark, clear conditions, M 39 can just be glimpsed by the naked eye, and it's likely that this cluster is the object recorded by Aristotle in 325 BC. Charles Messier independently re-discovered this object and logged it on the night of 24 October 1764.

M 39 is relatively easy to locate, lying about 9° ENE (east northeast) of Deneb. To locate M 39, just put your finder or binocular on Deneb and drift the field E until the m4 star 73-ϱ (rho) comes into view. That star is very prominent and difficult to mistake for any of the surrounding stars. M 39 lies 2.9° dead N of 73-ϱ (rho), and appears prominently in a 50mm binocular or finder.

Like many large, loose open clusters, M 39 looks better in a binocular or with very low magnification in a telescope than it does at higher magnification. With averted vision in our 50mm binoculars, we see two m7 stars that lie near the SE and NW edges of the cluster, and about two dozen m8 through m10 stars gathered in clumps and short chains. Although M 39 has only about 30 actual members, there are numerous unresolved dimmer foreground and background stars that present a faint nebulous haze surrounding the brighter stars in the cluster.

At 42X in our 10" reflector, M 39 begins to lose its character as a cluster and appears more as a loose, random collection of field stars. At higher magnification, the cluster nature of this object is lost entirely.

FIGURE 18-6.

NGC 7092 (M 39) (60' field width)

Image reproduced from Digitized Sky Survey courtesy Palomar Observatory and Space Telescope Science Institute

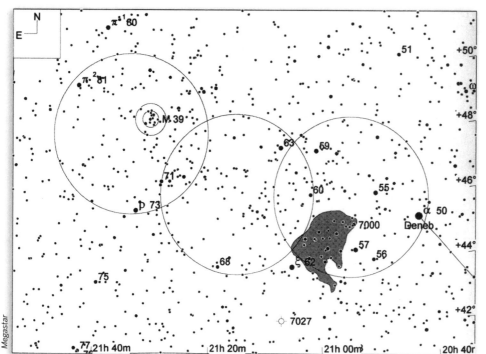

Megastar

NGC 6992/6995	★★★	✿✿✿✿	SR/EN	MBUDR
Chart 18-7	Figure 18-7, 18-8	m99.9, 80.0' x 26.0'	20h 57.0m	+31° 30'

NGC 6960	★★★	✿✿✿✿	SR/EN	MBUDR
Chart 18-7	Figure 18-9	m99.9, 60.0' x 9.0'	20h 45.9m	+30° 43'

NGC 6992, NGC 6995, and NGC 6960 all refer to parts of the same supernova remnant. These objects, separately or together, have several common names. NGC 6992/6995—sometimes NGC 6992 alone is used to refer to both of these objects—is sometimes called the Network Nebula and NGC 6960 the Filamentary Nebula, or vice versa. Sometimes one or the other of those names is assigned to the entire object, as are the names Bridal Veil Nebula and Cirrus Nebula. But the most commonly-used name for these objects is simply the Veil Nebula, usually with the two major segments differentiated as the Eastern Veil Nebula and Western Veil Nebula.

Whether you consider the Veil to be one, two, or three separate objects, it's easy to locate it/them. The Eastern Veil Nebula forms the SW apex of the flattened isosceles triangle whose base is the line between m3.2 64-ζ (zeta) and m2.5 53-ε (epsilon). Similarly, the Western Veil Nebula forms the SW apex of a right triangle formed by 64-ζ and 53-ε, and lies in contact with the m4.2 naked-eye star 52-Cygni. There's really no need to star-hop to locate these objects. Just place your finder or Telrad geometrically, and the object should be in the field of your low-power eyepiece.

The Veil Nebula is another object for which a narrowband filter

(or, better, an O-III filter) is essential. Although we have never tried it, some observers have reported seeing the brighter Eastern Veil segment naked-eye from very dark sites by using a hand-held O-III filter. With our 50mm binoculars, the Eastern Veil is faintly visible without filtration as a fishhook-shaped nebulosity with the shaft of the hook pointing directly toward 53-ε. We're unable to see the Western Veil at all with our 50mm binoculars, probably because the glare from 52-Cygni overwhelms the nebulosity.

More aperture reveals a great deal more detail in both sections of the Veil, but it's important to use the lowest-power, widest-field eyepiece you have access to because these objects are so large. At 42X in our 10" Dob, using a 40mm Pentax eyepiece and an O-III filter, NGC 6992/6995 appears as a beautiful swath of nebulosity extending 1.25° N-S and curving W at both ends. There's an immense amount of fine detail visible in the nebulosity, with some parts appearing lumpy or knotted and others resembling fine lace. The two westward prominences visible in Figure 18-8 are very faintly visible.

The Western Veil, NGC 6960, is fainter than the Eastern Veil, but is still spectacular with an O-III filter (and much less so unfiltered). The segment N of 52-Cygni extends about 20' and reveals considerable detail in its slender, twisted and knotted structure. The segment S of 52-Cygni is invisible to us without filtration, but with the O-III filter, reveals itself as a dimmer, looser nebulosity with a more granular appearance and some twisted filaments. In our 10" scope, we can't see the two southerly prongs visible in Figure 18-9, but we have seen them, very faintly and with great difficulty, in a friend's 17.5" scope

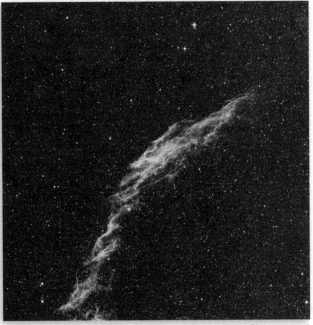

FIGURE 18-7.

NGC 6992 (Veil Nebula, eastern half, northern segment) (60' field width)

Image reproduced from Digitized Sky Survey courtesy Anglo-Australian Observatory and Space Telescope Science Institute

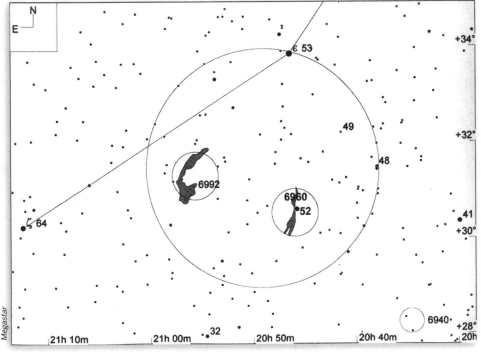

CHART 18-7.

NGC 6992/6995 (Veil Nebula eastern half) and NGC 6960 (Veil Nebula western half) (10° field width, 5.0° finder field, 1.0° eyepiece fields, LM 9.0)

FIGURE 18-8.

NGC 6995 (Veil Nebula, eastern half, southern segment) (60' field width)

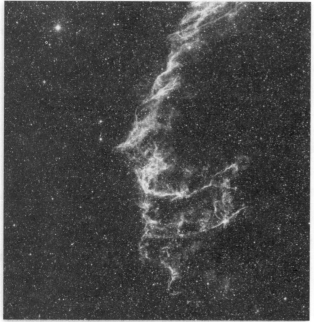

Image reproduced from Digitized Sky Survey courtesy Palomar Observatory and Space Telescope Science Institute

FIGURE 18-9

NGC 6960 (Veil Nebula, western half) (60' field width)

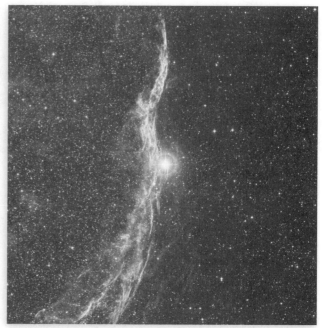

Image reproduced from Digitized Sky Survey courtesy Palomar Observatory and Space Telescope Science Institute

NGC 7063	★★	✸✸✸	OC	MBUDR
Chart 18-8	Figure 18-10	m7.0, 7.0'	21h 24.5m	+36° 30'

NGC 7063 is a moderately bright mid-size open cluster that is one of the objects required to complete the Astronomical League Deep-Sky Binocular Club list, but it's also worth observing with a telescope. To locate NGC 7063, place m3.2 64-ζ (zeta) at the S edge of your finder or binocular field and look for m4.4 66-υ (upsilon), which appears very prominently near the N edge of the field. Move the field NNW until 66-υ appears near the SW edge of the field and look for the two prominent stars m5.3 70-Cygni and m5.9 69-Cygni. NGC 7063 lies just 18' SW of 69-Cygni, where it is visible with averted vision as a small, relatively bright nebulous patch.

Using averted vision with our 50mm binoculars, NGC 7063 is visible as a relatively bright nebulous patch with half a dozen m9/10 stars embedded. At 90X in our 10" Dob, another half dozen stars are visible without nebulosity.

FIGURE 18-10.

NGC 7063 (60' field width)

Image reproduced from Digitized Sky Survey courtesy Palomar Observatory and Space Telescope Science Institute

ILLUSTRATED GUIDE TO ASTRONOMICAL WONDERS

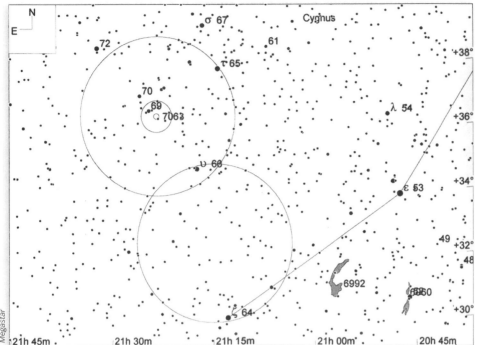

CHART 18-8.

NGC 7063 (15° field width, 5.0° finder fields, 1.0° eyepiece field, LM 9.0)

NGC 6819	★★★	✴✴✴	OC	MBUDR
Chart 18-9	Figure 18-11	m7.3, 5.0'	19h 41.3m	+40° 12'

Although the open cluster NGC 6819 is similar in size and magnitude to NGC 7063, it presents a distinctly difference appearance with both binocular and telescope.

To locate NGC 6819, place m3.9 21-η (eta) on the SE edge of your finder or binocular field, and look for the prominent arc of m5 stars, 25-, 22-, and 19-Cygni near the edge of the field in the NE quadrant. The westernmost of these stars, 19-Cygni, is easily identifiable because it has an m6.1 companion lying about 13' to the W. Place the 19-Cygni pair on the SE edge of your finder field to center NGC 6819 in the field of view.

Using averted vision with our 50mm binoculars, we see NGC 6819 as a moderately bright patch of nebulosity with two embedded m8/9 stars. At 90X in our 10" Dob, another 30+ stars down to m13 are visible, embedded in a faint nebulosity of unresolved dimmer stars.

FIGURE 18-11.

NGC 6819 (60' field width)

Image reproduced from Digitized Sky Survey courtesy Palomar Observatory and Space Telescope Science Institute

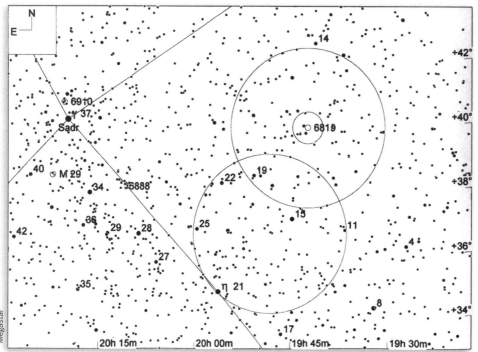

Megastar

CHART 18-9.

NGC 6819 (15° field width, 5.0° finder fields, 1.0° eyepiece field, LM 9.0)

NGC 6826	★★★	⬤⬤⬤	PN	MBUDR
Chart 18-10	Figure 18-12	m9.8, 0.6'	19h 44.8m	+50° 32'

NGC 6826, better known as the Blinking Planetary, is a beautiful planetary nebula that lies in the NW corner of Cygnus. At magnitude 10.6, the central star of this planetary is unusually bright, and easily seen even with scopes of moderate aperture.

To locate NGC 6826, extend the line from m3.8 1-ϰ (kappa) to m3.8 10-ι (iota) by 1.8° to locate m4.5 13-θ (theta) in your finder. Look for the m6.2 star SAO 31899, which lies 55' ENE of 13-θ, and appears very prominently in the finder. NGC 6826 lies 28' dead E of SAO 31899, where you will see it as an m10 star.

That m10 star may be NGC 6826 (the nebula) itself, or it may be the central star, depending on what you are seeing. (That sounds odd, but stay with us...) If you can't tell which is which, use higher magnification and glance back and forth between the nebula and the nearby bright star. As you look at the central star with direct vision, the nebula dims or even becomes invisible. When you shift to looking at the nearby bright star with direct vision, you're simultaneously using averted vision on the nebula, which pops into view. As you shift your gaze back to the central star, the nebula once again disappears, which is the source of its nickname, the Blinking Planetary.

At 125X in our 10" Dob (with averted vision, of course) NGC 6826 appears as a bright, slightly bluish-green featureless disc, about 20" in diameter.

FIGURE 18-12.

NGC 6826 (16-Cygni visible at right edge) (60' field width)

Image reproduced from Digitized Sky Survey courtesy Palomar Observatory and Space Telescope Science Institute

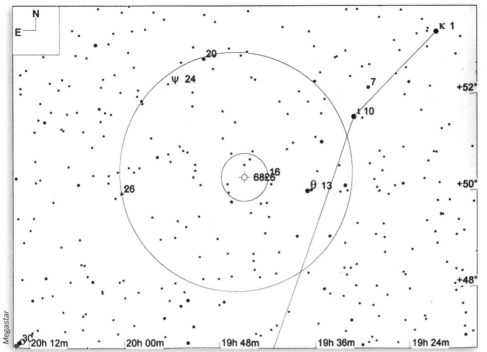

Megastar

Multiple Stars

6-beta (STFA 43Aa-B)	★★★★	๑๑๑๑		MS	UD
Chart 18-1 (off S edge)		m3.4/4.7, 34.7", PA 55° (2003)		19h 30.7m	+27° 57'

Albireo, which forms the head of the swan (or the foot of the Northern Cross) is the most spectacular double star visible in the night sky. In our 90mm refractor at 70X, the primary is a gorgeous golden-yellow, with a beautiful sapphire companion. Double stars simply don't get any better than this.

As is often true of double stars with saturated colors, using too much aperture is counterproductive because the additional brightness desaturates the colors. In our 10" Dob at 90X, for example, while the primary is still yellowish and the secondary bluish, those colors are a pale shadow of the deeply saturated colors visible in the 90mm refractor.

31-omicron1, 30 (STFA 50Aa-C)	★★★	๑๑๑๑		MS	UD
Chart 18-11		m3.9/7.0, 107.1", PA 173° (2000)		20h 13.6m	+46° 44'

31-omicron1, 30 (STFA 50Aa-D)	★★★	๑๑๑๑		MS	UD
Chart 18-11		m3.9/4.8, 330.7", PA 324° (1998)		20h 13.6m	+46° 44'

A very nice triple star system. To locate this system, put Deneb on the E edge of your finder and look for the m3.9/4.8 pair 31-/30-Cygni at or just beyond the NW edge of the field. In our 90mm refractor at 70X, the primary (31-o^1 (omicron^1)) is warm white, and the companion (30-Cygni) is a noticeably cooler white. The much dimmer tertiary has a distinct blue-white cast.

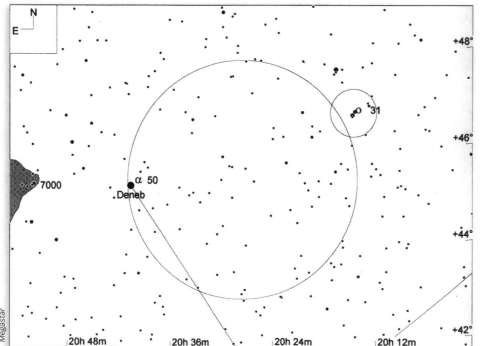

CHART 18-11.

*31-omicron1, 30 Cygni (10°
field width, 5.0° finder field, 1.0°
eyepiece field, LM 9.0)*

61 (STF 2758AB)	★★	◦◦		MS	UD
Chart 18-12		m5.4/6.1, 31.1", PA 151° (2006)		21h 06.9m	+38° 45'

61-Cygni is a rather ordinary double star. With our 90mm refractor at 70X, we see 61-Cygni as a bright warm white primary with a slightly dimmer secondary, also warm white, lying about 30" to the SSE.

61-Cygni is more interesting for another of its characteristics. It has the highest proper motion (motion relative to more distant background stars) of any star visible to the naked eye. 61-Cygni moves so quickly that its apparent position shifts by 30 arcminutes (the diameter of the full moon) in about 150 years, a gigantic leap in the blink of an eye in astronomical terms. The Italian astronomer Giuseppe Piazzi first determined and reported this high proper motion in 1792, and the star was subsequently labeled, rather irreverently, as Piazzi's Flying Star.

The very high proper motion of 61-Cygni results from three factors. First, the star actually moves very fast relative to us. Second, its motion is almost perpendicular to our line of sight. Third, 61-Cygni is very close to us. That last factor turned out to be very significant, because it allowed 61-Cygni to be the first star other than our own Sol for which an accurate distance could be determined.

In 1838, Friedrich Wilhelm Bessel was the first to determine the distance of 61-Cygni by using a technique called parallax, and arriving at a figure very close to the currently accepted value of 11.4 light years. Bessel made very precise measurements of the position of 61-Cygni relative to nearby background stars. Then, some months later, after Earth had moved a considerable distance in its orbit, Besell repeated those measurements. The distance the Earth had moved in the interim formed the baseline from which Bessel was able to determine the tiny apparent shift in position of 61-Cygni. From those values, he was able to calculate the angles of the triangle formed by 61-Cygni at one apex and Earth at the two other apices. From these angles, he calculated the distance of 61-Cygni.

61-Cygni lies 38° SE of Deneb, 8.7° E of Sadr, and 6.3° NE of m2.5 53-ε (epsilon) Cygni. You can locate it by star-hopping as shown in Chart 18-12, or simply by centering the m3.7 naked-eye star 65-τ (tau) in your finder and looking for 61-Cygni 1.7° to the WNW, where it appears prominently near the edge of the finder field.

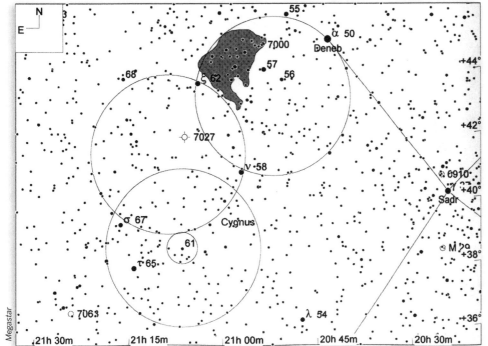

CHART 18-12.

61-Cygni (15° field width, 5.0° finder fields, 1.0° eyepiece field, LM 9.0)

19

Delphinus, The Dolphin

NAME: Delphinus (del-FEE-nus)

SEASON: Summer

CULMINATION: midnight, 31 July

ABBREVIATION: Del

GENITIVE: Delphini (del-FEE-nee)

NEIGHBORS: Aql, Aqr, Equ, Peg, Sge, Vul

BINOCULAR OBJECTS: NGC 6934

URBAN OBJECTS: NGC 6934, 12-gamma (STF 2727)

Delphinus is a small equatorial constellation, ranking 69th in size among the 88 constellations. It covers 189 square degrees of the celestial sphere, or about 0.5%. Despite the fact that the brightest stars in Delphinus are only of the fourth magnitude, Delphinus is quite prominent because it lies in a region of the sky that is devoid of bright stars.

Delphinus is an ancient constellation. In Greek mythology, Delphinus represents the dolphin that saved the poet Arion of Lesbos from drowning. As Arion was traveling by ship home from Tarentum to Corinth, the crew mutinied. As they were about to slay Arion, he begged to be allowed to sing a final dirge. The mutineers granted his request. Arion, knowing he could charm the sea creatures with his poetry and songs, threw himself overboard. His scheme worked, as when Arion jumped into the sea a dolphin immediately rescued him and carried him safely to shore. Zeus rewarded the dolphin by placing it in the heavens as the constellation Delphinus.

Despite lying on the fringes of the Milky Way, Delphinus lacks the open clusters and emission nebulae one expects to find in a Milky Way constellation (although it does host one rather ordinary globular cluster and a few planetary nebulae). Indeed, Delphinus is home to several galaxies—objects more commonly found in constellations that lie far outside the galactic plane.

Delphinus culminates at midnight on 31 July, and, for observers at mid-northern latitudes, is best placed for evening viewing during the early summer through mid-autumn months.

TABLE 19-1.

Featured star clusters, nebulae, and galaxies in Delphinus

Object	Type	Mv	Size	RA	Dec	M	B	U	D	R	Notes
NGC 6934	GC	8.9	7.1	20 34.2	+07 24			◉	◉		

TABLE 19-2.

Featured multiple stars in Delphinus

Object	Pair	M1	M2	Sep	PA	Year	RA	Dec	UO	DS	Notes
12-gamma*	STF 2727	4.4	5.0	9.2	266	2006	20 46.7	+16 07	◉	◉	

CHART 19-1.

The constellation Delphinus (field width 30°)

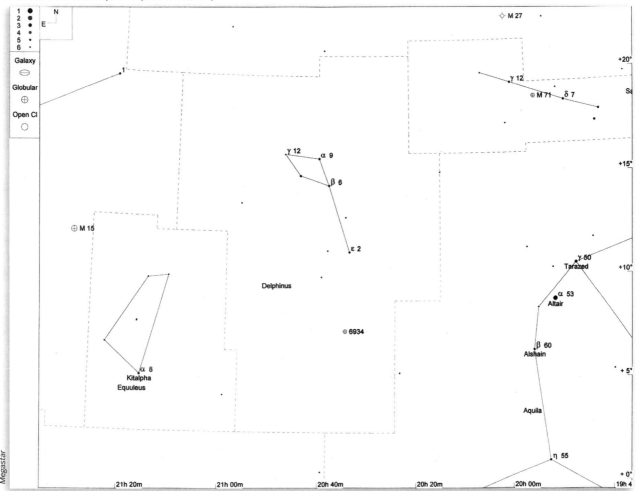

Clusters, Nebulae, and Galaxies

NGC 6934	★★	✷✷✷✷	GC	MBUDR
Chart 19-2	Figure 19-1	m8.9, 7.1'	20h 34.2m	+07° 24'

NGC 6934 is a moderately large, bright globular cluster that lies 3.9° S of m4.0 naked-eye star 2-ε (epsilon), the tail of the dolphin. NGC 6934 is easy to locate because it's visible in a 50mm binocular or finder. To locate NGC 6934, put 2-ε on the N edge of the field and look for a faint nebulosity 3.9° to the S.

Using averted vision with our 50mm binoculars, NGC 6934 is visible as a small, faint but distinct circular hazy patch lying about half a degree NW of an m6 star that appears prominently in the field of view. At 125X in our 10" reflector, the unresolved core is surrounded by a 3' halo with about 30 stars visible. A pair of m9 stars lies W of the halo, with the first just outside the halo and the second about 10' W. An m8 star appears prominently 24' ESE (east southeast) of the core, and an m9 star the same distance to the NE.

FIGURE 19-1.

NGC 6934 (60' field width)

Image reproduced from Digitized Sky Survey courtesy Palomar Observatory and Space Telescope Science Institute

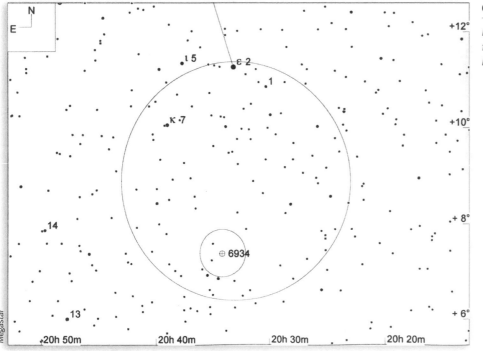

CHART 19-2.

NGC 6934 (10° field width, 5.0° finder field, 1.0° eyepiece field, LM 9.0)

Multiple Stars

12-gamma (STF 2727)	★★★	✧✧✧✧		MS	UD
Chart 19-1		m4.4/5.0, 9.2", PA 266° (2006)		20h 46.7m	+16° 07'

12-gamma, the nose of the dolphin, is a beautiful double star. At 125X in our 10" reflector, the primary appears yellow-white and the companion cool white. In our 90mm refractor at 100X, the primary is a more saturated yellow, and the secondary a striking blue-green shade.

20

Draco, The Dragon

NAME: Draco (DRAY-koe)

SEASON: Spring

CULMINATION: midnight, 24 May

ABBREVIATION: Dra

GENITIVE: Draconis (dray-KOE-nis)

NEIGHBORS: Boo, Cam, Cep, Cyg, Her, Lyr, UMa, UMi

BINOCULAR OBJECTS: NGC 5866 (M 102)

URBAN OBJECTS: none

Draco is a large, moderately dim, circumpolar constellation, ranking 8th in size among the 88 constellations. It covers 1,083 square degrees of the celestial sphere, or about 2.6%. Although Draco is home to one second magnitude star, five of third magnitude, and ten of fourth magnitude, those naked-eye stars are scattered across a sprawling constellation that extends across more than 100 degrees of sky.

Although Johann Bayer was never completely consistent in assigning letters to stars in order of their brightness, he outdid his usual inconsistency in Draco. Bayer assigned the brightest star in Draco, Etamin, the Bayer letter gamma, and Thuban, only the eighth-brightest star, the Bayer letter alpha. (Bayer may have shown favoritism for Thuban because it was the pole star 5,000 years ago, when the early Pharoahs were building the first pyramids.)

Draco is an ancient constellation, whom the Classical Greeks and Romans saw as a dragon, as did the Sumerians, Assyrians, and Egyptians before them. Many myths surround this constellation, including no less than three from ancient Greece. In one, Draco represents the dragon slain by the Hero Cadmus before he found the city of Thebes. In another, Draco is the dragon faced by Jason in his quest for the Golden Fleece. In the third and probably best known myth, Draco is the dragon with one hundred heads that guarded the Golden Apples of the Hesperides. In the eleventh of his twelve Labors, the Hero Heracles slew Draco with a poisoned arrow and made off with the Golden Apples.

Draco lies far outside the plane of the Milky Way, and so lacks the open clusters and emission nebulae common to Milky Way constellations. Draco is, however, rich in galaxies, three of which feature in the chapter, and is also home to a very bright planetary nebula, the Cat's Eye Nebula. Draco also has many interesting multiple stars, five of which are required objects for the Astronomical League Double Star club and are featured in this chapter.

Draco culminates at midnight on 24 May, and, for observers at mid-northern latitudes, is best placed for evening viewing during the early spring through mid-autumn months.

TABLE 20-1.

Featured star clusters, nebulae, and galaxies in Draco

Object	Type	Mv	Size	RA	Dec	M	B	U	D	R	Notes
NGC 5866	Gx	10.7	6.4 x 2.8	15 06.5	+55 46	◉	◉				M 102; Class SA0+ sp; SB 11.9
NGC 5907	Gx	11.1	12.9 x 1.3	15 15.9	+56 20					◉	Class SA(s)c: sp; SB 14.6
NGC 6503	Gx	10.9	7.1 x 2.4	17 49.5	+70 09					◉	Class SA(s)cd; SB 12.2
NGC 6543	PN	8.8	20"	17 58.6	+66 38					◉	Cat's Eye Nebula; Class 3a+2

TABLE 20-2.

Featured multiple stars in Draco

Object	Pair	M1	M2	Sep	PA	Year	RA	Dec	UO	DS	Notes
24-nu1, 25-nu2	STFA 35	4.9	4.9	63.4	311	2003	17 32.2	+55 10		◉	
21-mu*	STF 2130AB	5.7	5.7	2.3	11	2006	17 05.3	+54 28		◉	Alrakis
16/17	STF 2078AB	5.4	6.4	3.0	107	2003	16 36.2	+52 55		◉	
16/17	STFA 30AC	5.4	5.5	89.8	196	2003	16 36.2	+52 55		◉	
31-psi*	STF 2241AB	4.6	5.6	30.0	16	2006	17 41.9	+72 09		◉	
41/40	STF 2308AB	5.7	6.0	18.6	232	2003	18 00.1	+80 00		◉	

CHART 20-1.

The heart of the constellation Draco (field width 45°)

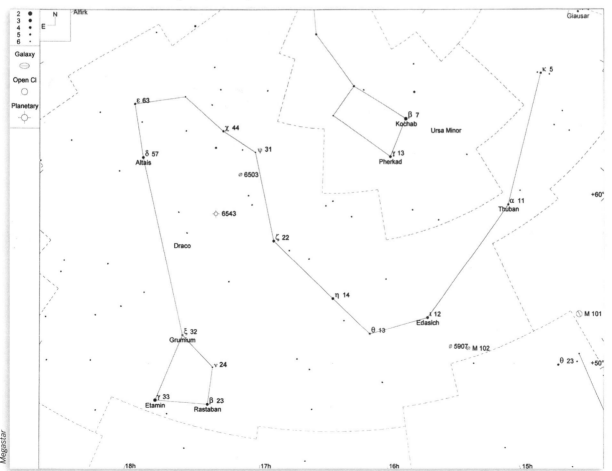

Clusters, Nebulae, and Galaxies

M 102 (NGC 5866)	★★	✹✹✹	GX	MBUDR
Chart 20-2	Figure 20-1	m10.7, 6.4' x 2.8'	15h 06.5m	+55° 46'

Although NGC 5866 is better known as Messier 102 or M 102, Messier may actually never have it. In the spring of 1781, Messier was working hard to meet the publication deadline for his catalog (we sympathize...). By mid-April, Messier had finished observing and cataloging 100 objects, many of which were his original discoveries and some of which he had observed after receiving reports of their locations from his friend and fellow comet enthusiast Pierre Méchain. With the deadline fast approaching, Messier added three objects—M 101, M 102, and M 103—that had been reported to him by Méchain but which he had not yet had an opportunity to observe for himself. (Messier noted that fact, so there was no dishonesty involved.)

Subsequent observations by other confirmed the identity and position of M 101 and M 103, but no one was able to locate the object that Messier had cataloged as M 102. Méchain later commented that he believed the object he had reported that was subsequently cataloged as M 102 was in fact a duplicate observation he'd made in error of M 101. Although that remains a possible explanation, most Messier enthusiasts now believe that Méchain (and later Messier himself) had actually observed NGC 5866, which lies about 9.1° E of M 101. Whatever the true explanation, NGC 5866 is now generally accepted as M 102.

M 102 is relatively easy to find. To locate it, place the m3.3 star 12-ι (iota), also known as Edasich, on the NE edge of your finder field, as shown in Chart 20-2, and look for a scalene triangle of m7 stars that appears near the SW edge of the finder field. M 102 lies about 10' NE of the E star in that triangle, and is clearly visible in a low-power eyepiece.

The Astronomical League Binocular Messier Club does not list M 102 as visible with 35/50mm binoculars, and lists it as a Tougher (their middle category) object for 80mm binoculars. We have never seen it with a 50mm binocular or finder, and have not attempted it with a larger binocular. At 125X in our 10" reflector, M 102 is visible as a bright oval extending 1' x 2.5' WNW-SSE (from west northwest to south southeast) that brightens smoothly to a non-stellar core.

FIGURE 20-1.

NGC 6910 (M 102) (60' field width)

Image reproduced from Digitized Sky Survey courtesy Palomar Observatory and Space Telescope Science Institute

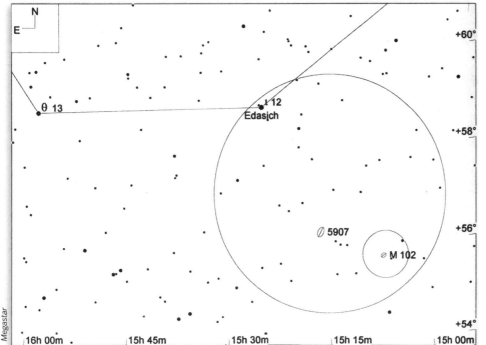

Megastar

NGC 5907	★★	◍◍◍	GX	MBUDR
Chart 20-2, 20-3	Figure 20-2	m11.1, 12.9' x 1.3'	15h 15.9m	+56° 20'

NGC 5907 is a rather dim galaxy that lies 1.4° ENE (east northeast) of M 102. If you're already viewing M 102, you can do a quick eyepiece star hop to NGC 5907, as shown in Chart 20-3. Otherwise, it's easiest to locate NGC 5907 by placing the m3.3 star 12-ι (iota), also known as Edasich, on the NE edge of your finder field, as shown in Chart 20-2, and locating the E-W chain of three m8 stars visible in Chart 20-2 just under the 5907 label. That chain points directly to NGC 5907, which lies just 23' NE of the E star in the chain and is faintly visible in a low-power eyepiece.

At 125X in our 10" reflector with averted vision, NGC 5907 appears as a moderately faint 0.5' x 7' streak of light extending NNW-SSE. A very thin brighter core occupies the central 2', with the outer halo tapering gradually down in brightness to invisibility.

FIGURE 20-2.

NGC 5907 (60' field width)

Image reproduced from Digitized Sky Survey courtesy Palomar Observatory and Space Telescope Science Institute

CHART 20-3.

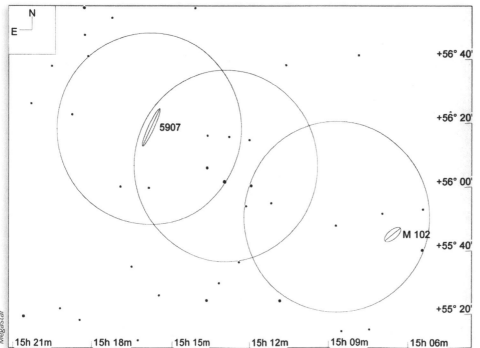

NGC 6503	★★	⚆⚆	GX	MBUDR
Chart 20-4	Figure 20-3	m10.9, 7.1' x 2.4'	17h 49.5m	+70° 09'

NGC 6503 is a relatively large, moderately bright galaxy. The easiest way to locate NGC 6503 is to orient yourself by locating the relatively bright naked-eye stars 14-η (eta), 22-ζ (zeta), and 44-χ (chi), which form a prominent 16° SW-NE line. Once you have located 44-χ (chi) at the NE end of that line, use it to put the dim naked-eye star 31-ψ (psi) on the NNW edge of your finder field, as shown in Chart 20-4. Look for the pattern formed by the prominent stars 27-, 28-, 37-, and 34-Dra and use that to center NGC 6503 in your finder field.

At 90X in our 10" reflector with averted vision, NGC 6503 is visible as a bright 1' x 4' thin oval halo extending NW-SE with some brightening to a poorly-defined central core.

FIGURE 20-3.

NGC 6503 (60' field width)

Image reproduced from Digitized Sky Survey courtesy Palomar Observatory and Space Telescope Science Institute

CHART 20-4.

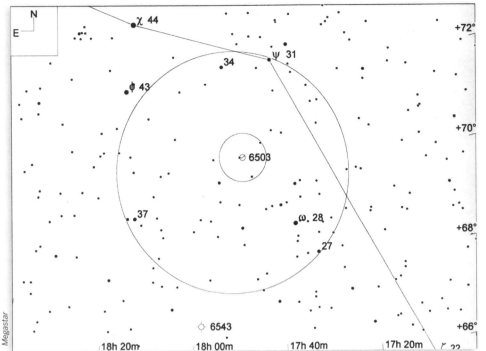

*NGC 6503 (10° field width, 5.0°
finder field, 1.0° eyepiece field,
LM 9.0)*

Megastar

NGC 6543	★★★	✿✿✿	PN	MBUDR
Chart 20-5	Figure 20-4	m8.8, 20"	17h 58.6m	+66° 38'

NGC 6543, better known as the Cat's Eye Nebula, is a small but very nice planetary nebula, and by far the best DSO in Draco. NGC 6543 lies 5.1° slightly N of E of m3.2 22-ζ (zeta) Draconis, where it can be reached by an easy hop. To locate NGC 6543, place 22-ζ (zeta) on the WSW edge of your finder field, as shown in Chart 20-5, with the m5 stars 27- and 28-Draconis near the N edge. Drift the finder E about half a field to center the Cat's Eye Nebula, which is visible in the finder but appears stellar. With our 5° finder field, we know we've gone slightly too far if the m5 stars 36- and 42-Draconis come into view on the E edge of the field; if your finder has a larger field of view, you can use these stars to center NGC 6543.

At 125X in our 10" reflector, the Cat's Eye is a small, round, bright disc with a blue-green tint. The m11 central star, which is invisible to us at lower magnification, is clearly visible at 125X. Using averted vision, some hints of the surrounding outer shell are faintly visible. Using an Ultrablock narrowband filter improves the contrast and visibility of the tendrils of the outer shell, but reveals no additional detail in the bright disc.

FIGURE 20-4.

NGC 6543 (60' field width)

Image reproduced from Digitized Sky Survey courtesy Palomar Observatory and Space Telescope Science Institute

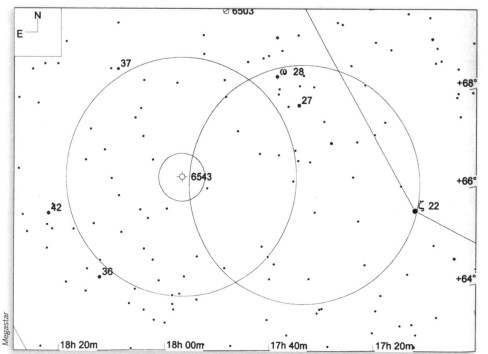

Multiple Stars

24-nu¹, 25-nu² (STFA 35)	★★	৬৬৬৬	MS	UD
Chart 20-1		m4.9/4.9, 63.4", PA 311° (2003)	17h 32.2m	+55° 10'

The double star 24-ν^1, 25-ν^2 Draconis (STFA 35) is a pretty, widely-spaced pair. This double is very easy to locate, lying just 2.9° N of the bright naked-eye star Rastaban, and itself faintly visible to the naked eye. At 90X in our 10" reflector or 100X in our 90mm refractor, this pair appear to be identical twins, both pure white and of the same brightness.

21-mu* (STF 2130AB)	★★	৬৬৬৬	MS	UD
Chart 20-6		m5.7/5.7, 2.3", PA 11° (2006)	17h 05.3m	+54° 28'

Like the preceding object, 21-μ Draconis (STF 2130AB), also known as Alrakis, is a pair of pure white stars of very similar brightness, but in this case the separation is much smaller. 21-μ is easy to locate. Place Rastaban on the SE edge of your finder and 25-ν on the NE edge and look for 21-μ on the W edge, where it appears much more prominent than any other star in the field of view.

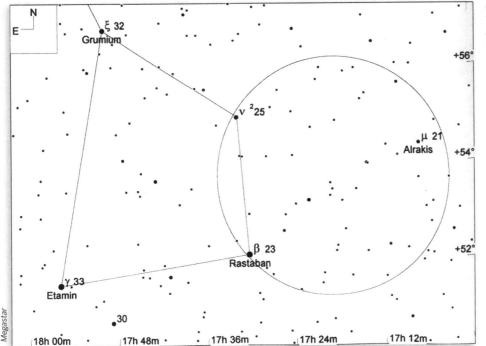

16/17 (STF 2078AB)	★★	❖❖❖	MS	UD
Chart 20-7		m5.4/6.4, 3.0", PA 107° (2003)	16h 36.2m	+52° 55'

16/17 (STFA 30AC)	★★	❖❖❖	MS	UD
Chart 20-7		m5.4/5.5, 89.8", PA 196° (2003)	16h 36.2m	+52° 55'

16/17-Draconis is a fine triple star, comprising a close m5.4/6.4 primary and companion, with an m5.5 tertiary much more widely separated from the primary-secondary pair. To locate 16/17-Draconis, locate Alrakis as described in the preceding section. Place Alrakis on the E edge of your finder field, and look for 16/17-Draconis on the WSW edge of the field, where it appears much more prominent than any of the surrounding stars.

At 70X in our 10" reflector, the primary and companion do not split cleanly. At 125X, the split is clean and all three stars appear a brilliant, pure white.

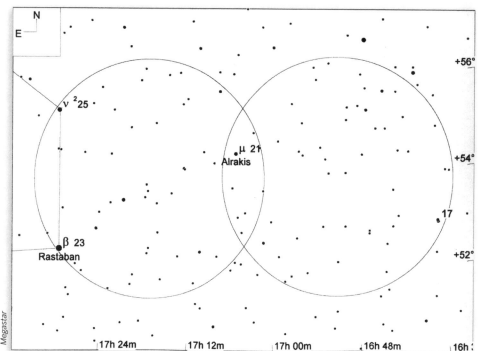

CHART 20-7.

16/17 Draconis (STF 2078AB and STFA 30AC) (10° field width, 5.0° finder fields, LM 9.0)

31-psi* (STF 2241AB)	★★	✹✹✹	MS	UD
Chart 20-1		m4.6/5.6, 30.0", PA 16° (2006)	17h 41.9m	+72° 09'

31-ψ (psi) Draconis (STF 2241AB) is a rather ordinary double star. The easiest way to locate 31-ψ is to orient yourself by locating the prominent 16° SW-NE line formed by the naked-eye stars 14-η (eta), 22-ζ (zeta), and 44-χ (chi). Once you locate 44-χ (chi) at the NE end of that line, it's easy to find the dim naked-eye star 31-ψ (psi), which lies 3.0° dead W.

At 90X in our 10" reflector, the primary appears warm white and the companion yellow white.

41/40 (STF 2308AB)	★★	✹✹	MS	UD
Chart 20-8		m5.7/6.0, 18.6", PA 232° (2003)	18h 00.1m	+80° 00'

41/40-Draconis (STF 2308AB) is another rather ordinary double star. To locate it, put 44-χ (chi) on the S edge of your finder field, as shown in Chart 20-8, and look for the stars 35- and 50-Draconis, which appear prominently near the NW and ENE edges of the field, respectively. Move the finder field to put 35-, 50-, and 59-Draconis on the edges of the field to confirm you have the correct stars, and then move the finder as shown to put 35-Draconis on the SSW edge of the field. 41/40-Draconis appears prominently in the N quadrant of the finder field.

At 90X in our 10" reflector, the primary and companion are of similar brightness and both are a warm white color.

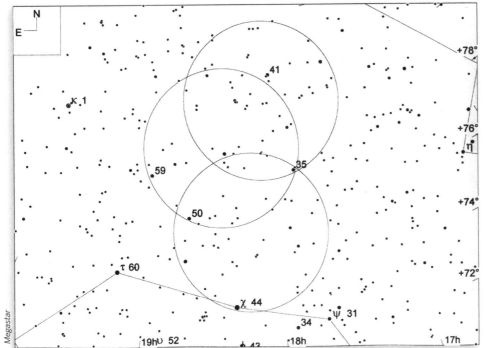

N
E

κ.1

41

59

35

50

τ 60

χ 44

50

ψ 31

34

19h υ 52

φ 43

18h

17h

+78°

+76°

η

+74°

+72°

Megastar

CHART 20-8.

41/40 Draconis (STF 2308AB)
(15° field width, 5.0° finder
fields, LM 9.0)

21

Eridanus, The River

NAME: Eridanus (air-uh-DAHN-us)	
SEASON: Autumn	
CULMINATION: midnight, 10 November	
ABBREVIATION: Eri	
GENITIVE: Eridani (air-uh-DAHN-ee)	
NEIGHBORS: Cae, Cet, For, Hor, Hyi, Lep, Ori, Phe, Tau	
BINOCULAR OBJECTS: none	
URBAN OBJECTS: none	

Eridanus is a dim, large, sprawling, southerly constellation, ranking 6th in size among the 88 constellations. It covers 1,138 square degrees of the celestial sphere, or about 2.8%. From its northern border on the celestial equator near Orion, Eridanus winds its way southwest, terminating near declination -57° at the 0th magnitude star Achernar, which means "the end of the river." The southerly section of Eridanus, including Achernar, never rises above the southern horizon for observers at mid-northerly latitudes.

Eridanus is an ancient constellation, known in Greek mythology as the river into which Phaëton plummeted from the chariot of his father Helios, the Sun God. Although he'd just gotten his learner's permit, Phaëton decided he could drive his father's chariot better than the old man could. Alas, his strength and skill was inadequate, so the chariot of the sun swerved through the heavens, creating deserts as it approached Earth too closely and arctic wastes

as it careened too far from Earth. Zeus, noticing the kid driving erratically, killed him with a thunderbolt. The smoking body of Phaëton plunged into the river Eridanus, thus ending his joyride.

Eridanus is relatively poor in deep-sky objects, but does have one interesting galaxy, a showpiece planetary nebula, and a pair of interesting double stars. Eridanus culminates at midnight on 10 November, and, for observers at mid-northern latitudes, is best placed for evening viewing from November through January.

TABLE 21-1.

Featured star clusters, nebulae, and galaxies in Eridanus

Object	Type	Mv	Size	RA	Dec	M	B	U	D	R	Notes
NGC 1232	Gx	10.5	7.4 x 6.4	03 09.8	-20 35					◉	Class SAB(rs)c; SB 13.7
NGC 1535	PN	9.6	1.0	04 14.3	-12 44					◉	Class 4+2c

TABLE 21-2.

Featured multiple stars in Eridanus

Object	Pair	M1	M2	Sep	PA	Year	RA	Dec	UO	DS	Notes
32	STF 470AB	4.8	5.9	6.8	348	2004	03 54.3	-02 57		◉	
55	STF 590	6.7	6.8	9.2	318	2004	04 43.6	-08 47		◉	

CHART 21-1.

The northern part of the constellation Eridanus (field width 40°)

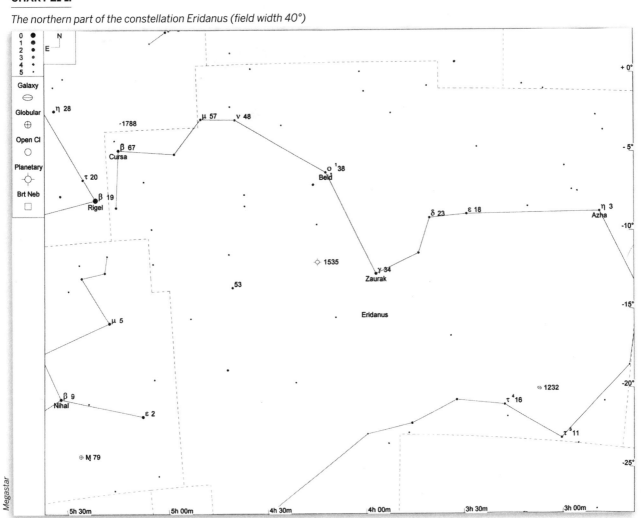

Megastar

Clusters, Nebulae, and Galaxies

NGC 1232	★★	◐◐	GX	MBUDR
Chart 21-2	Figure 21-1	m10.5, 7.4' x 6.4'	03h 09.8m	-20° 35'

NGC 1232 is a moderately large, rather pretty face-on spiral galaxy. Despite its relatively bright visual magnitude of 10.5, its 13.7 surface brightness makes it a difficult target in instruments smaller than 8" or 10".

NGC 1232 is rather difficult to locate. Begin by identifying the m3.0 star Zaurak, which lies 19.5° WSW of Rigel, and is the most prominent star in the vicinity. From Zaurack, look 12.3° SW to locate m3.7 16-τ^4 (tau^4)—which is the third and brightest of an E-W 10° chain of four m4 stars. From there, locate m4.1 11-τ^3 (tau^3), which lies 4.4° WSW (west southwest) of 16-τ^4. Having identified those stars, place 16-τ^4. on the E edge of your finder field and 11-τ^3 on the SW edge. Look in the NE quadrant of the finder field for a 1° arc of three m7 stars, which are quite prominent in the finder. NGC 35' lies dead W of the southernmost star in that arc, where it is visible in the same low-power eyepiece field about 7' W of an m10 star.

At 125X in our 10" Dob, NGC 1232 is visible with averted vision as a moderately faint 3.5' E-W halo surrounding a bright 1' oval core. Some mottling of the core is visible with difficulty.

FIGURE 21-1.

NGC 1232 (60' field width)

Image reproduced from Digitized Sky Survey courtesy Anglo-Australian Observatory and Space Telescope Science Institute

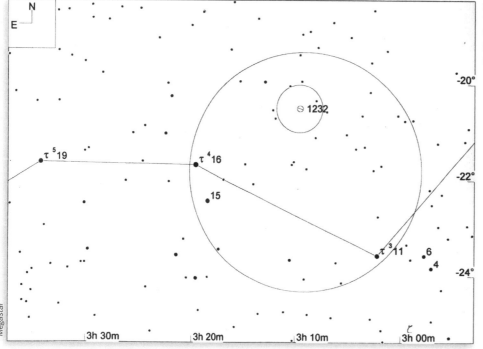

CHART 21-2.

NGC 1232 (10° field width, 5.0° finder field, 1.0° eyepiece field, LM 9.0)

Megastar

NGC 1535	★★★	❂❂❂	PN	MBUDR
Chart 21-3	Figure 21-2	m9.6, 1.0'	04h 14.3m	-12° 44'

NGC 1535 is bright, pretty planetary nebula, and by far the best DSO in Eridanus that's visible to mid-northerly observers. To locate NGC 1535, place the m3.0 star Zaurak on the SW edge of your finder field and look for the m4.9 star 39-Eridani, which appears prominently at or just beyond the NE edge of the finder field. Move the finder to place 39-Eridani on the N edge of the field. NGC 1535 lies 2.5° (about half a finder field) dead S of 39-Eridani.

At 125X in our 10" Dob with direct vision, NGC 1535 presents a bright bluish disc with good concentration toward the center. With averted vision, a hazy outer ring is visible although quite faint, and the central star is easily visible.

FIGURE 21-2.

NGC 1535 (60' field width)

Image reproduced from Digitized Sky Survey courtesy Anglo-Australian Observatory and Space Telescope Science Institute

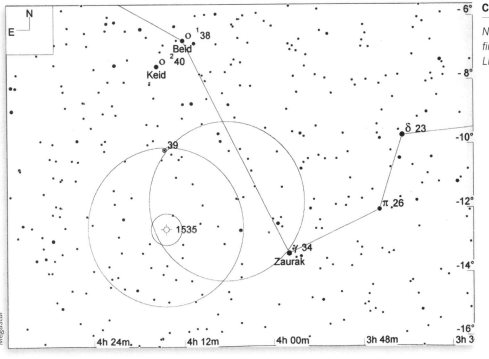

CHART 21-3.

NGC 1535 (15° field width, 5.0° finder fields, 1.0° eyepiece field, LM 9.0)

Multiple Stars

32 (STF 470AB)	★★	☼☼		MS	UD
Chart 21-4		m4.8/5.9, 6.8", PA 348° (2004)		03h 54.3m	-02° 57'

32-Eri is a rather ordinary double star. To locate 32-Eri, look 7.5° NNE (north northeast) of Zaurack for the m4 pair Beid and Keid. Place Beid on the ESE (east southeast) edge of your finder field and look near or just past the WNW edge of the field for the m5 star 30-Eri, which appears prominently in the finder. Place 30-Eri on the S edge of the finder field with the m5 star 35-Eri on the NE edge of the field, and look for m5 32-Eri near the center of the finder field.

At 90X in our 10" reflector, the primary has a golden yellow color and the companion appears cool white. (Some observers report the companion as greenish-white, but we see no green color cast.)

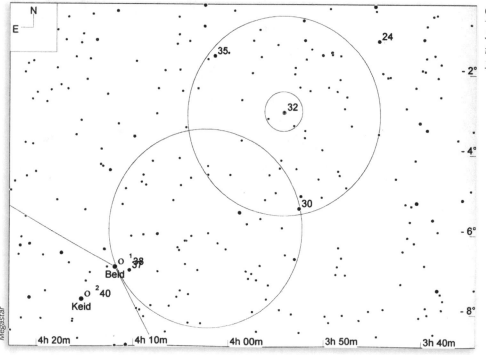

CHART 21-4.

32-Eridani (STF 470AB) (12° field width, 5.0° finder fields, 1.0° eyepiece field, LM 9.0)

55 (STF 590)	★★	◊◊◊		MS	UD
Chart 21-5		m6.7/6.8, 9.2", PA 318° (2004)		04h 43.6m	-08° 47'

55-Eri is another rather ordinary double star. To locate 55-Eri, place m2.8 Cursa on the NE edge of your finder field, with m4.4 61-ω (omega) on the NW edge and m4.8 65-ψ (psi) near the center of the field. Move the finder field SW until 61-ω is on the N edge of the field and 65-ψ on the E edge and look for a wide m6/7 double that appears prominently on the SW edge of the field. The southerly and dimmer member of that pair is 55-Eri.

At 90X in our 10" reflector, the primary is yellow-white and the companion warm white.

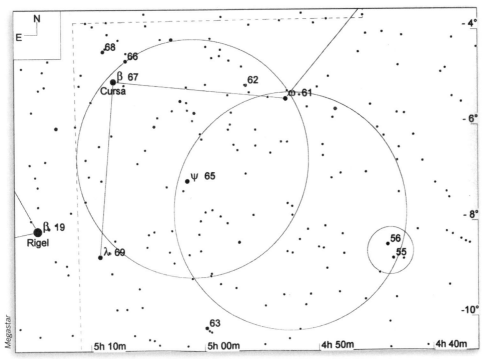

CHART 21-5.

55-Eridani (STF 590) (10° field width, 5.0° finder fields, 1.0° eyepiece field, LM 9.0)

22

Gemini, The Twins

NAME: Gemini (GEM-ih-nye)

SEASON: Winter

CULMINATION: midnight, 4 January

ABBREVIATION: Gem

GENITIVE: Geminorum (GEM-ih-NOR-um)

NEIGHBORS: Aur, CMi, Cnc, Lyn, Mon, Ori, Tau

BINOCULAR OBJECTS: NGC 2168 (M 35)

URBAN OBJECTS: NGC 2168 (M 35), NGC 2392

Gemini is a mid-size, prominent mid-northerly constellation, ranking 30th in size among the 88 constellations. It covers 514 square degrees of the celestial sphere, or about 1.2%.

Gemini is an ancient constellation that has always been associated with twins. In the most popular Greek myth, Gemini represents the twins Castor and Polydeuces (AKA Pollux) who, with their sister Helen (later of Troy), emerged from an egg laid by the human woman Leda, Queen of Sparta, after her dalliance with Zeus, who had assumed the guise of a swan to seduce her.

Gemini lies partially in the Milky Way, and so is home to several interesting open clusters. The northern half of Gemini, which includes the head, formed by the bright stars Castor and Pollux, lies outside the Milky Way, and contains many galaxies, most of which are far too dim to be visible with typical amateur instruments.

Gemini culminates at midnight on 4 January, and, for observers at mid-northern latitudes, is best placed for evening viewing during the late autumn through early spring months.

TABLE 22-1.

Featured star clusters, nebulae, and galaxies in Gemini

Object	Type	Mv	Size	RA	Dec	M	B	U	D	R	Notes
NGC 2168	OC	5.1	28.0	06 09.0	+24 21	◉	◉	◉			M 35; Mel 41; Class III 3 r
NGC 2371/2372	PN	13.0	55.0"	07 25.6	+29 29					◉	Class 3a+2
NGC 2392	PN	9.9	50.0"	07 29.2	+20 55			◉		◉	Class 3b+3b

TABLE 22-2.

Featured multiple stars in Gemini

Object	Pair	M1	M2	Sep	PA	Year	RA	Dec	UO	DS	Notes
55-delta*	STF 1066	3.6	8.2	5.7	226	2006	07 20.1	+21 59		◉	Wasat
66-alpha*	STF 1110AB	1.9	3.0	4.3	60	2006	07 34.6	+31 53		◉	Castor

CHART 22-1.

The constellation Gemini (field width 40°)

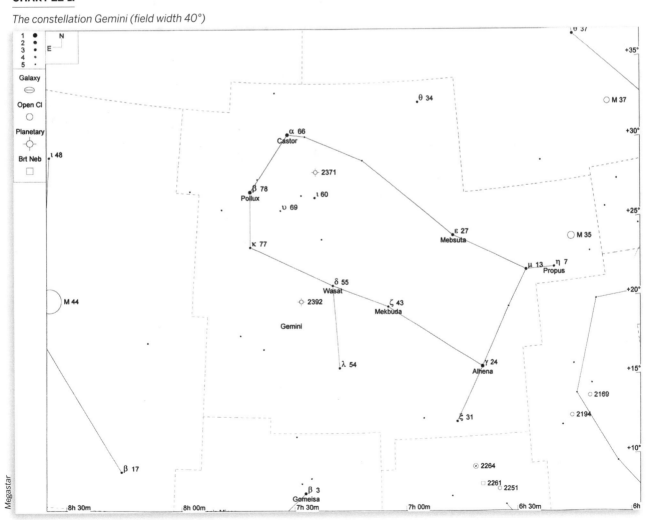

1	●
2	●
3	●
4	●
5	·

Galaxy
⊖

Open Cl
○

Planetary
◇

Brt Neb
□

N
E

θ 34
○ M 37
+35°

α 66
Castor
+30°

◇ 2371

ι 48

β 78
Pollux
ι 60

υ 69
+25°

ε 27
Mebsuta
○ M 35

κ 77
μ 13   η 7
Propus

δ 55
Wasat
+20°

◇ 2392
ζ 43
Mekbuda

Gemini

M 44

λ 54

γ 24
Alhena
+15°

○ 2169

ζ 31
○ 2194

+10°

⊙ 2264

β 17
□ 2261 ○ 2251

β 3
Gomeisa
+5°

8h 30m   8h 00m   7h 30m   7h 00m   6h 30m   6h

Megastar

θ 37

Clusters, Nebulae, and Galaxies

M 35 (NGC 2168)	★★★	✺✺✺	OC	MBUDR
Chart 22-2	Figure 22-1	m5.1, 28.0'	06h 09.0m	+24° 21'

NGC 2168, better known as Messier 35 or M 35, is a bright open cluster about the size of the full moon. Under very dark conditions, it can just be glimpsed by the naked eye with averted vision.

M 35 was discovered by Philippe Loys de Chéseaux in 1745 or 1746, and independently re-discovered by English astronomer John Bevis before the 1750 publication of his star atlas, *Uranographia Britannica*. (Incidentally, there is absolutely no truth to the rumor we started that John Bevis had a German assistant named Arschkopf (Butthead).) Charles Messier, who attributed the discovery to Bevis, observed M 35 on the night of 30 August 1764 and described it as a "cluster of very small stars, near the left foot of Castor, at a little distance from the stars Mu & Eta of that constellation."

Messier's description remains the best way to locate M 35. At the NW corner of Gemini, look for the naked-eye stars m2.9 13-μ (mu) and m3.3 7-η (eta), also called Propus, which form a prominent 1.9° E-W pair. M 35 lies 2.2° NW of Propus, and is easily visible in a 50mm binocular or finder.

With our 50mm binoculars, M 35 is visible as a bright hazy patch with numerous embedded stars. Half a dozen m7/8 stars are visible with another dozen down to m10. At 42X in our 10" Newtonian, M 35 is a magnificent sight, with 100+ stars visible down to m13 within a half-degree circle. It's very difficult to determine where the edges of the cluster lie because the cluster merges into the very rich background of Milky Way field stars of similar magnitude.

When you view M 35, take a moment to look for the much dimmer cluster NGC 2158, which lies about half a degree to the SW, and is visible at the lower right of Figure 22-1. NGC 2158 lies about six times the distance from us of M 35. At the same distance, it would be one of the most magnificent open clusters in the night sky. At low magnification in our 10" scope, NGC 2158 is just a dim glow. At 125X, it reveals itself as a 4' open cluster with many dim stars. We think it looks like a smaller, dimmer version of M 37 in Auriga.

FIGURE 22-1.

NGC 2168 (M 35) with NGC 2158 to lower right (60' field width)

Image reproduced from Digitized Sky Survey courtesy Palomar Observatory and Space Telescope Science Institute

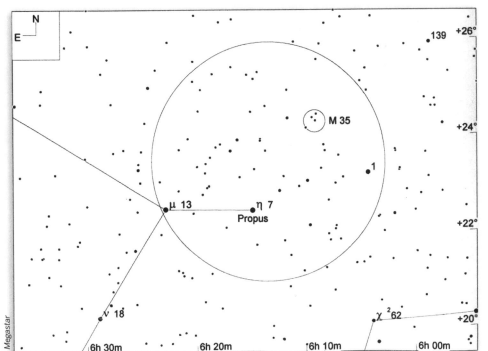

CHART 22-2.

NGC 2168 (M 35) (10° field width, 5.0° finder field, LM 9.0)

NGC 2371/2372	★	⊙	PN	MBUDR
Chart 22-3	Figure 22-2	m13.0, 55.0"	07h 25.6m	+29° 29'

NGC 2371/2372 is a rather ordinary planetary nebula. It's assigned two NGC numbers because in images it appears to be split into NW and SE segments.

To locate NGC 2371/2372, place Castor on the N edge of your finder field, with Pollux on the SE edge. To center NGC 2371/2372 in the finder, move the finder field about half a field dead W, as shown in Chart 22-3, until the bright stars 60-, 62-, 64-, and 65-Gem are placed as shown in the chart.

We rate this object as very difficult to find because we had a very hard time finding it. It's easy enough to get into the general vicinity of the object, but identifying the object with certainty is quite difficult, or at least we found it so.

At 42X in our 10" Dob, this planetary appears as just another dim star. We attempted to confirm the object by "blinking" it (moving the filter between our eye and the eyepiece) with our Ultrablock narrowband filter, but the object is so faint that it disappeared entirely along with the background stars we'd intended to block. Shifting to higher magnification allowed us to see the object as non-stellar. At 125X, it appears as a slightly fuzzy dim star. At 240X with averted vision, we were finally able to discern a very faint disc. We could see no hint of surface detail.

FIGURE 22-2.

NGC 2371/2372 (60' field width)

Image reproduced from Digitized Sky Survey courtesy Palomar Observatory and Space Telescope Science Institute

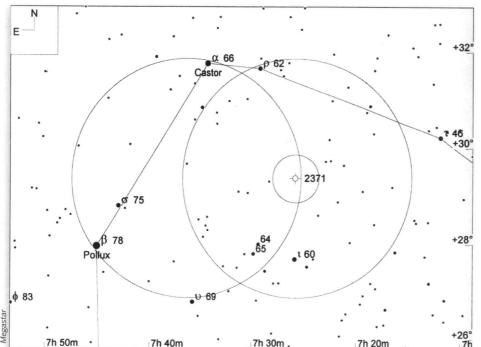

NGC 2392	★★★	◊◊◊◊	PN	MBUDR
Chart 22-4	Figure 22-3	m9.9, 50.0"	07h 29.2m	+20° 55'

NGC 2392, better known as the Eskimo Nebula or the Clown Face Nebula, is one of the finest planetary nebulae in the sky. To locate it, place the m3.5 star 55-δ (delta), also known as Wasat, on the WNW (west northwest) edge of your finder field and look for the stars 61- and 63-Gem, which appear prominently near the center of the finder field. NGC 2392 forms a triangle with those two stars, as shown in Chart 22-4, and will be located at or near the center of your finder field. With averted vision, we can actually make out NGC 2392 in our 50mm finder, although it appears as just another very dim star.

At 42X in our 10" Dob, NGC 2392 looks like a fuzzy star. Boosting magnification to 240X and using averted vision reveals considerable detail. The disc is well defined, and is a pretty pale, pure blue rather than the bluish-green color we see in most bright planetaries. The central star is easily visible. It's easy to "blink" this nebula by staring at the central star with direct vision until the disc fades from view and then looking to one side suddenly, whence the disc again pops into view.

FIGURE 22-3.

NGC 2392 (60' field width)

Image reproduced from Digitized Sky Survey courtesy Palomar Observatory and Space Telescope Science Institute

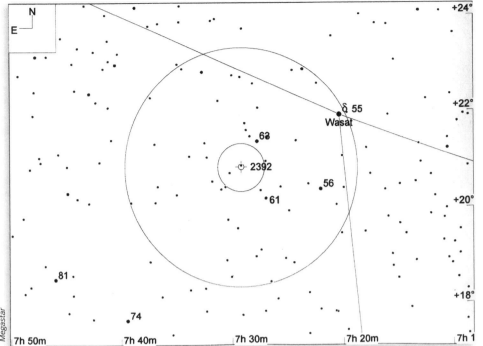

Multiple Stars

55-delta* (STF 1066)	★★★	✴✴✴✴		MS	UD
Chart 22-1		m3.6/8.2, 5.7", PA 226° (2006)		07h 20.1m	+21° 59'

55-δ (delta) Geminorum (Wasat) is a very pretty double star, easily located because Wasat is a naked-eye star. At 125X in our 10" reflector, the primary is a bright yellow-white, and the companion a pale blue-white.

66-alpha* (STF 1110AB)	★★	✴✴✴✴		MS	UD
Chart 22-1		m1.9/3.0, 4.3", PA 60° (2006)		07h 34.6m	+31° 53'

66-α (alpha) Geminorum (Castor) is another pretty double star. At 125X in our 10" reflector, the primary and its companion are both a brilliant, blazing white.

23

Hercules, The Hero

NAME:	Hercules (HUR-cue-leez)
SEASON:	Early summer
CULMINATION:	midnight, 13 June
ABBREVIATION:	Her
GENITIVE:	Herculis (HUR-cue-lis)
NEIGHBORS:	Aql, Boo, CrB, Dra, Lyr, Oph, Ser, Sge, Vul
BINOCULAR OBJECTS:	NGC 6205 (M 13), NGC 6341 (M 92)
URBAN OBJECTS:	NGC 6205 (M 13), NGC 6210, NGC 6341 (M 92)

Hercules is a large, bright, mid-northerly constellation, ranking 5th in size among the 88 constellations. It covers 1,225 square degrees of the celestial sphere, or about 3.0%.

Hercules is the Roman name for Herakles, the greatest Hero of Greek mythology. The son of Zeus and the mortal woman Alcmena, Herakles drew the ire of the jealous goddess Hera, wife of Zeus. When Herakles was still an infant, Hera sent two venomous serpents to bite him in his crib. Hera's scheme failed, and Herakles was found alive and well in his crib, holding a strangled serpent in each hand.

Herakles grew to manhood, married, and had children, at which time a vengeful Hera again intervened. She caused Herakles to slay his wife and children in a fit of madness. When the madness passed, an inconsolable Herakles became a hermit and lived alone in the wilderness. Finally found by his cousin Theseus, Herakles allowed himself to be persuaded to visit the Oracle at Delphi, who assigned him as penance for the murder of his wife and children a series of twelve impossible tasks, the Twelve Labors of Herakles. Impossible they may have been, but Herakles completed those tasks and by doing so Herakles made the world safe for mankind and assumed the mantle of the greatest Hero of Greek mythology.

Hercules is home to several interesting multiple stars, a nice planetary nebula (NGC 6210), and two of the most magnificent globular clusters visible to mid-northerly observers, M 13 and M 92. Hercules culminates at midnight on 13 June, and, for observers at mid-northern latitudes, is best placed for evening viewing from mid-spring until early autumn.

TABLE 23-1.

Featured star clusters, nebulae, and galaxies in Hercules

Object	Type	Mv	Size	RA	Dec	M	B	U	D	R	Notes
NGC 6205	GC	5.8	20.0	16 41.7	+36 28	◉	◉	◉			M 13
NGC 6341	GC	6.5	14.0	17 17.1	+43 08	◉	◉	◉			M 92
NGC 6210	PN	9.3	30.0"	16 44.5	+23 48			◉		◉	Class 2+3b

TABLE 23-2.

Featured multiple stars in Hercules

Object	Pair	M1	M2	Sep	PA	Year	RA	Dec	UO	DS	Notes
7-kappa	STF 2010AB	5.1	6.2	27.4	13	2003	16 08.1	+17 02		◉	
64-alpha*	STF 2140Aa-B	3.5	5.4	4.6	104	2006	17 14.6	+14 23		◉	Rasalgethi
65-delta	STF 3127Aa-B	3.1	8.3	11.0	282	2001	17 15.0	+24 50		◉	
75-rho	STF 2161Aa-B	4.5	5.4	4.1	319	2004	17 23.7	+37 08		◉	
95	STF 2264	4.9	5.2	6.3	257	2003	18 01.5	+21 35		◉	

CHART 23-1.

The constellation Hercules (field width 50°)

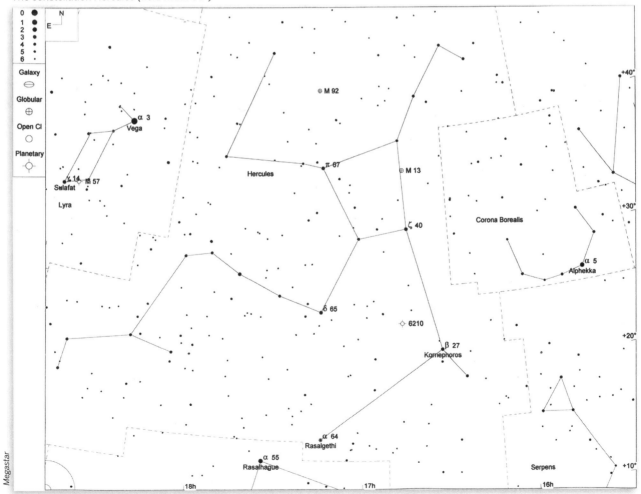

Clusters, Nebulae, and Galaxies

M 13 (NGC 6205)	★★★★	✦✦✦✦	GC	MBUDR
Chart 23-2	Figure 23-1	m5.8, 20.0'	16h 41.7m	+36° 28'

NGC 6205, better known as Messier 13, M 13, or the Great Hercules Cluster, is the most magnificent globular cluster visible to mid-northerly observers. With even the slightest optical aid, such as an opera glass, M 13 is a prominent object. On a very dark, very clear night when M 13 is near zenith, it is sometimes just possible to glimpse it as a very faint patch of haze using only the naked eye with averted vision.

In fact, Edmond Halley (of Halley's Comet fame) was the first person to report sighting M 13, which he first observed in 1714 without optical aid. Charles Messier first observed this cluster on the night of 1 June 1764, and logged it as the thirteenth object in his catalog.

M 13 is trivially easy to locate. It lies 2.5° dead S of m3.5 44-η (eta) Herculis, the NW star in the prominent Keystone of Hercules, and is prominently visible in a 50mm binocular or finder.

With our 50mm binocular, M 13 is visible as a very bright 10' patch of nebulosity without any stars resolved. At 90X in our 10" Newtonian, M 13 is a brilliant, dense, glittering 6' spherical core embedded in a dimmer halo extending irregularly out to 12', with chains of stars extending out around the periphery. At 180X, the globular is resolved as hundreds of individual stars all the way to the core. A pair of m6/m7 stars frame M 13 to the SSW and E.

FIGURE 23-1.

NGC 6205 (M 13) (60' field width)

Image reproduced from Digitized Sky Survey courtesy Palomar Observatory and Space Telescope Science Institute

CHART 23-2.

NGC 6205 (M 13) (15° field width, 5.0° finder field, 1.0° eyepiece field, LM 9.0)

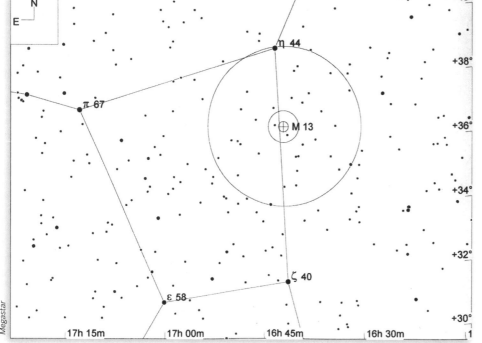

ILLUSTRATED GUIDE TO ASTRONOMICAL WONDERS

M 92 (NGC 6341)	★★★★	✦✦✦	GC	MBUDR
Chart 23-3	Figure 23-2	m6.5, 14.0'	17h 17.1m	+43° 08'

If M 13 didn't exist, M 92 would be famous as the best globular cluster visible to mid-northerly observers. As it is, M 92 is often overlooked in favor of the nearby M 13. That's unfortunate, because M 92 is a spectacular object in its own right.

Johann Elert Bode discovered M 92 on the night of 27 December 1777, but made little effort to publicize his discovery. Charles Messier independently re-discovered M 92 on his "big night" of 18 March 1781, the same night he discovered or re-discovered no less than nine galaxies in Virgo and Coma Berenices.

One reason that M 92 is often overlooked is that it's harder to find than M 13, but it's not really that much harder. To locate M 92, place m3.1 67-π (pi) Herculis on the S edge of your finder field and drift the finder field N by half a field or so until M 92 comes into view. In a 50mm binocular or finder, M 92 looks like a smaller, dimmer version of M 13. But it's smaller and dimmer only in a relative sense. It still appears as a relatively large, relatively bright patch of haze with no stars resolved.

At 125X in our 10" Newtonian, M 92 is a bright, densely concentrated 3' spherical core surrounded by a dimmer halo extending irregularly out to 8'. Numerous chains of stars extend from the core to the edges of the outer halo.

FIGURE 23-2.

NGC 6341 (M 92) (60' field width)

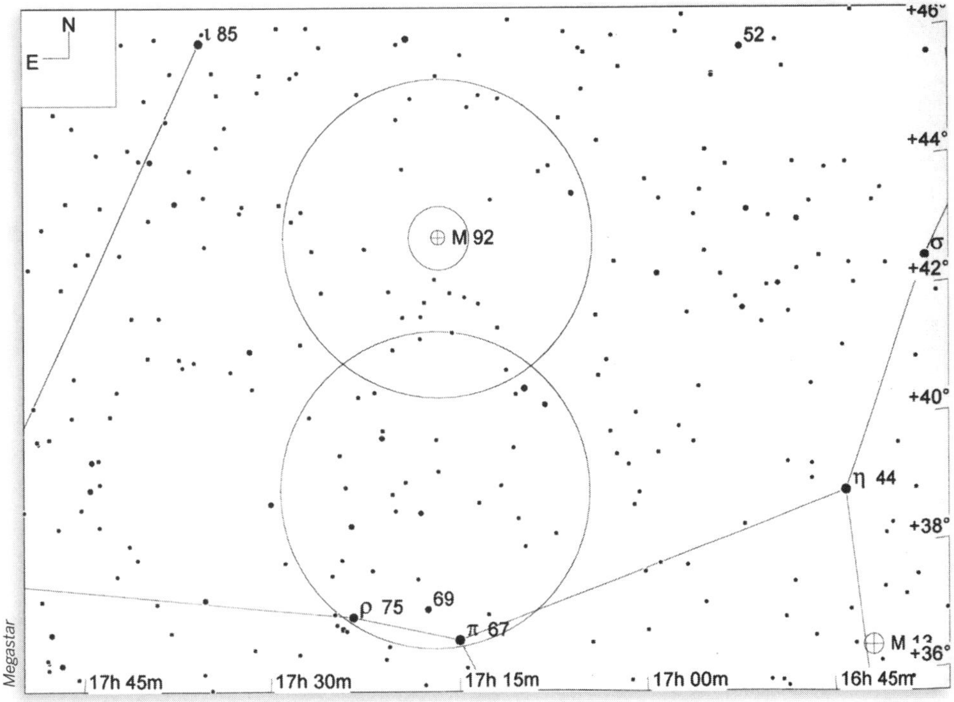

CHART 23-3.

NGC 6341 (M 92) (15° field width, 5.0° finder fields, 1.0° eyepiece field, LM 9.0)

NGC 6210	★★	✸✸✸	PN	MBUDR
Chart 23-4	Figure 23-3	m9.3, 30.0"	16h 44.5m	+23° 48'

NGC 6210 is a rather pedestrian planetary nebula that is easily located 3.8° NE of the m2.8 star 27-β (beta) Herculis in the southern part of the constellation.

To locate NGC 6210, place 27-β (beta) on the SW edge of your finder field and drift the finder NE until the m5 star 51-Her comes into view on the NE edge of the field. NGC 6210 lies exactly on and one third of the way along the line from 51-Her to 27-β. We can't see NGC 6210 in our 50mm finder, but it forms the N apex of a 16' isosceles triangle with two m7 stars that are prominently visible in the finder field (and in Figure 23-3).

At 42X in our 10" Dob, NGC 6210 is visible as a slightly fuzzy m9 star. At 90X, the object has a bluish coloration and is obviously non-stellar, but no surface detail is visible. Using 180X reveals a featureless blue 20" x 12" E-W oval disc with no central star visible. A narrowband or O-III filter improves the contrast but reveals no additional detail.

FIGURE 23-3.

NGC 6210 (60' field width)

Image reproduced from Digitized Sky Survey courtesy Palomar Observatory and Space Telescope Science Institute

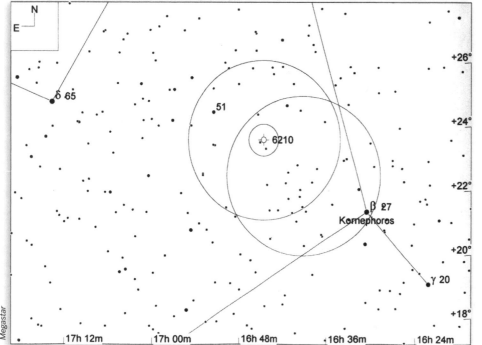

CHART 23-4.

NGC 6210 (15° field width, 5.0° finder fields, 1.0° eyepiece field, LM 9.0)

Multiple Stars

7-kappa (STF 2010AB)	★★	◔◔◔		MS	UD
Chart 23-5		m5.1/6.2, 27.4", PA 13° (2003)		16h 08.1m	+17° 02'

To locate 7-ϰ (kappa), put m2.8 Komephoros on the NE edge of your finder field and look for m3.7 20-γ (gamma) and m5.7 16-Herculis, which appear prominently in the SW quadrant of the finder. Drift the finder SW to place 20-γ on the NE edge of the finder field, and look for a bright pair of stars that appear prominently in the SW quadrant of the finder field. The brighter of that pair is m5 7-ϰ.

In our 10" reflector at 90X, the primary and companion are both yellow-white.

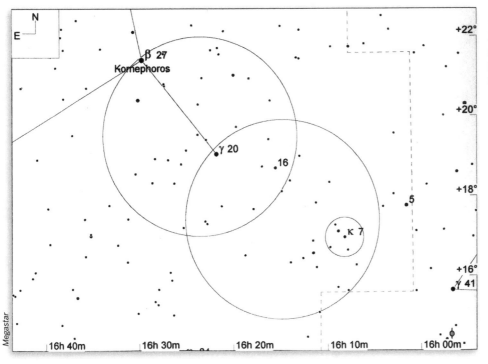

CHART 23-5.

7-kappa Herculis (STF 2010AB) (12° field width, 5.0° finder fields, 1.0° eyepiece field, LM 9.0)

64-alpha* (STF 2140Aa-B)	★★	◔◔◔◔		MS	UD
Chart 23-1		m3.5/5.4, 4.6", PA 104° (2006)		17h 14.6m	+14° 23'

64-α (alpha) Herculis, better known as Rasalgethi, is a bright naked-eye star that lies on the S border of Hercules with Ophiuchus, just 5.3° WNW of second magnitude Rasalhague (55-α Ophiuchi), with which it forms a prominent naked-eye pair. In our 10" reflector at 125X, the primary is yellow and the companion is cool white.

65-delta (STF 3127Aa-B)	★★	❀❀❀❀		MS	UD
Chart 23-1		m3.1/8.3, 11", PA 282° (2001)		17h 15.0m	+24° 50'

65-δ (delta) Herculis is the bright naked eye star that lies 7° SSE of the Keystone of Hercules. In our 10" reflector at 125X, the primary is brilliant white and the companion blue-white.

75-rho (STF 2161Aa-B)	★★	❀❀❀❀		MS	UD
Chart 23-6		m4.5/5.4, 4.1", PA 319° (2004)		17h 23.7m	+37° 08'

75-ϱ (rho) lies just 1.8° ENE of 67-π (pi), the star that forms the NE corner of the Keystone of Hercules. In our 10" reflector at 125X, the primary and the companion are both a brilliant, pure white.

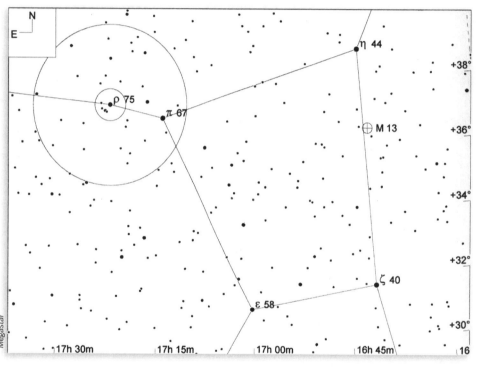

CHART 23-6.

75-rho Herculis (STF 2161Aa-B) (15° field width, 5.0° finder field, 1.0° eyepiece field, LM 9.0)

95 (STF 2264)	★★	❀❀		MS	UD
Chart 23-7, 23-8		m4.9/5.2, 6.3", PA 257° (2003)		18h 01.5m	+21° 35'

The easiest way to locate 95-Herculis is to follow the 12.5° chain of m3 and m4 stars that leads E from m3.1 65-δ (delta) to m3.8 103-o (omicron). Once you have identified 103-o, look 10° SE for a 10° E-W arc of four m4 stars, which from E to W are 111-, 110-, 109-, and 102-Herculis. Place 102-Herculis on the ESE edge of your finder field, as shown in Chart 23-8, and look for the distinctive arc of stars formed by 97-, 98-, 102-, and 101-Herculis, with 96- and 95-Herculis lying to the W near the center of the finder field. In our 10" reflector at 125X, the primary and the companion are both warm white.

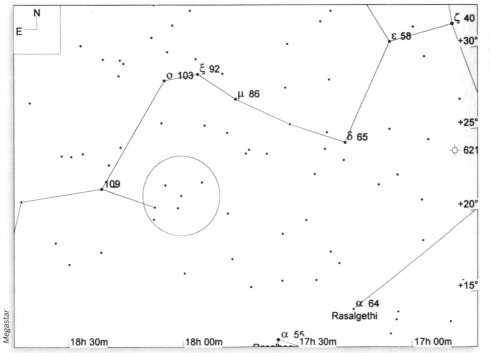

CHART 23-7.

95-Herculis (STF 2264) overview (30° field width, 5.0° finder field, LM 6.0)

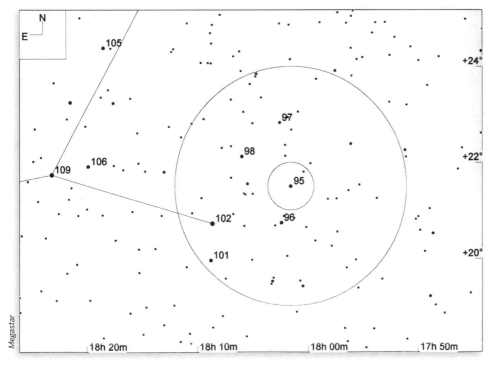

CHART 23-8.

95-Herculis (STF 2264) detail (10° field width, 5.0° finder field, 1.0° eyepiece field, LM 9.0)

24
Hydra, The Water Snake

NAME: Hydra (HYE-druh)

SEASON: Winter

CULMINATION: midnight, 9 February

ABBREVIATION: Hya

GENITIVE: Hydrae (HYE-dry)

NEIGHBORS: Ant, Cen, CMi, Cnc, Crt, Crv, Leo, Lib, Lup, Mon, Pup, Pyx, Sex, Vir

BINOCULAR OBJECTS: NGC 2548 (M 48), NGC 4590 (M 68), NGC 5236 (M 83)

URBAN OBJECTS: NGC 2548 (M 48), NGC 3242

Hydra is the largest constellation, just edging out second-place Virgo. Hydra covers 1,303 square degrees of the celestial sphere, or about 3.2%. Despite its size, Hydra has few bright stars. At magnitude 1.99, 30-α (alpha) Hydrae, also known as Alphard, is Hydra's brightest star. Hydra is home to only six third magnitude stars, scattered across its huge extent. All other stars in Hydra are fourth magnitude or dimmer.

In Greek mythology, Hydra is seen as a water snake. In one Greek myth, Apollo sent his servant, Corvus the Crow, to fetch water for him in Crater the Cup. Corvus returned with an empty cup but presented Apollo with the water snake as evidence that he had searched for water. Angered by the failure of his servant, Apollo cast all of them—Corvus, Crater, and Hydra—up into the heavens, where they remain in close proximity. In another Greek myth, the Hero Herakles faced Hydra during one of his Twelve Labors. Each time Herakles cut off one of Hydra's heads, two more grew in its place. Herakles finally solved the problem by burning off the heads instead of cutting them off.

Although you might expect a constellation the size of Hydra to have many interesting deep-sky objects, Hydra is nearly as poor in bright DSOs as it is in bright stars. Lying far off the Milky Way, Hydra does have numerous galaxies, but most of these are too dim to be interesting in any but the largest amateur instruments. There are, however, three bright Messier objects in Hydra, including the open cluster M 48, the globular cluster M 68, and the galaxy M 83. Hydra is also home to NGC 3242, one of the better planetary nebulae.

Because Hydra covers nearly seven hours of right ascension, the best time to observe it depends on whether the object you're looking for is in eastern Hydra or western Hydra. For example, M 48 (on the W edge of Hydra) culminates at midnight local time during late January, but M 83, more than five hours of right ascension to the east, doesn't culminate at midnight until early May.

TABLE 24-1.

Featured star clusters, nebulae, and galaxies in Hydra

Object	Type	Mv	Size	RA	Dec	M	B	U	D	R	Notes
NGC 2548	OC	5.8	54.0	08 13.7	-05 45	◉	◉	◉			M 48; Class I 3 r
NGC 3242	PN	8.6	75"	10 24.7	-18 39			◉		◉	Ghost of Jupiter; Class 4+3b
NGC 4590	GC	7.3	11.0	12 39.5	-26 45	◉	◉				M 68; Class X
NGC 5236	Gx	8.2	12.8 x 11.4	13 37.1	-29 52	◉	◉				M 83; Class SAB(s)c; SB (high) ???

TABLE 24-2.

Featured multiple stars in Hydra

Object	Pair	M1	M2	Sep	PA	Year	RA	Dec	UO	DS	Notes
N	H 96	5.6	5.7	9.4	210	2003	11 32.3	-29 15		◉	

CHART 24-1.

The constellation Hydra (field width 90°)

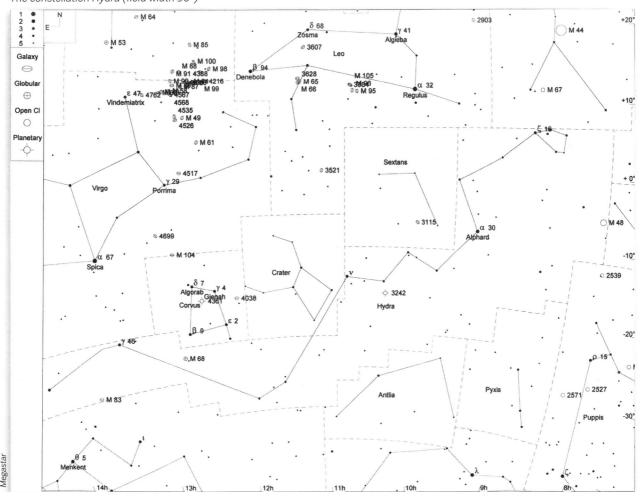

Clusters, Nebulae, and Galaxies

M 48 (NGC 2548)	★★★	✹✹✹	OC	MBUDR
Chart 24-2	Figure 24-1	m5.8, 54.0'	08h 13.7m	-05° 45'

NGC 2548, better known as Messier 48 or M 48, is a large, very bright open cluster. Charles Messier discovered and logged M 48 on the night of 19 February 1771, but he recorded its position incorrectly, leading to M 48 being listed as a "missing Messier" for more than 150 years. M 48 was re-discovered independently by Johann Elert Bode some time before 1783, and by Caroline Herschel (sister of William Herschel, and a respected observer in her own right) in 1783.

Although you can locate M 48 as shown in Chart 24-2 by star hopping from Procyon and 13-ζ (zeta) in Canis Minor to 28- and 29-Monocerotis and then over the boundary to M 48, it's much easier just to eyeball the 4° NW-SE line formed by Gomeisa to Procyon, extending it another 11° until you come to m4.4 29-ζ (zeta) Monocerotis, which is visible to the naked eye. Once you locate that star by eye, put it on the NW edge of your binocular or finder field and look 2.9° SSE for M 48, which is conspicuous in a 50mm finder or binocular.

Some observers have reported seeing M 48 naked-eye on very dark, transparent nights. M 48 is a prominent object with even the slightest optical aid, such as an opera glass. With our 50mm binocular and direct vision, M 48 is a very bright 30' nebulous patch with half a dozen m8 stars resolved and 15+ dimmer stars visible with averted vision. At 42X in our 10" reflector, M 48 is spectacular. About 40 stars m8 and dimmer are visible in the denser central 30' core with two dozen more scattered around the periphery. The brighter stars in the denser central region range from m8 to m10 and are arranged in chains, clumps, and knots, with one prominent N-S chain of m8/9 stars nearly at the center of the cluster. At 90X in our 10" Dob with our 14mm Pentax eyepiece, only the central 45' of the cluster lies within the field of view, but many dimmer stars are visible, nearly doubling the star count. The brighter stars are so prominent that the dimmer stars appear to recede, giving the cluster an almost three-dimensional appearance.

FIGURE 24-1.

NGC 2548 (M 48) (60' field width)

Image reproduced from Digitized Sky Survey courtesy Anglo-Australian Observatory and Space Telescope Science Institute

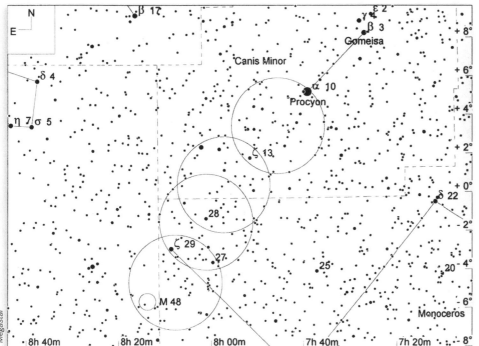

CHART 24-2.

NGC 2548 (M 48) (25° field width; 5° finder circles; LM 9.0)

Megastar

NGC 3242	★★★	◦◦	PN	MBUDR
Chart 24-3, 24-4	Figure 24-2	m8.6, 75"	10h 24.7m	-18° 39'

NGC 3242, better known as the Ghost of Jupiter, is a beautiful planetary nebula. William Herschel discovered this planetary nebula in 1785. It came to be known as the Ghost of Jupiter for its similarity in appearance to the planet Jupiter. In fact, it was after observing this object and the similar Saturn Nebula that Herschel named this class of objects planetary nebulae.

To locate NGC 3242, begin by identifying the m2.0 star Alphard, otherwise known as 30-α (alpha) Hydrae. Look 8.5° SE for the m4.1 star 39-υ¹ (upsilon¹) Hydrae, and then another 8.5° ESE for m3.8 42-μ (mu) Hya. Place 42-μ near the N edge of your finder field, and look for an m6 star that appears prominently in the SW quadrant of the finder field and place your finder cross hairs just over halfway on a line from 42-μ to that m6 star. NGC 3242 should be visible in your low-power eyepiece.

At 42X in our 10" reflector, NGC 3242 is visible as a fuzzy m9 star. We confirmed its identity by "blinking" it (moving the filter between our eye and the eyepiece) with our Ultrablock narrowband filter, which dims field stars to invisibility and leaves the planetary brightly visible. At 125X, NGC 3242 shows a bright, bluish 30" disc, slightly extended E-W (from east to west), with a prominent m12 central star. At 240X with our Ultrablock narrowband filter, considerable additional detail is visible. The contrast is greatly enhanced, revealing a relatively bright double inner shell with a very faint outer shell lying a few seconds outside the brighter inner shells.

FIGURE 24-2.

NGC 3242 (60' field width)

Image reproduced from Digitized Sky Survey courtesy Anglo-Australian Observatory and Space Telescope Science Institute

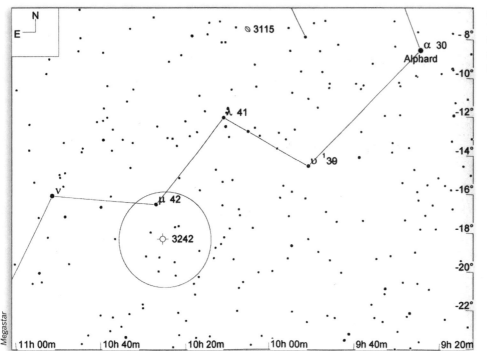

CHART 24-3.

NGC 3242 (Ghost of Jupiter) overview (25° field width; 5° finder circle; LM 7.0)

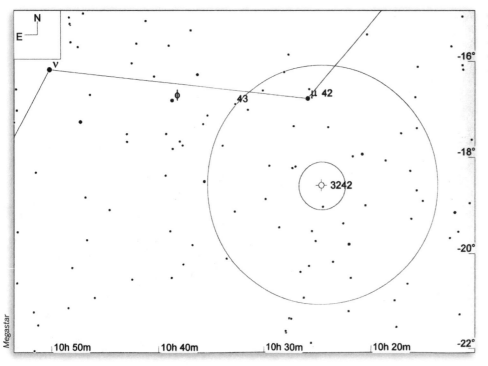

CHART 24-4.

NGC 3242 (Ghost of Jupiter) detail (10° field width; 5° finder circle; 1° eyepiece circle; LM 9.0)

M 68 (NGC 4590)	★★★	✿✿✿✿	GC	MBUDR
Chart 24-5	Figure 24-3	m7.3, 11.0'	12h 39.5m	-26° 45'

NGC 4590, better known as Messier 68 or M 68, is a pretty globular cluster. Charles Messier discovered M 68 on the night of 9 April 1780.

M 68 is easy to find, lying just 3.5° SSE (south southeast) of m2.7 9-β (beta) Corvi (the SW star of the five bright stars that make up the trapezoid pattern of Corvus). To locate M 68, place 9-β Corvi on the NW edge of your finder field and look for an m5.4 star that appears prominently near the S edge of the field. M 68 lies half a degree NE of that star, and is very faintly visible with averted vision in a 50mm finder or binocular on a clear, dark night.

At 125X in our 10" reflector, M 68 shows a bright, tight 2.5' central core with distinct mottling that cannot be resolved into individual stars. The core is surrounded by a moderately bright, loose 10' halo with many individual stars resolvable.

FIGURE 24-3.

NGC 4590 (M 68) (60' field width)

Image reproduced from Digitized Sky Survey courtesy Anglo-Australian Observatory and Space Telescope Science Institute

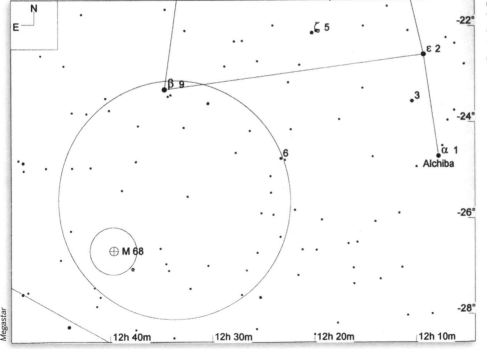

CHART 24-5.

NGC 4590 (M 68) (10° field width; 5° finder circle; 1° eyepiece circle; LM 9.0)

M 83 (NGC 5236)	★★★	♁♁♁	GX	MBUDR
Chart 24-6		m8.2, 12.8' x 11.4'	13h 37.1m	-29° 52'

NGC 5236, better known as Messier 83, M 83, or the Southern Pinwheel Galaxy (compare with M 33 in Triangulum), is a bright, beautiful face-on spiral galaxy. M 83 was first observed by Nicholas Louis de la Caille on 23 February 1752. Charles Messier, unaware of that discovery, independently re-discovered M 83 on the night of 17 February 1781.

In Messier's original description, he stated that it was only with the greatest difficulty that he could see M 83 at all. That had less to do with the object itself, and more to do with Messier's location. From his observatory in Paris at 49° N latitude, M 83 never rose higher than about 11° above his southern horizon. At declination -29° 52', M 83 is the most southerly galaxy on Messier's list. M 83 lies only 5° N of Messier's most southerly object, the bright open cluster M 7. All eight of the Messier objects that lie south of M 83 are bright globular clusters or open clusters.

To locate M 83, first identify m2.1 5-θ (theta) Centauri, also known as Menkent. Place Menkent on the SE edge of your binocular or finder field, and look for m4.3 2-Centauri 4° WNW. Move the finder slightly N until the arc of three m4/5 stars 1-, 3-, and 4-Centauri come into view, and then place them along the SE edge of your field. M 83 lies near the NW edge of the field about half a degree SW of an m5.8 star that appears prominent in the finder. M 83 is faintly visible with averted vision in a 50mm finder or binocular.

At 90X in our 10" Dob, M 83 is a beautiful sight. A small, very bright circular core is embedded in the center of a bright 3' NE-SW bar that lies within a fainter 15' halo. With averted vision, both spiral arms are visible. A broad spiral arm extends from the SW end of the bar, curling N and then W, and showing a distinct mottled texture. A narrower spiral arm extends from the NE end of the bar, curling SW. Several dim stars are very faintly visible within the halo, both between the bar and the spirals and outside the spirals.

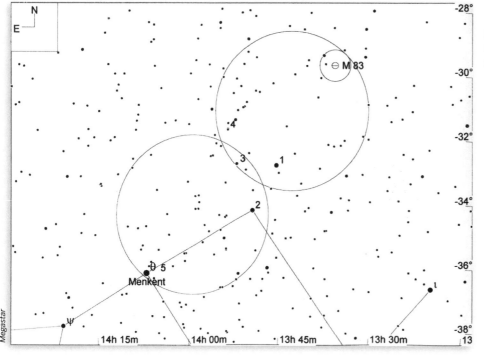

CHART 24-6.

NGC 5236 (M 83) (15° field width; 5° finder circle; 1° eyepiece circle; LM 9.0)

Multiple Stars

N (H 96)	★★	⊙⊙		MS	UD
Chart 24-7, 24-8		m5.6/5.7, 9.4", PA 210° (2003)		11h 32.3m	-29° 15'

To locate N-Hydrae, follow the line that forms the W edge of the trapezoid pattern of Corvus about 10° SSW to locate the m4.3 star β (beta) Hydrae. Look 4.6° WNW to locate the m3.5 star ξ (xi) Hydrae. Put ξ (xi) Hya on the S edge of your finder field and look for an m5.6 star that appears very prominently near the center of the field. That star is N-Hydrae.

At 125X in our 10" Dob, it's difficult to tell the primary from the companion. Both are of very similar brightness, and both are the same very pale yellow color.

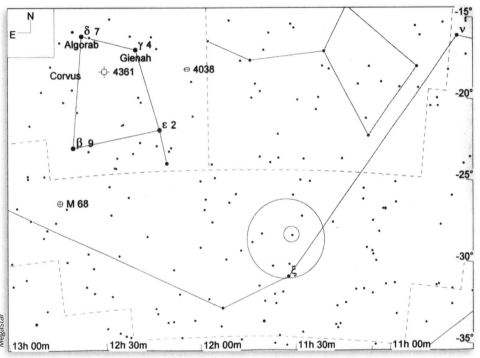

CHART 24-7.

N-Hydrae overview (30° field width; 5° finder circle; 1° eyepiece circle; LM 7.0)

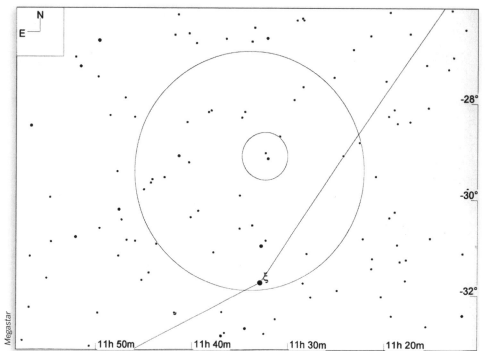

CHART 24-8.

N-Hydrae detail (10° field width; 5° finder circle; 1° eyepiece circle; LM 9.0)

Megastar

25

Lacerta, The Lizard

NAME:	Lacerta (lay-CERT-uh)
SEASON:	Summer
CULMINATION:	midnight, 28 August
ABBREVIATION:	Lac
GENITIVE:	Lacertae (lay-CERT-eye)
NEIGHBORS:	And, Cas, Cep, Cyg, Peg
BINOCULAR OBJECTS:	NGC 7209, NGC 7243
URBAN OBJECTS:	NGC 7209, NGC 7243

Lacerta is a small, dim, inconspicuous mid-northerly constellation that has the misfortune to lie surrounded by the large, prominent constellations Andromeda, Cassiopeia, Cepheus, Cygnus, and Pegasus. Lacerta ranks 68th in size among the 88 constellations. It covers only 201 square degrees of the celestial sphere, or about 0.5%. Although its brightest star is only fourth magnitude, Lacerta presents a striking pattern once you know where to look for it. To us, the zig-zag pattern of Lacerta looks just like a much dimmer, much smaller version of Cassiopeia. Lacerta is one of the "modern" constellations, created in the late 17th century by Johannes Hevelius, so it has no mythology associated with it.

Poor little Lacerta. No bright stars, ignored by the ancient myth weavers, and—you guessed it—not much in the way of interesting DSOs. The best Lacerta has to offer is a pair of ordinary open clusters, NGC 7209 and NGC 7243, and one rather pedestrian double star.

Lacerta culminates at midnight on 28 August, and, for observers at mid-northern latitudes, is best placed for evening viewing from the early summer through early winter months.

TABLE 25-1.

Featured star clusters, nebulae, and galaxies in Lacerta

Object	Type	Mv	Size	RA	Dec	M	B	U	D	R	Notes
NGC 7243	OC	6.4	21.0	22 15.3	+49 53			●	●		Cr 448; Mel 240; Class II 2 m
NGC 7209	OC	7.7	24.0	22 05.2	+46 30			●	●		Cr 444; Mel 238; Class III 1 m

TABLE 25-2.

Featured multiple stars in Lacerta

Object	Pair	M1	M2	Sep	PA	Year	RA	Dec	UO	DS	Notes
8	STF 2922Aa-B	5.7	6.3	22.2	185	2004	22 35.9	+39 38		●	

CHART 25-1.

The constellation Lacerta (field width 35°, North to the right)

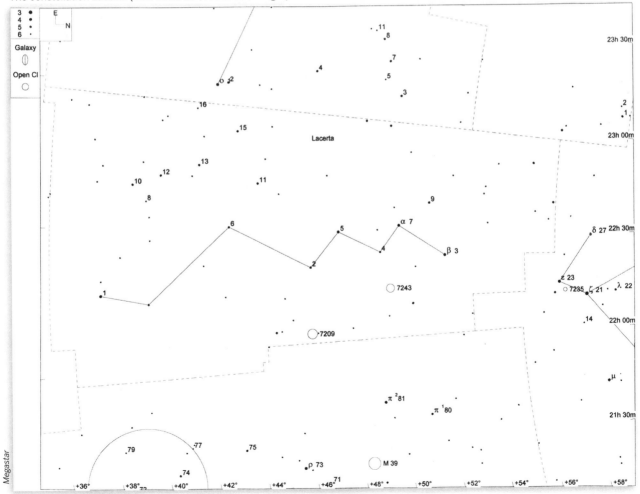

Clusters, Nebulae, and Galaxies

NGC 7243	★★	✦✦✦	OC	MBUDR
Chart 25-2, 25-3	Figure 25-1	m6.4, 21.0'	22h 15.3m	+49° 53'

NGC 7243 is a large, bright, but very loose and scattered open cluster. To locate NGC 7243, begin at the SW corner of the Cepheus "ice cream cone" as shown in Chart 25-2, and work your way S through the zig-zag pattern of Lacerta until you have 3-β (beta) and 7-α (alpha) on the NNE and E edges, respectively, of your finder or binocular field. NGC 7243 lies near the center of the field, and is visible in a 50mm finder or binocular.

With our 50mm binoculars and averted vision, NGC 7243 is visible as a relatively bright hazy patch with three embedded m8/9 stars. At 90X in our 10" Dob, NGC 7243 is bright, loose, and coarse, with about three dozen stars visible, grouped into chains and lumps. Despite its looseness and irregular shape, it is well detached from the surrounding star field.

FIGURE 25-1.

NGC 7243 (60' field width)

Image reproduced from Digitized Sky Survey courtesy Palomar Observatory and Space Telescope Science Institute

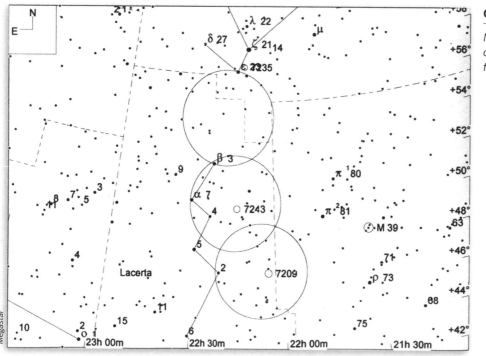

CHART 25-2.

NGC 7243 and NGC 7209 overview (25° field width; 5° finder circles; LM 7.0)

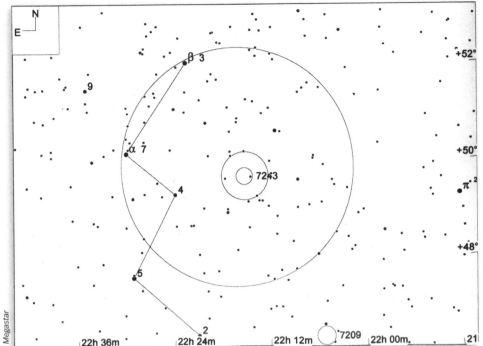

Megastar

NGC 7209	★★★★★	✸✸✸✸✸	OC	MBUDR
Chart 25-2, 25-4	Figure 25-2	m7.7, 24.0'	22h 05.2m	+46° 30'

NGC 7209 is another large, loose, scattered open cluster, about the same size as 7243, but noticeably dimmer. To locate NGC 7209, continue working your way S along the zig-zag pattern of Lacerta, as shown in Chart 25-2, until you arrive at m4.6 2-Lacertae. Place that star on the E edge of your finder or binocular field, and look for NGC 7209 near the center of the field, where it is faintly visible with averted vision just S of an m6 field star.

With our 50mm binoculars and averted vision, NGC 7209 is visible as a fairly large, fairly dim hazy patch. Three embedded m8/9 stars are easily visible, with a dozen or so dimmer stars fading in and out of visibility. At 90X in our 10" Dob, NGC 7209 is very loose and scattered across a 25' extent, with 70+ stars visible from m9 to m12.

FIGURE 25-2.

NGC 7209 (60' field width)

Image reproduced from Digitized Sky Survey courtesy Palomar Observatory and Space Telescope Science Institute

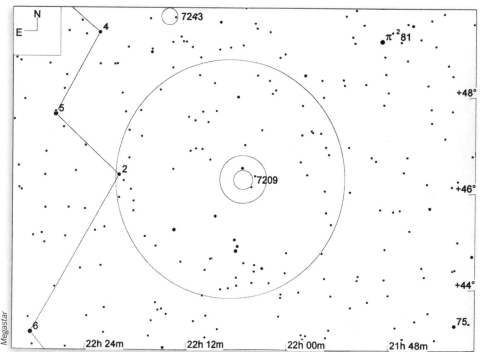

CHART 25-4.

NGC 7209 (10° field width; 5° finder circle; 1° eyepiece circle; LM 9.0)

Multiple Stars

8 (STF 2922Aa-B)	★★	⬤⬤	MS	UD
Chart 25-5		m5.7/6.3, 22.2", PA 185° (2004)	22h 35.9m	+39° 38'

To locate 8-Lacertae, follow the zig-zag pattern of the constellation S from Cepheus, as shown in Chart 25-1 and Chart 25-2, until you come to m4.5 6-Lacertae. Place that star at the N edge of your finder field and look for a prominent N-S (from north to south) line of three m5 stars, 13-, 12-, and 10-Lacertae, on the E edge of the finder field, as shown in Chart 25-5. 8-Lacertae appears prominently in the finder about one degree NW of 10-Lacertae.

The main problem with 8-Lacertae is figuring out which two stars make up STF 2922. There are enough relatively bright stars in the vicinity that this object looks more like a tiny open cluster than a double star. The actual double star STF 2922 is the closely separated m6 pair. At 125X in our 10" reflector, the primary and companion both appear as blazing blue-white.

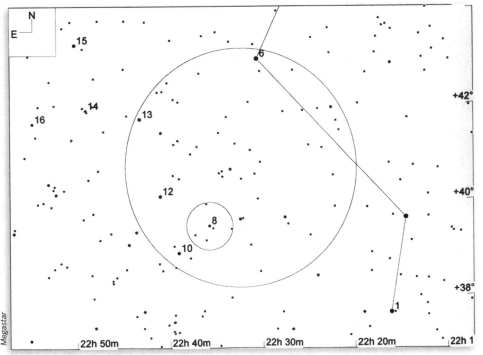

CHART 25-5.

8-Lacertae (STF 2922Aa-B)
(10° field width; 5° finder circle;
1° eyepiece circle; LM 9.0)

26

Leo, The Lion

NAME: Leo (LEE-oh)

SEASON: Winter

CULMINATION: midnight, 1 March

ABBREVIATION: Leo

GENITIVE: Leonis (LEE-own-is)

NEIGHBORS: Cnc, Com, Crt, Hya, LMi, Lyn, Sex, UMa, Vir

BINOCULAR OBJECTS: NGC 3351 (M 95), NGC 3368 (M 96), NGC 3379 (M 105), NGC 3623 (M 65), NGC 3627 (M 66)

URBAN OBJECTS: 41-gamma (STF 1424)

Leo is a large, bright, prominent equatorial constellation that ranks 12th in size among the 88 constellations. Leo covers 947 square degrees of the celestial sphere, or about 2.3%.

Greek mythology identifies Leo as representing the Nemean Lion that Herakles killed during one of his Twelve Labors. The easily recognized asterism that forms the Sickle of Leo is anchored by the bright stars Regulus and Algieba and represents the head and mane of Leo, while the bright star Denebola represents his tail. In ancient times, Leo was larger than it is today. Ptolemy considered the modern-day constellation Coma Berenices a part of Leo, forming the tuft on the end of his tail.

Like the adjacent large constellations Coma Berenices, Virgo, and Ursa Major, Leo is the realm of galaxies. Leo is home to scores of galaxies that are visible in amateur instruments, including no less than five bright Messier galaxies, and possesses several fine galaxies that are bright enough to be glimpsed with even a binocular. Several of these bright galaxies are tightly clustered, making it possible to observe multiple bright galaxies in the same low-power eyepiece field.

Leo culminates at midnight on 1 March, and, for observers at mid-northern latitudes, is best placed for evening viewing from the mid-winter through late spring months.

TABLE 26-1.

Featured star clusters, nebulae, and galaxies in Leo

Object	Type	Mv	Size	RA	Dec	M	B	U	D	R	Notes
NGC 2903	Gx	9.7	12.6 x 6.0	09 32.2	+21 30					◉	Class SAB(rs)bc; SB 12.4
NGC 3351	Gx	10.5	7.5 x 5.0	10 44.0	+11 42	◉	◉				M 95; Class SB(r)b; SB 12.7
NGC 3368	Gx	10.1	7.6 x 5.2	10 46.8	+11 49	◉	◉				M 96; Class SAB(rs)ab; SB 12.5
NGC 3379	Gx	10.2	5.4 x 4.8	10 47.8	+12 35	◉	◉				M 105; Class E1; SB 11.3
NGC 3384	Gx	10.9	5.5 x 2.5	10 48.3	+12 38					◉	Class SB(s)0-:; SB 11.0
NGC 3521	Gx	9.8	11.0 x 7.1	11 05.8	+00 02					◉	Class SAB(rs)bc; SB 11.8
NGC 3607	Gx	9.9	5.5 x 5.0	11 16.9	+18 03					◉	Class SA(s)0^:; SB 11.0 (???)
NGC 3623	Gx	10.3	9.8 x 2.8	11 18.9	+13 06	◉	◉				M 65; Class SAB(rs)a; SB 12.8
NGC 3627	Gx	9.7	9.1 x 4.1	11 20.3	+12 59	◉	◉				M 66; Class SAB(s)b; SB 11.9
NGC 3628	Gx	10.3	14.8 x 2.9	11 20.3	+13 35					◉	Class Sb pec sp; SB 12.4 (???)

TABLE 26-2.

Featured multiple stars in Leo

Object	Pair	M1	M2	Sep	PA	Year	RA	Dec	UO	DS	Notes
32-alpha	STF 6AB	1.4	8.2	176.0	308	2000	10 08.4	+11 58		◉	Regulus
41-gamma*	STF 1424AB	2.4	3.6	4.4	125	2006	10 20.0	+19 50	◉	◉	Algieba
54	STF 1487	4.5	6.3	6.6	111	2003	10 55.6	+24 44		◉	

CHART 26-1.

The constellation Leo (field width 60°)

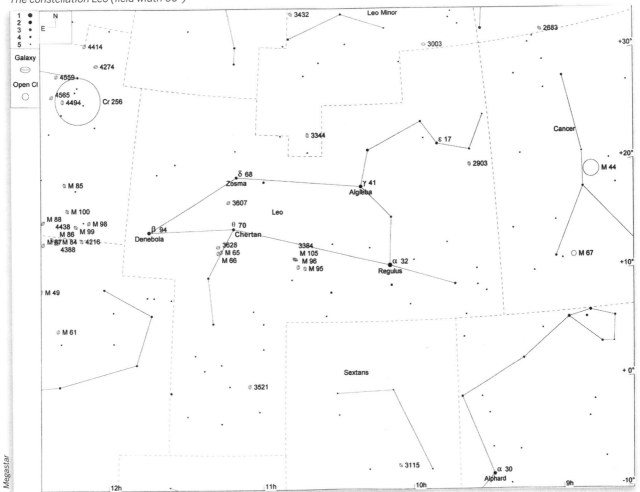

Clusters, Nebulae, and Galaxies

NGC 2903	★★★	✹✹✹✹	GX	MBUDR
Chart 26-2	Figure 26-1	m9.7, 12.6' x 6.0'	09h 32.2m	+21° 30'

NGC 2903 is a fine, bright spiral galaxy that we see nearly face-on. It's one of the best non-Messier galaxies, as good as most of Messier's galaxies and better than many. We can only assume that Charles Messier missed this one because he never got around to scanning this area of the sky.

NGC 2903 is easy to find. Start from the bright star m3.0 17-ε (epsilon) in the Sickle of Leo, and look 3.3° WSW (west southwest) for m4.3 4-λ (lambda). With 4-λ near the N center of your finder field, NGC 2903 is about centered in the field, forming the S apex of a triangular pattern with two m7 stars that appear prominently in the finder (and are visible in Figure 26-1).

At 125X in our 10" reflector, NGC 2903 shows a large extended core that brightens to a stellar nucleus. The core is embedded in a moderately bright 8' x 4' NNE-SSW (from north northeast to south southwest) halo that shows distinct knots and mottling. This is a beautiful galaxy.

FIGURE 26-1.

NGC 2903 (60' field width)

Image reproduced from Digitized Sky Survey courtesy Palomar Observatory and Space Telescope Science Institute

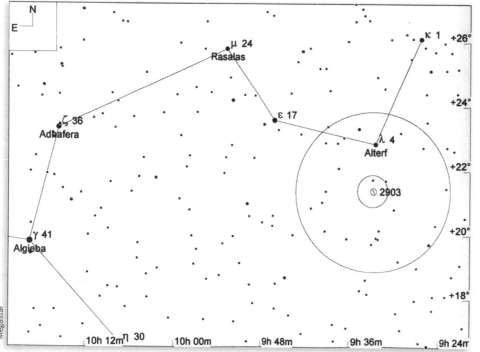

CHART 26-2.

NGC 2903 (15° field width; 5° finder circle; 1° eyepiece circle; LM 9.0)

M 95 (NGC 3351)	★★★	✿✿✿	GX	MBUDR
Chart 26-3	Figure 26-2	m10.5, 7.5' x 5.0'	10h 44.0m	+11° 42'

M 96 (NGC 3368)	★★★	✿✿✿	GX	MBUDR
Chart 26-3	Figure 26-2	m10.1, 7.6' x 5.2'	10h 46.8m	+11° 49'

NGC 3351, better known as Messier 95 or M 95, and NGC 3368 (Messier 96 or M 96) are bright spiral galaxies, both of which we see nearly face-on. Pierre Méchain discovered M 95 and M 96 on 20 March 1781, and communicated his discovery to his friend and colleague Charles Messier. Messier observed M 95 and M 96 four nights later, on 24 March 1781, and added them to his catalog.

M 95 and M 96 are relatively difficult to locate by star hopping because they lie far from any bright stars. Fortunately, it's easy to find them by applying a little field geometry. Begin by identifying the bright stars Chertan (m3.3) and Regulus (m1.4). Not quite halfway between them, nearer Chertan, there is an m5.5 star called 52-Leonis. That star may be just barely visible to the naked eye from a dark site, but even if it's not you can locate 52-Leonis simply by pointing your finder just under halfway on the line from Chertan to Regulus and looking for a star that's prominent in the finder. Once you locate 52-Leonis, put it near the N edge of your finder and use your low-power eyepiece. M 95 and/or M 96 should be visible in that eyepiece. If not, scan around a little until you locate the unmistakable pair of M 95 and M 96.

The Astronomical League Binocular Messier Club lists M 95 and M 96 as Challenge Objects (its most difficult category) with an 80mm binocular, and our experience confirms that. We have never seen M 95 or M 96 in any binocular or finder. At 90X in our 10" Dob, M 95 and M 96 are both visible in the same eyepiece field of our 14mm Pentax, with M 95 slightly fainter than M 96. M 95 shows a circular 3' halo that gradually brightens to bright core with a stellar nucleus. No mottling or other detail is visible in the halo. M 96 shows a moderately bright 3' x 5' halo extending WNW-ESE (from west northwest to east southeast) with a bright extended core and stellar nucleus.

FIGURE 26-2.

NGC 3351 (M 95, right) and NGC 3368 (M 96) (60' field width)

Image reproduced from Digitized Sky Survey courtesy Palomar Observatory and Space Telescope Science Institute

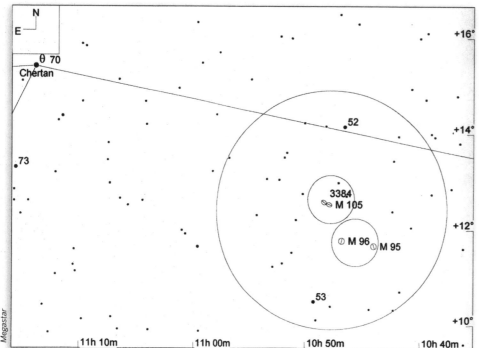

CHART 26-3.

NGC 3351 (M 95), NGC 3368 (M 96), NGC 3379 (M 105), and NGC 3384 (10° field width; 5° finder circle; 1° eyepiece circle; LM 9.0)

M 105 (NGC 3379)	★★★	✪✪✪	GX	MBUDR
Chart 26-3	Figure 26-3	m10.2, 5.4' x 4.8'	10h 47.8m	+12° 35'

NGC 3384	★★	✪✪✪	GX	MBUDR
Chart 26-3	Figure 26-3	m10.9, 5.5' x 2.5'	10h 48.3m	+12° 38'

NGC 3379, better known as Messier 105 or M 105, and NGC 3384 are a pair of bright galaxies that, like M 95 and M 96, are close enough to fit into one eyepiece field. Pierre Méchain discovered M 105 on 24 March 1781. The fact that Méchain did not also see NGC 3384—which lies only 7' ENE (east northeast) of M 105—that night is a good indication of the poor light gathering ability of even the better telescopes of that time. The discovery of NGC 3384 had to wait for William Herschel, who used a much larger telescope.

For unknown reasons, Charles Messier did not include M 105 in the final published version of his catalog, which contained only the first 103 of the 109 or 110 objects that are currently considered to be Messier Objects. M 105 was among the six or seven objects that were later added to Messier's catalog based on at least some evidence that Charles Messier had in fact observed them but had for some reason not included them in his catalog.

The M 105 and NGC 3384 pair are easy to locate if you start from M 95 and M 96. The center of this pair lies about 50' NNE of M 96,

and it is just possible to get all three of these galaxies into one 1° eyepiece field.

The Astronomical League Binocular Messier Club lists M 105 as a Challenge Object (its most difficult category) with an 80mm binocular. We have never seen either of these objects in a 50mm binocular or finder. We've never attempted either object with an 80mm binocular, but we found M 105 very difficult with an 80mm short tube refractor, and NGC 3384 impossible.

At 90X in our 10" Newtonian, both galaxies are visible in the same field of view with our 14mm Pentax eyepiece. M 105 shows a bright 2.5' circular halo that brightens gradually to a well-defined core with a stellar nucleus. NGC 3384 is noticeably fainter than M 105, but shows a distinct 1' x 3' halo extending NE-SW. The halo brightens sharply to a bright, distinct circular core with a stellar nucleus.

FIGURE 26-3.

NGC 3379 (M 105, right) and NGC 3384, with the faint NGC 3389 to the lower left (60' field width)

Image reproduced from Digitized Sky Survey courtesy Palomar Observatory and Space Telescope Science Institute

NGC 3521	★★★	◎◎	GX	MBUDR
Chart 26-4	Figure 26-4	m9.8, 11.0' x 7.1'	11h 05.8m	+00° 02'

NGC 3521 is a fine, bright galaxy that lies a bit off the beaten path. To locate NGC 3521, begin at m3.3 Chertan in the tail of Leo. Look 5.4° SSE of Chertan for m4 78-ι (iota) and then 4.5° S of 78-ι for m4 77-σ (sigma). Place 77-σ on the N edge of your finder field and locate the m5 stars 75- and 79-Leonis, which appear prominently near the S edge of the finder field. Place 79-Leonis on the E edge of the finder field, and locate the m5 stars 65- and 69-Leonis, as shown in Chart 26-4. With 65- and 69-Leonis as the 2.7° baseline, NGC 3521 forms the SW apex of an isosceles triangle with 2° sides. Near that apex is the m5.9 star 62-Leonis, which appears prominently in the finder. NGC 3521 lies 33' dead E of that star.

At 125X in our 10" reflector, NGC 3521 is visible as a bright oval core and bright stellar nucleus lying off-center nearer the W edge of a much fainter 5' x 2' NNW-SSE diffuse halo. Some mottling is visible in the core and, with extreme difficulty, in the brighter areas of the halo near the core.

FIGURE 26-4.

NGC 3521 (60' field width)

Image reproduced from Digitized Sky Survey courtesy Palomar Observatory and Space Telescope Science Institute

CHART 26-4.

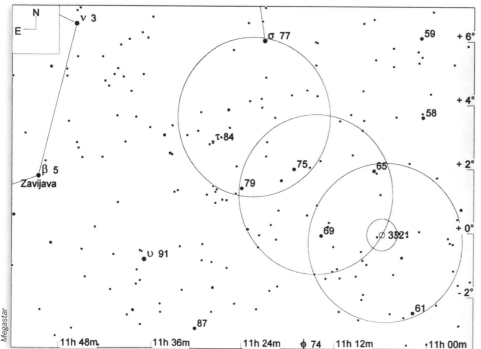

*NGC 3521 (15° field width; 5°
finder circles; 1° eyepiece circle;
LM 9.0)*

NGC 3607	★★	◍◍◍	GX	MBUDR
Chart 26-5	Figure 26-5	m9.9, 5.5' x 5.0'	11h 16.9m	+18° 03'

NGC 3607 is a rather ordinary galaxy, but has the advantage of
being very easy to find. To locate it put m2.6 Zosma on the N edge
of your finder and m3.3 Chertan on the S edge. NGC 3607 lies 40' E
of the center of that line, where you can center it in the cross hairs
of your finder just by drifting the finder field less than a degree E.

At 125X in our 10" reflector, NGC 3607 shows a bright circular
core with a stellar nucleus embedded in a fainter 1' halo. No detail
is visible in the core or halo.

FIGURE 26-5.

*NGC 3607, with NGC 3605 to its lower
right, NGC 3608 above, and NGC 3599
near right edge (60' field width)*

Image reproduced from Digitized Sky Survey courtesy Palomar Observatory and Space
Telescope Science Institute

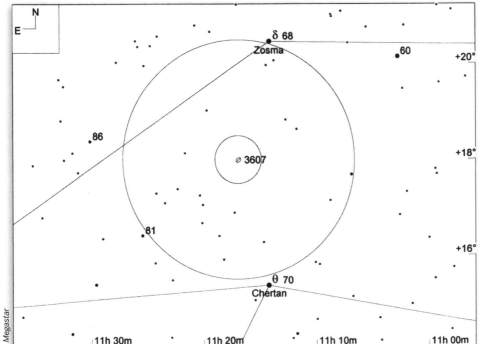

Megastar

M 65 (NGC 3623)	★★★	✧✧✧	GX	MBUDR
Chart 26-6	Figure 26-6	m10.3, 9.8' x 2.8'	11h 18.9m	+13° 06'

M 66 (NGC 3627)	★★★	✧✧✧	GX	MBUDR
Chart 26-6	Figure 26-6	m9.7, 9.1' x 4.1'	11h 20.3m	+12° 59'

NGC 3628	★★	✧✧✧	GX	MBUDR
Chart 26-6	Figure 26-6	m10.3, 14.8' x 2.9'	11h 20.3m	+13° 35'

NGC 3623, better known as Messier 65 or M 65, and NGC 3627 (Messier 66 or M 66) form, with NGC 3628, the famous Leo Trio of galaxies. Although there are many areas in Coma and Virgo where a dozen or more galaxies are visible in one eyepiece field, nowhere in the sky can you find three such bright galaxies in such close proximity as the Leo Trio.

Charles Messier discovered M 65 and M 66 on the night of 1 March 1780. Messier notes that he missed these two objects when he observed this area of the sky on the nights of 1 and 2 November 1773. On those evenings, he was observing a comet that lay between M 65 and M 66. Either Messier was so enamored of the comet that he failed to notice M 65 and M 66, or perhaps the light from the bright comet caused him to overlook them. Many modern sources attribute the discovery of M 65 and M 66 to Messier's friend and colleague Pierre Méchain, but we think that's unlikely. In every other instance, Messier gave Méchain proper credit for the initial discovery of objects that Messier later observed. Messier does not mention Méchain in his observing notes for M 65 and M 66, so we think it's likely that Messier did discover these objects himself, or at least re-discovered them independently without help from Méchain.

To locate these three galaxies, place m3.3 Chertan on the NW edge of your finder field. The galaxy group will be centered in your finder, about 2.7° E of a prominent m5 star, and surrounding an m7 star that's visible near the center of the finder field.

The Astronomical League Binocular Messier Club lists M 65 and M 66 as Challenge Objects (its most difficult category) with a 35mm or 50mm binocular, and as Tougher Objects (its middle category) with an 80mm binocular. We have never seen either object with our 50mm binocular or finder, although we find both to be relatively easy with an 80mm short-tube refractor.

At 42X in our 10" reflector, all three galaxies lie within the eyepiece field and are visible with direct vision, with M 66 the brightest, M 65 somewhat dimmer, and NGC 3628 considerably fainter. At 125X, M 65 shows a very bright, irregular 2' x 3' N-S (from north to south) core with a very bright, fuzzy nucleus embedded in a much fainter, uneven halo elongated 7' N-S. Significant mottling is visible in the halo, particularly on the N side. M 66 shows a large, very bright nucleus embedded in a patchy, mottled halo that extends 2' x 5' N-S. NGC 3268 is much fainter than either of the Messier galaxies. At 125X, it shows as a very thin, moderately bright streak of light. A needle-thin 3' core without a visible nucleus is centered in an extremely elongated 1' x 10' E-W halo. No mottling or other structural detail is visible.

FIGURE 26-6.

*NGC 3623 (M 65) at the lower right,
NGC 3627 (M 66) at the lower left, and
NGC 3628 (60' field width)*

Image reproduced from Digitized Sky Survey courtesy Palomar Observatory and Space
Telescope Science Institute

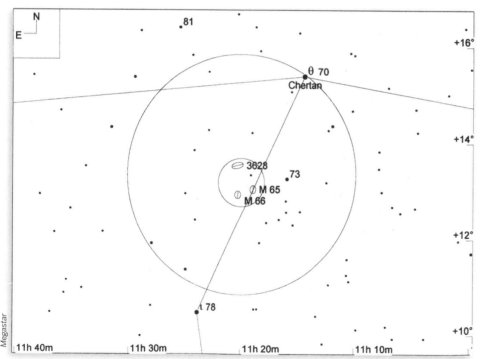

CHART 26-6.

*NGC 3623 (M 65), NGC 3627
(M 66), and NGC 3628 (10°
field width; 5° finder circle; 1°
eyepiece circle; LM 9.0)*

Multiple Stars

32-alpha (STF 6AB)	★★	✹✹✹✹		MS	UD
Chart 26-1		m1.4/8.2, 176.0", PA 308° (2000)		10h 08.4m	+11° 58'

32-α (alpha) Leonis is better known as Regulus, the brightest star in Leo. At 125X in our 10" reflector, Regulus is a brilliant, glaring white, with its much dimmer companion showing a warm white color.

41-gamma (STF 1424AB)	★★	✹✹✹✹		MS	UD
Chart 26-1		m2.4/3.6, 4.4", PA 125° (2006)		10h 20.0m	+19° 50'

41-γ (gamma) Leonis, sometimes called Algieba, is the middle star in the Sickle of Leo. At 125X in our 10" reflector, Algieba is a yellow-white with a very pale yellow companion.

54 (STF 1487)	★★★	✹✹✹		MS	UD
Chart 26-7		m4.5/6.3, 6.6", PA 111° (2003)		10h 55.6m	+24° 44'

54-Leonis is a pretty double that offers some color contrast. To locate 54-Leonis, begin at m2.6 Zosma. Locate m4.4 60-Leonis 2.8° W of Zosma, and put that star on the S edge of your finder field. 54-Leonis lies 4.8° NNW of 60-Leonis, and will be prominently visible near the NW edge of your finder field. At 125X in our 10" reflector, the primary is yellow-white and the secondary cool white with a distinct bluish tinge.

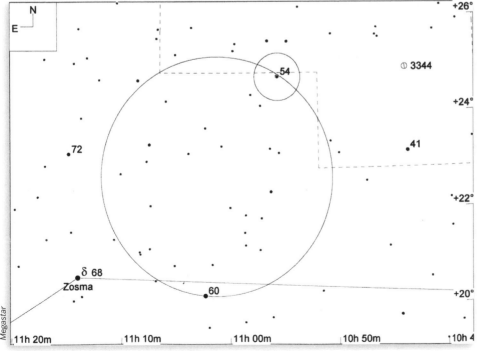

CHART 26-7.

54-Leonis (STF 1487) (10° field width; 5° finder circle; 1° eyepiece circle; LM 9.0)

27

Leo Minor, The Small Lion

NAME: Leo Minor (LEE-oh MYE-nur)

SEASON: Winter

CULMINATION: midnight, 24 February

ABBREVIATION: LMi

GENITIVE: Leonis Minoris (lee-OWN-is min-OR-is)

NEIGHBORS: Cnc, Leo, Lyn, UMa

BINOCULAR OBJECTS: none

URBAN OBJECTS: none

Leo Minor is a small, dim, mid-northerly constellation that ranks 64th in size among the 88 constellations. Leo Minor covers 232 square degrees of the celestial sphere, or about 0.6%. Leo Minor is the dark area that lies between the bright constellations Ursa Major to the north and Leo to the south. The brightest star in Leo Minor is fourth magnitude, and the faint triangle of stars that makes up its constellation outline is just visible to the naked eye on a dark night.

Leo Minor is one of the "modern" constellations, defined by Johannes Hevelius in 1687. No mythology is associated with Leo Minor, because the ancients considered it just a dark area of the sky and did not associate it with any of the surrounding constellations. There's not much more there for modern astronomers. In fact, Leo Minor would not have appeared in this book except that it is home to three rather ordinary galaxies that are required to complete the RASC list.

Leo Minor culminates at midnight on 24 February, and, for observers at mid-northern latitudes, is best placed for evening viewing from the mid-winter through early summer months.

TABLE 27-1.

Featured star clusters, nebulae, and galaxies in Leo Minor

Object	Type	Mv	Size	RA	Dec	M	B	U	D	R	Notes
NGC 3344	Gx	10.5	7.3 x 6.4	10 43.5	+24 55					◉	Class (R)SAB(r)bc; SB 13.3
NGC 3003	Gx	12.3	5.9 x 1.3	09 48.6	+33 25					◉	Class Sbc?; SB 13.7
NGC 3432	Gx	11.7	6.8 x 1.4	10 52.5	+36 37					◉	Class SB(s)m sp; SB 13.7

CHART 27-1.

The constellation Leo Minor (field width 30°)

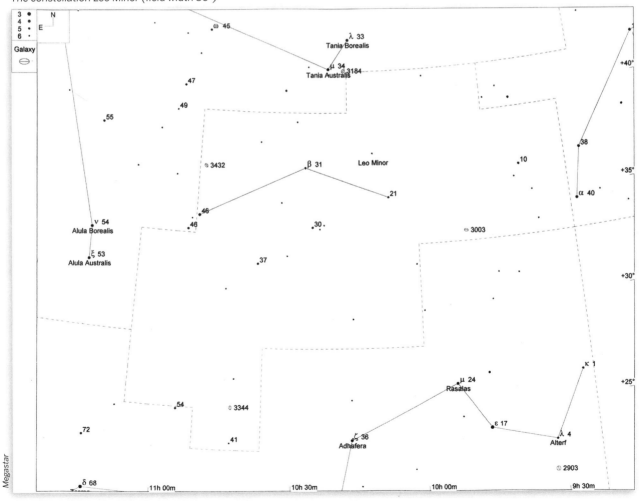

Clusters, Nebulae, and Galaxies

NGC 3344	★★	◆◆◆	GX	MBUDR
Chart 26-7, 27-2	Figure 27-1	m10.5, 7.3' x 6.4'	10h 43.5m	+24° 55'

NGC 3344 is a moderately bright spiral galaxy that we see almost face-on. To locate NGC 3344, begin at m2.6 Zosma in Leo. Locate m4.4 60-Leonis 2.8° W of Zosma, and put that star on the S edge of your finder field. 54-Leonis lies 4.8° NNW of 60-Leonis, and is prominently visible near or just past the NW edge of your finder field. Place 54-Leonis on the E edge of the finder field and look for the two stars m5.5 40-LMi and m5.1 LMi, both of which are shown in Chart 27-2 and appear very prominently in the finder. NGC 3344 lies near the halfway point between those stars, just slightly closer to 40-LMi.

At 125X in our 10" reflector, NGC 3607 shows a 4' diffuse halo brightening gradually to a dense core with a stellar nucleus. No detail is visible in the core or halo.

FIGURE 27-1.

NGC 3344 (60' field width)

Image reproduced from Digitized Sky Survey courtesy Palomar Observatory and Space Telescope Science Institute

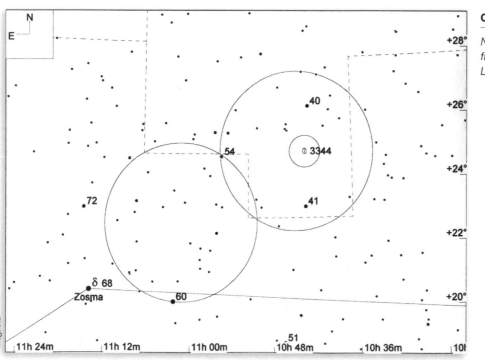

CHART 27-2.

NGC 3344 (15° field width; 5° finder circles; 1° eyepiece circle; LM 9.0)

NGC 3003	★	◦◦	GX	MBUDR
Chart 27-3, 27-4, 27-5	Figure 27-2	m12.3, 5.9' x 1.3'	09h 48.6m	+33° 25'

NGC 3003 is a small, faint, edge-on galaxy. To locate NGC 3003, identify the star 21-Leonis Minoris, which is the W star in the triangle of Leo Minor. Put 21-LMi on the NE edge of your finder field, and look for m5.4 20-LMi which appears very prominently near the S edge of the finder. NGC 3003 lies near the WSW edge of the finder field, surrounded by an arc of three m7/8 stars, as shown in Chart 27-5.

At 125X in our 10" reflector, NGC 3003 is a moderately faint 0.5' x 3' WNW-ESE (from west northwest to east southeast) streak of light with some brightening along the axis that suggests a thin core with an extremely faint halo. No other detail is visible.

FIGURE 27-2.

NGC 3003 (60' field width)

Image reproduced from Digitized Sky Survey courtesy Palomar Observatory and Space Telescope Science Institute

CHART 27-3.

Locating Leo Minor (45° field width; LM 5.0)

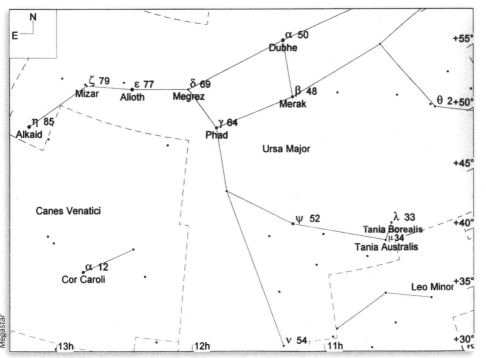

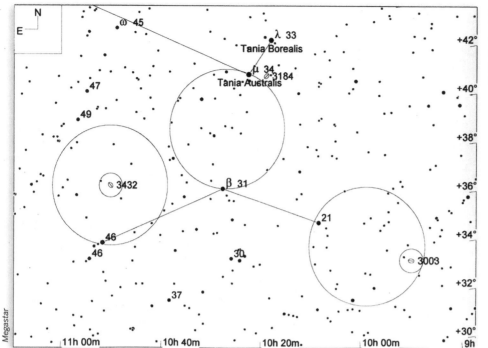

CHART 27-4.

NGC 3003 and NGC 3432 overview (20° field width; 5° finder circles; 1° eyepiece circles; LM 9.0)

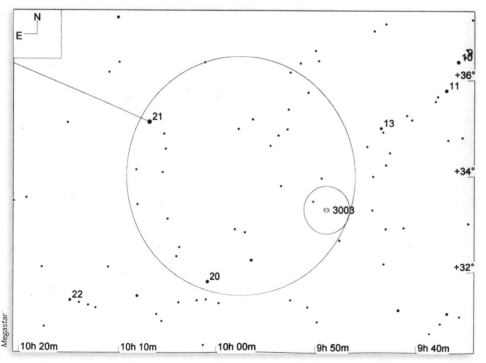

CHART 27-5.

NGC 3003 (10° field width; 5° finder circle; 1° eyepiece circle; LM 9.0)

NGC 3432	★★	✧✧✧	GX	MBUDR
Chart 27-4, 27-6	Figure 27-3	m11.7, 6.8' x 1.4'	10h 52.5m	+36° 37'

NGC 3432 is another small, faint, edge-on galaxy. To locate NGC 3432, identify the star 46-Leonis Minoris, which is the E star in the triangle of Leo Minor. Put 46-LMi on the S edge of your finder field to center NGC 3432 in the finder.

At 125X in our 10" reflector, NGC 3432 shows a bright, needle-thin 2' core extending NE-SW, centered in and surrounded by an uneven 1' x 4' halo. No detail is visible in the core or halo.

FIGURE 27-3.

NGC 3432 (60' field width)

Image reproduced from Digitized Sky Survey courtesy Palomar Observatory and Space Telescope Science Institute

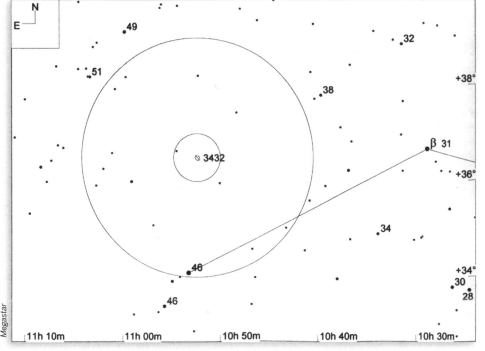

CHART 27-6.

NGC 3432 (10° field width; 5° finder circle; 1° eyepiece circle; LM 9.0)

Multiple Stars

There are no featured multiple stars in Leo Minor.

28

Lepus, The Hare

Lepus is a mid-size, south equatorial constellation that ranks 51st in size among the 88 constellations. Lepus covers 290 square degrees of the celestial sphere, or about 0.7%. Lepus is easy to overlook, lying as is does between the two bright constellations Orion and Canis Major, at the feet of Orion and to the west of Orion's hunting dog Canis.

Lepus is an ancient constellation, but has little mythology associated with it. To the Greeks and Romans of antiquity, Lepus was the hare pursued by Orion the Hunter, nestled cringing at his feet, having been brought to bay by Orion's two hunting dogs, Canis Major and Canis Minor.

With the exception of the Messier globular cluster M 79, Lepus has few DSOs of interest to amateur astronomers. Lepus is home to hundreds of very faint galaxies, but all but the brightest of those are out of reach of any but the largest amateur instruments.

Lepus culminates at midnight on 13 December, and, for observers at mid-northern latitudes, is best placed for evening viewing from the late autumn through late winter months.

TABLE 28-1.

Featured star clusters, nebulae, and galaxies in Lepus

Object	Type	Mv	Size	RA	Dec	M	B	U	D	R	Notes
NGC 1904	GC	7.7	9.6	05 24.2	-24 31	◉	◉				M 79; Class V

TABLE 28-2.

Featured multiple stars in Lepus

Object	Pair	M1	M2	Sep	PA	Year	RA	Dec	UO	DS	Notes
13-gamma	H 40AB	3.6	6.3	96.9	350	1999	05 44.4	-22 26		◉	

CHART 28-1.

The constellation Lepus (field width 30°)

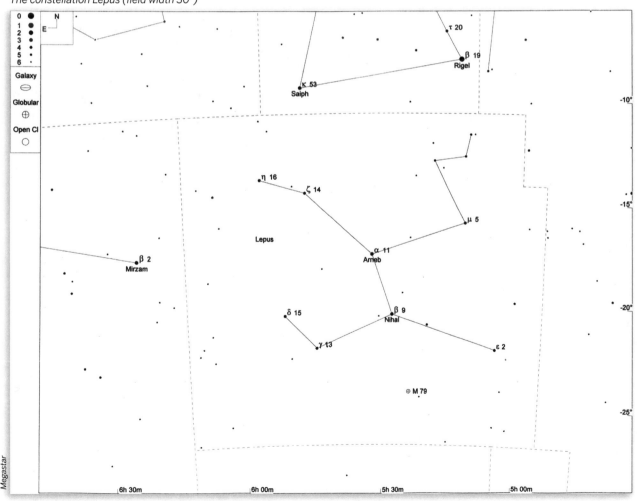

Clusters, Nebulae, and Galaxies

M79 (NGC 1904)	★★	✹✹✹	GC	MBUDR
Chart 28-2	Figure 28-1	m7.7, 9.6'	05h 24.2m	-24° 31'

NGC 1904, better known as Messier 79 or M 79, is a globular cluster that shouldn't be where it is. Pierre Méchain discovered M 79 on 26 October 1780 and reported his find to Charles Messier, who observed it on the night of 17 December 1780 and subsequently added it to his catalog.

Nearly all other known globular clusters lie in the summer constellations, which is to say in the direction of the galactic center. M 79 lies in the opposite direction, in the winter constellation Lepus, at about three times our 20,000 light year distance from the galactic center. For many years, astronomers were mildly puzzled by the location of M 79, which simply shouldn't be where it is. In 2003, a possible answer was discovered. It's now thought that M 79 is an intergalactic immigrant. M 79 appears to belong to the Canis Major Dwarf Galaxy, which is colliding with and being subsumed by our Milky Way Galaxy. Along with M 79, three other "misplaced" globular clusters—NGC 1851, NGC 2298, and NGC 2808—appear also to be members of the Canis Major Dwarf Galaxy.

Although you can locate M 79 with a tedious star hop, it's easier to locate it geometrically. Identify the stars m3.6 13-γ (gamma) and m3.2 2-ε (epsilon), which together with Nihal form a prominent triangle pattern on the S side of Lepus. Draw an imaginary baseline from 13-γ to 2-ε and place your finder so that the N edge of the field about touches that line near the center, but just slightly to the 2-ε side. M 79 should be about centered in your finder.

In a 50mm finder or binocular, M 79 is visible with averted vision as a fuzzy star. At 125X in our 10" reflector, M 79 shows a large, bright, densely concentrated core with a fainter 4' halo surrounding it. No stars can be resolved at the core.

FIGURE 28-1.

NGC 1904 (M 79) (60' field width)

Image reproduced from Digitized Sky Survey courtesy Anglo-Australian Observatory and Space Telescope Science Institute

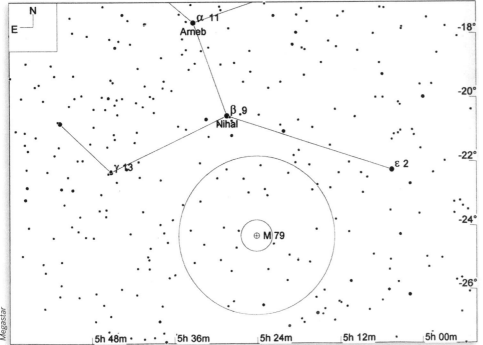

Multiple Stars

13-gamma (H 40AB)	★★	◍◍◍◍	MS	UD
Chart 28-2		m3.6/6.3, 96.9", PA 350° (1999)	05h 44.4m	-22° 26'

13-γ (gamma) Leporis is the SE apex of the triangle shown in Chart 28-2. At 90X in our 10" reflector, the primary appears yellow-white and the secondary a more saturated yellow.

29
Libra, The Scales

NAME: Libra (LEE-bruh)	
SEASON: Spring	
CULMINATION: midnight, 9 May	
ABBREVIATION: Lib	
GENITIVE: Librae (LEE-brye)	
NEIGHBORS: Cen, Hya, Lup, Oph, Sco, Ser, Vir	
BINOCULAR OBJECTS: none	
URBAN OBJECTS: none	

Libra is a mid-size, dim south equatorial constellation that ranks 29th in size among the 88 constellations. Libra covers 528 square degrees of the celestial sphere, or about 1.3%. The brightest stars in Libra are only third magnitude, but it is still relatively easy to discern the pattern of Libra, lying as it does in an area between Scorpius and Virgo that's relatively devoid of bright stars.

In early constellation mythology, Libra formed part of the claws of Scorpius to its southeast. In later Greek mythology, Libra was seen as a set of scales belonging to Virgo the Virgin, who was identified with Astraea, the goddess of justice.

Libra culminates at midnight on 9 May, and, for observers at mid-northern latitudes, is best placed for evening viewing from the late spring through the summer months.

TABLE 29-1.

Featured multiple stars in Libra

Object	Pair	M1	M2	Sep	PA	Year	RA	Dec	UO	DS	Notes
9-alpha2, 8-alpha1	SHJ 186AB	2.7	5.2	231.1	315	2002	14 50.9	-16 02		◉	Zubenelgenubi

Clusters, Nebulae, and Galaxies

There are no featured clusters, nebulae, or galaxies in Libra.

Multiple Stars

9-alpha2, 8-alpha1 (SHJ 186AB)	★★	◐◐◐◐		MS	UD
Chart 29-1		m2.7/5.2, 231.1", PA 315° (2002)		14h 50.9m	-16° 02'

9-α^2 (alpha^2), 8-α^1, better known as Zubenelgenubi, is easily seen in a 50mm binocular as a widely-separated double star with a white primary and yellow secondary. At 90X in our 10" reflector, both stars appear pure white.

CHART 29-1.

The constellation Libra (field width 45°)

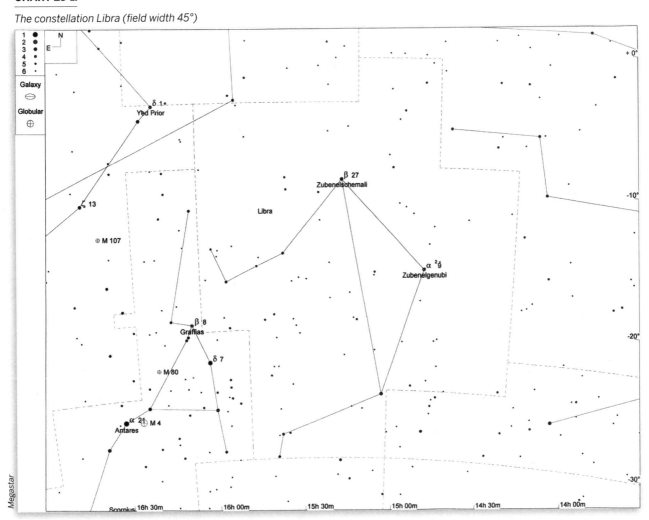

30

Lynx, The Lynx

NAME:	Lynx (LINKS)
SEASON:	Winter
CULMINATION:	midnight, 20 January
ABBREVIATION:	Lyn
GENITIVE:	Lyncis (LINN-cis)
NEIGHBORS:	Aur, Cam, Cnc, Gem, Leo, LMi, UMa
BINOCULAR OBJECTS:	none
URBAN OBJECTS:	none

Lynx is a very faint, mid-size, mid-northerly constellation that ranks 28th in size among the 88 constellations. Lynx covers 545 square degrees of the celestial sphere, or about 1.3%. Lynx is the dark area that lies between the bright constellations Ursa Major to the northwest and Auriga to the southwest. Lynx has one third magnitude star, and only five of the fourth magnitude.

Lynx is one of the "modern" constellations, defined by Johannes Hevelius in 1687. No mythology is associated with Lynx, because the ancients considered it just a dark area of the sky and did not associate it with any of the surrounding constellations. Hevelius simply wanted to fill in the dark area between Ursa Major and Auriga, so he created a new constellation there. Hevelius is said to have named the constellation Lynx because the constellation is so dim that it requires the eyes of a cat to see. On a very dark night, Lynx is faintly visible as a zig-zag line of dim stars that extends from the boundary of Camelopardalis in the northwest to the boundary of Leo in the southeast.

Lynx culminates at midnight on 20 January, and, for observers at mid-northern latitudes, is best placed for evening viewing from early winter through early summer.

If you're sharp-eyed, you may notice that the star with the Flamsteed number 10 in Chart 30-1 is out of sequence with the Flamsteed numbers of the other stars in Lynx. That's because that star originally belonged to the constellation Ursa Major, where it was known as 10-UMa. When the modern constellations were defined in the 1930's, that section of the sky was reassigned to the constellation Lynx, but for some reason the Flamsteed number assigned to that star was not officially changed. Technically, that star should still be labeled 10-UMa even though it now lies within the boundaries of Lynx, but most star charts assign it the designation 10-Lyncis. (And, no, there isn't another 10-Lyncis in the western part of Lynx, where you'd expect to find it.) To balance things out, Lynx also lost a star to Ursa Major. The former 41-Lyncis now lies with the boundaries of Ursa Major.

TABLE 30-1.

Featured star clusters, nebulae, and galaxies in Lynx

Object	Type	Mv	Size	RA	Dec	M	B	U	D	R	Notes
NGC 2683	Gx	10.6	10.5 x 2.5	08 52.7	+33 25					◉	Class SA(rs)b; SB 12.4

TABLE 30-2.

Featured multiple stars in Lynx

Object	Pair	M1	M2	Sep	PA	Year	RA	Dec	UO	DS	Notes
12	STF 948AC	5.4	7.1	8.7	309	2004	06 46.2	+59 26		◉	
19	STF 1062AB	5.8	6.7	14.8	315	2004	07 22.9	+55 16		◉	
38	STF 1334A-Bb	3.9	6.1	2.6	226	2004	09 18.8	+36 48		◉	

CHART 30-1.

The constellation Lynx (field width 45°)

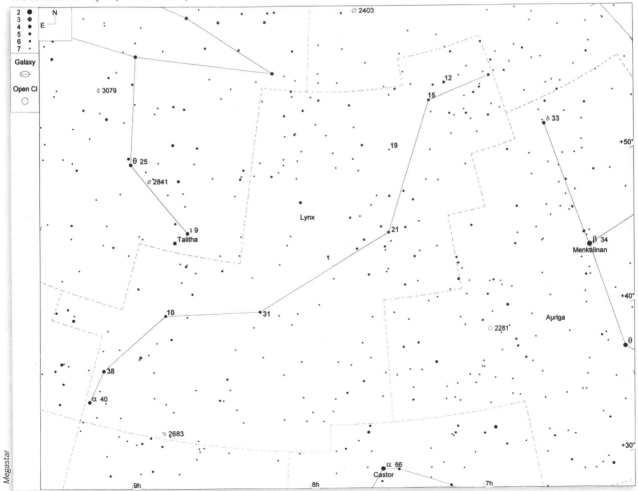

Megastar

Clusters, Nebulae, and Galaxies

NGC 2683	★★	◐◐	GX		MBUDR
Chart 30-2, 30-3	Figure 30-1	m10.6, 10.5' x 2.5'	08h 52.7m		+33° 25'

NGC 2683 is by far the finest DSO in Lynx. It's a moderately bright edge-on galaxy that takes some effort to find. Fortunately, NGC 2683 lies not far from m3.1 40-α (alpha) Lyncis, which is the brightest star in Lynx and relatively easy to identify with the naked eye. To locate 40-α, look about 15° NNE of the Sickle of Leo. 40-α is the brightest star in the vicinity, and appears prominent to the naked eye. Once you've identified 40-α, center it in your finder and look for m3.9 38-Lyn on the N edge of your finder field. Then place 40- and 38-Lyn on the E edge of your finder field and drift the finder field most of a field WSW (west southwest), and place the prominent arc of m5/6 stars as shown in Chart 30-3. NGC 2683 lies about 30' ENE (east northeast) of a prominent m6 star that's nearly centered in the finder.

At 125X in our 10" reflector, NGC 2683 is visible as a moderately bright 1.5' x 6' NE-SW (from northeast to southwest) halo that brightens somewhat to a very thin, dense core. Using averted vision, some extremely faint mottling is visible in the halo adjacent to the core.

FIGURE 30-1.

NGC 2683 (60' field width)

Image reproduced from Digitized Sky Survey courtesy Palomar Observatory and Space Telescope Science Institute

CHART 30-2.

NGC 2683 overview (30° field width; 5° finder circle; LM 7.0)

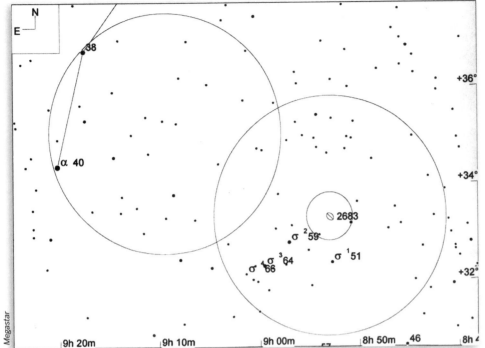

Megastar

Multiple Stars

12 (STF 948AC)	★★	◐◐	MS	UD
Chart 30-4		m5.4/7.1, 8.7", PA 309° (2004)	06h 46.2m	+59° 26'

Although 12-Lyn lies in Lynx, the easiest way to locate it is to start from m3.7 33-δ (delta) Aurigae, using the star hop shown in Chart 30-4. At 125X in our 10" reflector, the primary is pure white and the companion a pretty straw yellow.

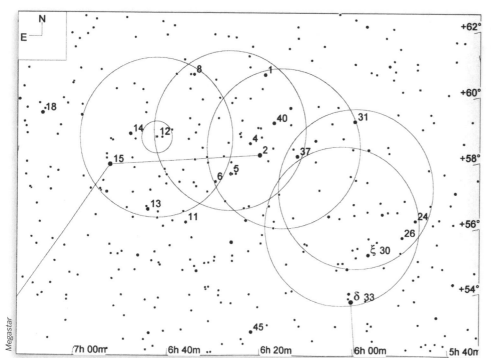

CHART 30-4.

*12-Lyncis (15° field width; 5°
finder circles; 1° eyepiece circle;
LM 9.0)*

Megastar

19 (STF 1062AB)	★	⚭	MS	UD
Chart 30-5		m5.8/6.7, 14.8", PA 315° (2004)	07h 22.9m	+55° 16'

19-Lyncis is an unremarkable double star, best located by using
the star hop shown in Chart 30-5. At 125X in our 10" reflector, the
primary and its slightly dimmer companion are both a cool white
color.

38 (STF 1334A-Bb)	★★	⚭⚭	MS	UD
Chart 30-2, 30-3		m3.9/6.1, 2.6", PA 226° (2004)	09h 18.8m	+36° 48'

38-Lyncis is a pretty double, easily located by following the
directions for NGC 2683 earlier in this chapter. At 125X in our 10"
reflector, the primary is a sparkling cool white and the noticeably
dimmer secondary yellow-orange.

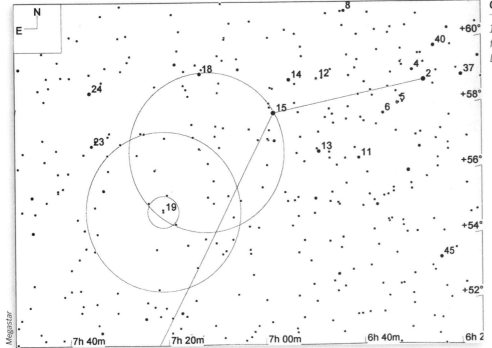

CHART 30-5.

19-Lyncis (15° field width; 5° finder circles; 1° eyepiece circle; LM 9.0)

31

Lyra, The Lyre

NAME: Lyra (LEER-uh)	
SEASON: Summer	
CULMINATION: midnight, 2 July	
ABBREVIATION: Lyr	
GENITIVE: Lyrae (LEER-eye)	
NEIGHBORS: Cyg, Dra, Her, Vul	
BINOCULAR OBJECTS: NGC 6779 (M 56)	
URBAN OBJECTS: NGC 6720 (M 57)	

Lyra is a mid-size, mid-northerly constellation that ranks 52nd in size among the 88 constellations. Lyra covers 286 square degrees of the celestial sphere, or about 0.7%. Although Lyra has only one bright star—0th magnitude Vega, which is the northwestern apex of the Summer Triangle asterism with Deneb and Altair—it is easy to recognize the constellation because of the pattern of 3rd and 4th magnitude stars that forms a prominent parallelogram pattern just to the southeast of Vega.

Lyra is an ancient constellation, recognized as such since at least Babylonian and Assyrian times. In early Greek mythology, Lyra represented a vulture associated with the Stymphalian Birds, one of the Twelve Labors of Herakles. In later Greek mythology, Lyra came to be associated with the lyre created by the young god Hermes.

Lyra is relatively poor in interesting DSOs, but it is home to two bright Messier objects, the globular cluster M 56 and the planetary

TABLE 31-1.

Featured star clusters, nebulae, and galaxies in Lyra

Object	Type	Mv	Size	RA	Dec	M	B	U	D	R	Notes
NGC 6720	PN	8.8	1.8 x 1.4	18 53.6	+33 02	◉		◉			M 57; Class 4+3
NGC 6779	GC	8.4	8.8	19 16.6	+30 11	◉	◉				M 56; Class X

TABLE 31-2.

Featured multiple stars in Lyra

Object	Pair	M1	M2	Sep	PA	Year	RA	Dec	UO	DS	Notes
5-epsilon2*	STF 2383Cc-D	5.3	5.4	2.4	80	2006	18 44.3	+39 40		◉	Double
4-epsilon1*	STF 2382AB	5.0	6.1	2.4	349	2006	18 44.3	+39 40		◉	Double
4-epsilon1, 5-epsilon2	STFA 37AB-CD	5.0	5.3	210.5	174	1998	18 44.3	+39 40		◉	Double Double
6-zeta1, 7-zeta2	STFA 38AD	4.3	5.6	43.8	150	2003	18 44.8	+37 36		◉	
10-beta	STFA 39AB	3.6	6.7	46.0	150	2002	18 50.1	+33 21		◉	Sheliak
"Otto Struve 525"	SHJ 282AC	6.1	7.6	45.1	349	2004	18 54.9	+33 58		◉	

nebula M 57. M 56 is an excellent example of a very loose (Class X) globular cluster, while M 57 is generally considered to be the finest planetary nebula in the sky. Lyra is also home to the famous Double Double, or Epsilon Lyrae, a wide double star each of whose components is itself a close double star.

Lyra culminates at midnight on 2 July, and, for observers at mid-northern latitudes, is best placed for evening viewing from the mid-spring through mid-autumn months.

CHART 31-1.

The constellation Lyra (field width 15°)

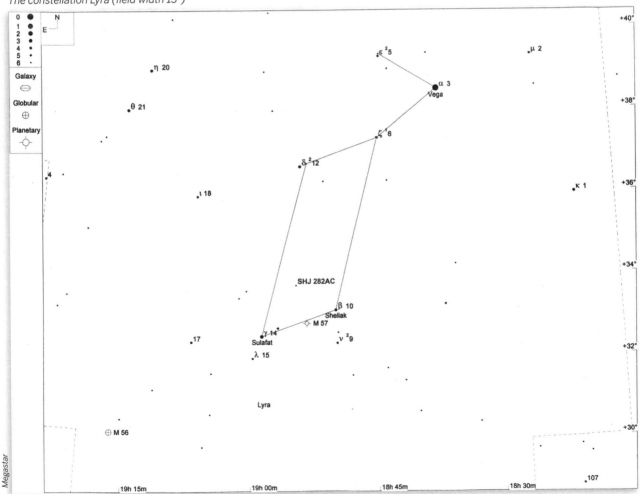

Clusters, Nebulae, and Galaxies

M 57 (NGC 6720)	★★★	✺✺✺✺	PN	MBUDR
Chart 31-1, 31-2	Figure 31-1	m8.8, 1.8' x 1.4'	18h 53.6m	+33° 02'

NGC 6720, better known as Messier 57, M 57, or the Ring Nebula, is considered by many observers to be the archetypical planetary nebula. M 57 was discovered by Antoine Darquier de Pellepoix while he was observing the Comet of 1779. Charles Messier independently re-discovered M 57 on the night of 31 January 1779 as he observed the same comet.

Crediting de Pellepoix with the original discovery, Messier recorded M 57 as a "cluster of light between Gamma & Beta Lyrae, discovered when looking for the Comet of 1779, which has passed it very close: it seems that this patch of light, which is round, must be composed of very small stars: with the best telescopes it is impossible to distinguish them; there stays only a suspicion that they are there. M. Messier reported this patch of light on the Chart of the Comet of 1779."

M 57 is easy to find. It lies not quite halfway on the 1.9° line from 10-β (beta) Lyrae (Sheliak) to 14-γ (gamma) Lyrae (Sulafat), which form the S end of the parallelogram. At 42X in our 10" Dob, M 57 is a tiny, delicate celestial smoke ring. At 180X, M 57 appears as a bright, subtly blue-green, nearly circular 1' disc that is slightly elongated ENE-WSW. The bright annulus surrounds a noticeably darker, although not black, central area.

No hint of the central star is visible. The m13 star that lies just E of the ring is a field star. Glimpsing the central star of M 57 requires excellent conditions and large aperture. We have seen the central star, but only in 17.5" and 20" instruments.

FIGURE 31-1.

NGC 6720 (M 57) (60' field width)

CHART 31-2.

NGC 6720 (M 57) and double star SHJ 282AC (7.5° field width; 5° finder circle; 1° eyepiece circles; LM 9.0)

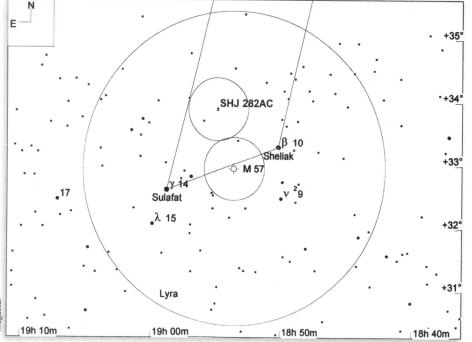

ILLUSTRATED GUIDE TO ASTRONOMICAL WONDERS

M 56 (NGC 6779)	★★	⊙⊙⊙	GC	MBUDR
Chart 31-3	Figure 31-2	m8.4, 8.8'	19h 16.6m	+30° 11'

NGC 6779, better known as Messier 56 or M 56, is a small, dim, loose globular cluster. Charles Messier discovered M 56 on the night of 19 January 1779, the same night he discovered the Comet of 1779.

M 56 is relatively easy to find. It lies not quite halfway on the 8.3° line from 6-β (beta) Cygni (Albireo) to 14-γ (gamma) Lyrae (Sulafat). To locate M 56, place Albireo on the SE edge of your binocular or finder field. Look for the m5 star 2-Cygni, which appears prominently near the center of the field, and then for an m6 star that appears prominently in the NW quadrant of the field. M 56 lies 25' SE of that m6 star.

The Astronomical League Binocular Messier Club lists M 56 as a Challenge Object (the most difficult category) with a 35mm or 50mm binocular, and as a Tougher Object (the middle category) with an 80mm binocular. We have logged M 56 with our 50mm binoculars, but it's visible only as a dim, slightly fuzzy star. If we didn't know it was there, we'd just have passed it by.

At 125X in our 10" Dob, M 56 shows a bright, very loose 2' core surrounded by a halo of easily resolved stars. Half a dozen stars can be resolved in the core, embedded in a nebulous glow. This is one of the poor Messier globulars.

FIGURE 31-2.

NGC 6779 (M 56) (60' field width)

Image reproduced from Digitized Sky Survey courtesy Palomar Observatory and Space Telescope Science Institute

CHART 31-3.

NGC 6779 (M 56) (10° field width; 5° finder circle; 1° eyepiece circle; LM 9.0)

Multiple Stars

5-epsilon2 (STF 2383Cc-D)	★★★★	७७७७		MS	UD
Chart 31-1, 31-4		m5.3/5.4, 2.4", PA 80° (2006)		18h 44.3m	+39° 40'

4-epsilon1 (STF 2382AB)	★★★★	७७७७		MS	UD
Chart 31-1, 31-4		m5.0/6.1, 2.4", PA 349° (2006)		18h 44.3m	+39° 40'

4-epsilon1, 5-epsilon2 (STFA 37AB-CD)	★★★★	७७७७		MS	UD
Chart 31-1, 31-4		m5.0/5.3, 210.5", PA 174° (1998)		18h 44.3m	+39° 40'

Epsilon Lyrae is our favorite multiple star after Albireo. The stars themselves are unexceptional. At 125X in our 10" Dob, all are of similar brightness, and all are pure white or slightly warm white. What is extraordinary about this multiple star is the placement of the components. The AB-CD pair is very widely separated, and can be split with the slightest optical aid. It's visible as a double in even a 30mm finder. The A-B and C-D pairs are much tighter, and require medium to high magnification to split. The components of the N (northernmost) pair, 4-ϵ^1 (epsilon^1) are aligned perpendicularly to the components of the S (southernmost) pair, 5-ϵ^2.

To locate epsilon Lyrae, center Vega in your finder and look 2.2° NNE (north northeast).

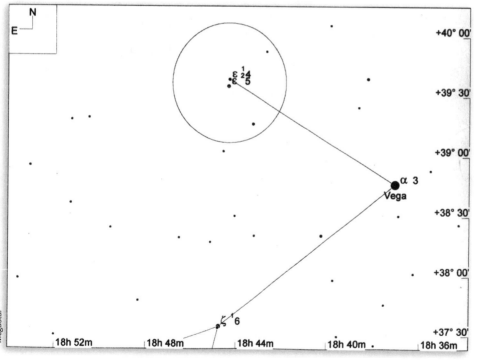

CHART 31-4.

Epsilon Lyrae (The Double Double) (4° field width; 1° eyepiece circle; LM 9.0)

6-zeta1, 7-zeta2 (STFA 38AD)	★★	ଡ଼ଡ଼ଡ଼ଡ଼		MS	UD
Chart 31-1		m4.3/5.6, 43.8", PA 150° (2003)		18h 44.8m	+37° 36'

6,7-ζ (zeta) Lyrae is a rather ordinary double star. It's easy enough to find, as it forms the NW corner of Lyra's parallelogram pattern. At 125X in our 10" Dob, the primary and companion are both cool white.

10-beta (STFA 39AB)	★★	ଡ଼ଡ଼ଡ଼ଡ଼		MS	UD
Chart 31-1		m3.6/6.7, 46.0", PA 150° (2002)		18h 50.1m	+33° 21'

10-β (beta) Lyrae (Sheliak) is another double star of rather pedestrian appearance. It's easy enough to find, as it forms the SW corner of Lyra's parallelogram pattern. At 125X in our 10" Dob, the primary and companion are both blue-white.

SHJ 282AC	★★★	ଡ଼ଡ଼ଡ଼ଡ଼		MS	UD
Chart 31-1, 31-2		m6.1/7.6, 45.1", PA 349° (2004)		18h 54.9m	+33° 58'

This multiple star system is misidentified in Astronomical League Double Star Club list as "Otto Struve 525". The actual pair required for the AL DS club is SAO 67566 and SAO 67565, which is properly identified as SHJ 282AC. STT 525 is a much closer double, separated by less than 2", and comprises the m6.1 primary and an m9.1 companion lying at PA 128°.

SHJ 282AC is easy to find. It forms the N apex of a triangle with Sulafat and Sheliak, and is by far the most prominent star in the near vicinity. At 125X in our 10" Dob, the primary is yellow-white and the companion blue-white.

32

Monoceros, The Unicorn

NAME: Monoceros (MON-oh-SAIR-ose)

SEASON: Winter

CULMINATION: midnight, 5 January

ABBREVIATION: Mon

GENITIVE: Monocerotis (MON-oh-sur-OH-tis)

NEIGHBORS: CMa, CMi, Gem, Hya, Lep, Ori, Pup

BINOCULAR OBJECTS: NGC 2232, NGC 2244, NGC 2251, NGC 2264, NGC 2301, NGC 2323 (M 50), NGC 2343

URBAN OBJECTS: NGC 2232, NGC 2244, NGC 2264, NGC 2301, NGC 2323 (M 50), 11-beta (STF 919)

Monoceros is a dim, mid-size equatorial constellation that ranks 35th in size among the 88 constellations. Monoceros covers 482 square degrees of the celestial sphere, or about 1.2%. Monoceros is one of the faintest of the constellations. It's brightest star is only fourth magnitude, and the fact that it is surrounded by the bright constellations Orion, Gemini, Canis Minor, and Canis Major makes it even less prominent than it otherwise would be. The ancients generally ignored the area of sky that is now Monoceros, considering it to be just a dark area of sky devoid of interest.

Monoceros is one of the "modern" constellations, and so has little or no mythology associated with it. The Dutch astronomer Petrus Plancius named Monoceros in 1613, when he proposed it as a new constellation. The German astronomer Jakob Bartsch charted Monoceros in 1624, but under the name Unicornus.

Despite its faintness, Monoceros has a fair number of interesting DSOs. Most of those are open clusters, including Messier 50. Monoceros is also home to the Rosette Nebula, a very nice emission nebula.

Monoceros culminates at midnight on 5 January, and, for observers at mid-northern latitudes, is best placed for evening viewing from the early winter through mid-spring months.

TABLE 32-1.

Featured star clusters, nebulae, and galaxies in Monoceros

Object	Type	Mv	Size	RA	Dec	M	B	U	D	R	Notes
NGC 2244	OC	4.8	23.0	06 32.3	+04 51			●	●		Mel 47; Class II 3 r n
NGC 2237	EN	99.9	80.0 x 60.0	06 31.7	+05 04					●	Rosette Nebula; Class E
NGC 2251	OC	7.3	10.0	06 34.7	+08 22				●		Cr 101; Class III 2 m
NGC 2261	EN/RN	var	2.0 x 1.7	06 39.2	+08 45					●	Hubble's Variable Nebula; Class E+R
NGC 2264	OC/EN	4.1	20.0	06 41.0	+09 54			●	●		Cr 112; Mel 49; Class III 3 m n
NGC 2301	OC	6.0	12.0	06 51.8	+00 28			●	●		Cr 119; Mel 54; Class I 3 r
NGC 2323	OC	5.9	16.0	07 02.8	-08 23	●	●	●			M 50; Cr 124; Class II 3 r
NGC 2343	OC	6.7	6.0	07 08.1	-10 37				●		Cr 128; Class II 2 p n
NGC 2232	OC	4.2	29.0	06 28.0	-04 51			●	●		Cr 93; Class III 2 p

TABLE 32-2.

Featured multiple stars in Monoceros

Object	Pair	M1	M2	Sep	PA	Year	RA	Dec	UO	DS	Notes
8-epsilon	STF 900AB	4.4	6.6	12.1	29	2004	06 23.8	+04 35		◉	
11-beta	STF 919AB	4.6	5.0	7.1	133	2002	06 28.8	-07 01	◉	◉	

CHART 32-1.

The constellation Monoceros (field width 35°)

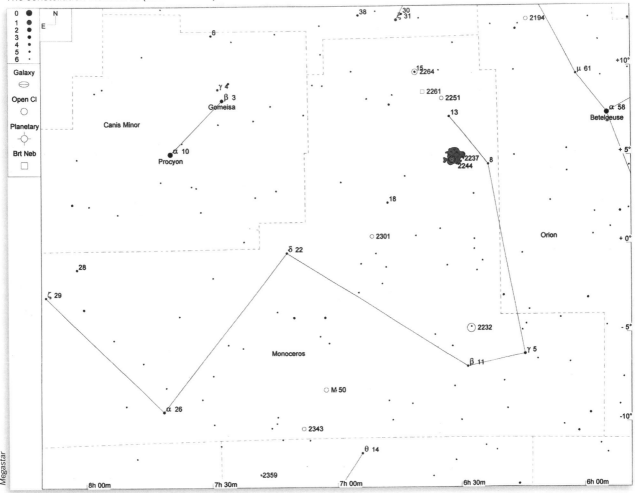

Clusters, Nebulae, and Galaxies

NGC 2244	★★	✹✹✹	OC	MBUDR
Chart 32-2, 32-3	Figure 32-1	m4.8, 23.0'	06h 32.3m	+04° 51'

NGC 2237	★★★	✹✹✹	EN	MBUDR
Chart 32-2, 32-3	Figure 32-1	m99.9, 80.0' x 60.0'	06h 31.7m	+05° 04'

NGC 2244 is an open cluster embedded in the emission nebula NGC 2237, better known as the Rosette Nebula. When William Herschel discovered the Rosette Nebula, he saw the brighter portions individually as a group of separate, disconnected objects, and assigned a separate Herschel number to each of them. When the New General Catalog (NGC) was compiled, those parts of the Rosette Nebula, which was later determined to be a single object, were assigned the NGC numbers 2237, 2238, 2239, and 2246. NGC 2237 is now used to represent the entire Rosette Nebula, and the other NGC numbers are no longer used.

To locate NGC 2244 and NGC 2237, place Betelgeuse in Orion on the WNW (west northwest) edge of your finder field and drift the finder field ESE (east southeast) on the line from Meissa to Betelgeuse by 7.6° until m4.4 8-ε (epsilon) Monocerotis comes into view. 8-ε appears very prominently in the finder, and is much brighter than any other star in the near vicinity. Place 8-ε on the W edge of the finder field to center NGC 2244/2237 in the finder field.

With our 50mm binoculars, NGC 2244 is visible as a small, faint hazy patch extending NW of yellowish m6 12-Monocerotis, with another m6 star lying on the NE edge of the cluster and half a dozen m7 through m9 stars filling out the cluster. At 90X in our 10" reflector, NGC 2244 is a very bright, loose cluster with 30+ m6 through m12 stars surrounded by tendrils of much fainter nebulosity. Using our Ultrablock narrowband filter makes all but the brightest stars in the cluster invisible, but enhances the contrast to reveal significant detail in the surrounding nebulosity.

The Rosette Nebula is a huge object, about one degree in diameter, or about four times the size of the full moon. Although it has notoriously low surface brightness, some observers have reported seeing the Rosette Nebula naked-eye with the aid of a narrowband or O-III filter. We have never managed that feat, but we have seen the Rosette clearly with our 50mm binoculars by using our Ultrablock narrowband filter between our eyes and the binocular eyepiece.

At 42X in our 10" reflector with the Ultrablock filter in place, the Rosette Nebula is visible in context as an irregular doughnut shape, much brighter on the E and W edges, but with tendrils faintly visible on the N and S edge of the doughnut hole. Using higher magnification to examine individual parts of the nebula reveals significant additional detail. Under higher magnification, the Rosette reminds us of a much fainter version of M 42, the Great Orion Nebula, in its delicacy and detail.

FIGURE 32-1.

NGC 2244 and NGC 2237 (60' field width)

Image reproduced from Digitized Sky Survey courtesy Palomar Observatory and Space Telescope Science Institute

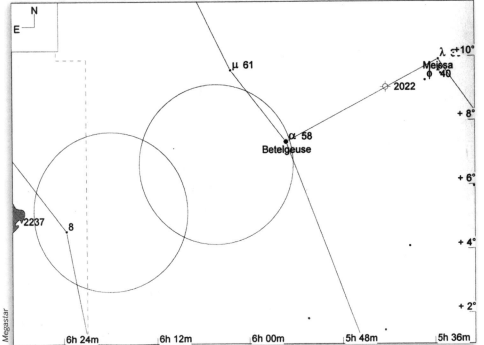

CHART 32-2.

Locating 8-Monocerotis (15° field width; 5° finder circles; LM 5.0)

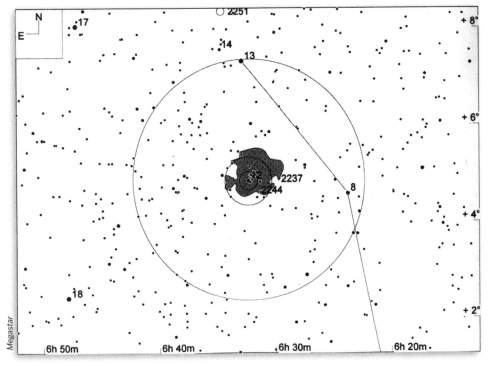

CHART 32-3.

NGC 2244 and NGC 2237 (Rosette Nebula) (10° field width; 5° finder circle; 1° eyepiece circle; LM 9.0)

NGC 2251	★★	✧✧✧	OC	MBUDr
Chart 32-4	Figure 32-2	m7.3, 10.0'	06h 34.7m	+08° 22'

NGC 2251 is a rather ordinary open cluster that's a required object to complete the Astronomical League Deep-Sky Binocular club list. To locate NGC 2251, place m4.4 8-ε (epsilon) Monocerotis on the SW edge of your binocular or finder field and look 3.5° NE for m4.5 13-Mon. NGC 2251 lies 1.2° NNE (north northeast) of 13-Mon, where it is visible in a 50mm finder or binocular as a small, moderately bright hazy patch with one m9 star embedded in the nebulosity.

At 90X in our 10" reflector, NGC 2251 is a moderately bright, very loose cluster of unusual appearance. Rather than the typical roundish shape, this cluster comprises three distinct groups of stars that together form a 10' x 5' NW-SE (from northwest to southeast) oval. At the N and NW edge of the cluster is an 5' E-W arc of three m10 stars. At the cluster's center is a pair of m9/10 stars, with a second m10 pair lying on the SE edge of the cluster. About two dozen much dimmer stars fill out the cluster.

FIGURE 32-2.

NGC 2251 (60' field width)

Image reproduced from Digitized Sky Survey courtesy Palomar Observatory and Space Telescope Science Institute

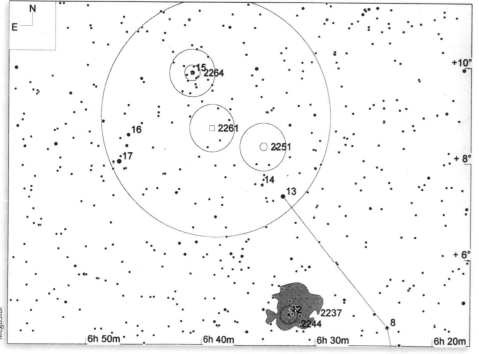

CHART 32-4.

NGC 2251, NGC 2261 (Hubble's Variable Nebula), and NGC 2264 (10° field width; 5° finder circle; 1° eyepiece circles; LM 9.0)

NGC 2261	★★★	✹✹	EN/RN	MBUDR	
Chart 32-4	Figure 32-3	m (variable), 2.0' x 1.7'	06h 39.2m	+08° 45'	

NGC 2261, better known as Hubble's Variable Nebula, looks more like a comet than any other object we have observed. If the great comet hunter Charles Messier had ever seen this object, he'd have been convinced he'd found yet another new comet, at least until he realized it wasn't moving against the background stars. The variability of this nebula is caused by the star R-Monocerotis, which lies at the southern tip of the nebula and illuminates it.

NGC 2261 lies 1.2° ENE (east northeast), and is easiest to find by placing NGC 2251 on the WSW (west southwest) edge of your low-power eyepiece field and drifting the eyepiece field ENE until NGC 2261 comes into view. (With our 10" Dob, our low-power eyepiece has a true field of nearly 2°, so both objects are visible in the same field of view.)

At 125X in our 10" reflector, NGC 2261 appears strikingly like a comet. The star R-Monocerotis forms the nucleus of the "comet", with the nebula forming the tail. The nebula extends 1.5' N from R-Monocerotis and curves W to form a comma shape. Using our Ultrablock narrowband filter increases the contrast of the nebula, but does not increase its visible extent. Some very subtle faint mottling is visible in the nebulosity with the Ultrablock in place, but it requires a larger scope and higher magnification to see much additional detail in the nebula. Although we have observed this object only once, other observers tell us that the visible structural details of the nebula change from session to session.

FIGURE 32-3.

NGC 2261 (60' field width)

Image reproduced from Digitized Sky Survey courtesy Palomar Observatory and Space Telescope Science Institute

NGC 2264	★★★	✹✹✹	OC/EN	MBUDR	
Chart 32-4	Figure 32-4	m4.1, 20.0'	06h 41.0m	+09° 54'	

NGC 2264, better known as the Christmas Tree Cluster, is one of the best non-Messier open clusters. It's easily located, lying 3.2° NE of 13-Mon. In fact, this cluster is bright enough to be a naked-eye object under very transparent conditions if you use averted vision from a dark site when the object is at high elevation. NGC 2264 is very prominent with even the slightest optical aid.

With our 50mm binoculars, NGC 2264 is a large, very bright open cluster with a dozen bright stars and two dozen dimmer ones. The m4.7 star 15-Mon at the S end of the cluster forms the base of the Christmas tree, which appears upside-down in a binocular or correct-image finder. At 90X in our 10" reflector, the Christmas tree is right-side up. The tree is outlined by m8/9 stars, with an m6 star representing the star on top of the tree, and dozens of dimmer stars scattered throughout the cluster. A small patch of extremely faint nebulosity is visible just SW of 15-Mon. Using our Ultrablock narrowband filter increases the contrast and visible extent of the nebulosity, but reveals little additional detail. This is a gorgeous open cluster, but the emission nebula component is unimpressive to say the least.

FIGURE 32-4.

NGC 2264 (60' field width)

NGC 2301	★★★	◑◑◑	OC	MBUDR
Chart 32-5	Figure 32-5	m6.0, 12.0'	06h 51.8m	+00° 28'

NGC 2301 is a bright, pretty open cluster. It's relatively easy to locate by placing 8- and 13-Mon on the W and N edges of your binocular or finder field, respectively, and panning a bit over one full field SE until NGC 2301 comes into view, about 2.2° SSE of the m4.5 star 18-Mon, which appears very prominently in the finder.

With our 50mm binoculars, NGC 2301 is a small, relatively bright patch of haze centered on an m7 star, with half a dozen other stars visible down to m10. At 90X in our 10" Newtonian, 50+ stars are visible from m9 through m11, tightly concentrated into an 8' core, with a 7' arc of four m8/9 stars extending S from the SW edge of the cluster.

FIGURE 32-5.

NGC 2301 (60' field width)

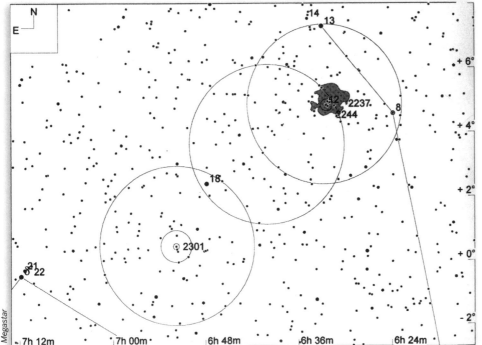

Megastar

CHART 32-5.

NGC 2301 (15° field width; 5° finder circles; 1° eyepiece circle; LM 9.0)

M 50 (NGC 2323)	★★★	✿✿✿	OC	MBUDR
Chart 32-6	Figure 32-6	m5.9, 16.0'	07h 02.8m	-08° 23'

NGC 2323, better known as Messier 50 or M 50, is the finest open cluster in Monoceros. M 50 was first logged by Giovanni Cassini at some time prior to 1711, and independently re-discovered by Charles Messier on the night of 5 April 1772. Messier logged M 50 as a "Cluster of small stars, more or less brilliant, above the right loins of the Unicorn, above the star Theta of the ear of Canis Major, & near a star of 7th magnitude. It was while observing the Comet of 1772 that M. Messier observed this cluster."

If you've just finished observing NGC 2301, you can locate M 50 by moving your finder as shown in Chart 32-6. Otherwise, it's easiest to locate M 50 from Canis Major. To do so, locate m4.1 14-θ (theta) CMa, which lies 5.1° NNE of Sirius and is clearly visible to the naked eye. Place 14-θ on the SW edge of your finder or binocular field, and look near the N edge of the field for M 50, which appears prominently.

With our 50mm binoculars, M 50 appears as a large, bright patch of haze with a dozen or so m8 through m10 stars embedded. With our 10" reflector at 90X, M 50 has a dozen m8 through m10 stars scattered among 50+ dimmer stars down to m12. Although many observers report that the dense central core resembles a heart or a blunt arrowhead, we've never been able to see such a pattern. To us, the central concentration looks more like a fan with the narrow end pointing N.

FIGURE 32-6.

NGC 2323 (60' field width)

Image reproduced from Digitized Sky Survey courtesy Anglo-Australian Observatory and Space Telescope Science Institute

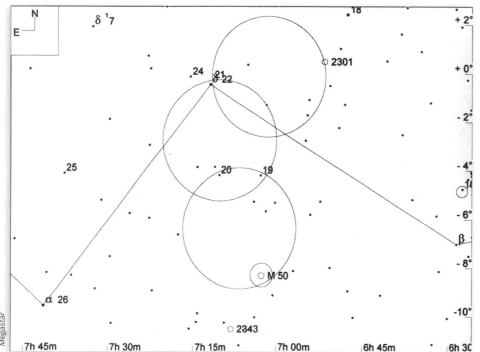

CHART 32-6.

*NGC 2323 (M 50) (20° field
width; 5° finder circles; 1°
eyepiece circle; LM 7.0)*

NGC 2343	★★	❂❂❂	OC	MBUDR
Chart 32-7	Figure 32-7	m6.7, 6.0'	07h 08.1m	-10° 37'

NGC 2343 is a rather ordinary open cluster. If you've just finished observing M 50, you can locate NGC 2343 by moving your finder as shown in Chart 32-7. Otherwise, it's easiest to locate NGC 2343 from Canis Major. To do so, locate m4.1 14-θ (theta) CMa, which lies 5.1° NNE of Sirius and is clearly visible to the naked eye. Place 14-θ on the SW edge of your finder or binocular field, and look near the E edge of the field for NGC 2343, which appears as a small, relatively bright patch of nebulosity.

With our 50mm binocular, NGC 2343 is a 5' hazy patch lying W and NW of an m9.4 yellow-orange star on the E edge of the cluster. With averted vision, half a dozen or so other stars flicker just on the edge of visibility. At 125X in our 10" Dob, about two dozen scattered stars from m9 through m13 are visible.

FIGURE 32-7.

NGC 2343 (60' field width)

Image reproduced from Digitized Sky Survey courtesy Anglo-Australian Observatory and Space Telescope Science Institute

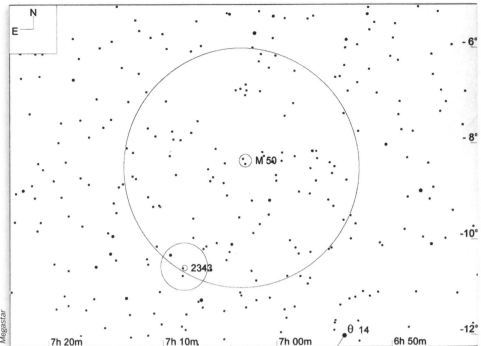

NGC 2232	★★	✥✥✥	OC	MBUDR
Chart 32-8	Figure 32-8	m4.2, 29.0'	06h 28.0m	-04° 51'

NGC 2232 is a very large, very bright, very loose open cluster, centered on the m5.1 star 10-Mon. If you've just finished observing NGC 2343, you can locate NGC 2232 by moving your finder as shown in Chart 32-8. Otherwise, the easiest way to locate NGC 2232 is to extend the line formed by the belt stars in Orion ESE (east southeast) 9.5° to locate the m4.0 star 5-γ (gamma) Monocerotis, which is the brightest naked-eye star in the immediate vicinity. Place 5-γ on the W edge of your binocular or finder and look for NGC 2232 in the NE quadrant of the field, where it appears prominently.

With our 50mm binoculars, NGC 2232 is a large, very bright hazy patch with a dozen or so embedded stars from m5 down to m10. At 90X with our 10" reflector, about three dozen stars down to m13 are visible within a 30' area, with 10-Mon near the center of the cluster. An m8 star lies on the NE edge of the cluster and an m7 star on the S edge. The prominent 35' arc of m7 stars that curves S and SW from m6.5 9-Mon is not a part of the cluster.

FIGURE 32-8.

NGC 2232 (60' field width)

Image reproduced from Digitized Sky Survey courtesy Anglo-Australian Observatory and Space Telescope Science Institute

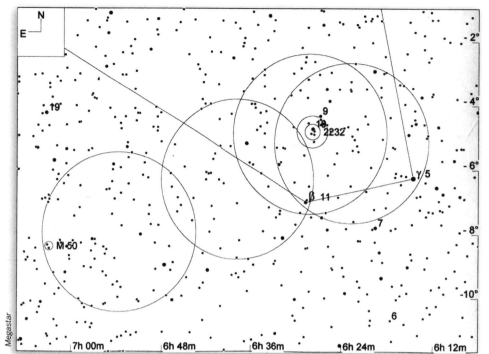

Megastar

Multiple Stars

8-epsilon (STF 900AB)	★★	✺✺✺		MS	UD
Chart 32-1, 32-2		m4.4/6.6, 12.1", PA 29° (2004)		06h 23.8m	+04° 35'

8-ε (epsilon) Monocerotis is a pretty double star set in a rich Milky Way star field. To locate 8-ε, drift your finder ESE on the line from Meissa to Betelgeuse in Orion by 7.6° until m4.4 8-ε comes into view. 8-ε appears very prominently in the finder, and is much brighter than any other star in the near vicinity. At 125X in our 10" Dob, the primary and companion are both yellow-white.

11-beta (STF 919AB)	★★★	✺✺✺		MS	UD
Chart 32-1, 32-8		m4.6/5.0, 7.1", PA 133° (2002)		06h 28.8m	-07° 01'

Beta Monocerotis is one of the most famous multiple stars. Although we've rated β-Mon as relatively easy to find, much depends on the darkness of your observing site. From a dark-sky site, β-Mon is a dim naked-eye star, easily confirmed by looking 3.6° WNW for 5-γ (gamma), which at m4.0 is nearly twice as bright as β-Mon.

From an urban site with brightly-lit skies, it's a bit more difficult. We generally start by putting Alnitak in Orion on the W edge of our finder field and looking for the m4.5 star SAO 132732 near the E edge of the finder field. That star is much brighter than any other star in the immediate vicinity, and therefore difficult to mistake for another star. Put SAO 132732 on the NW edge of the finder field and look near the SE edge of the field for m4.0 5-γ, which is blazingly bright in the finder. Then move the finder to put 5-γ on the W edge, and look for β-Mon 3.6° to the ESE. Alternatively, if you've just logged NGC 2232, you can hop to β-Mon from there, as shown in Chart 32-8.

Beta Monocerotis is certainly the most famous triple star. Most triples have one or two bright stars with the companion or companions much dimmer. Beta Monocerotis has three stars of similar brightness and color, arranged in a flattened triangle. At 125X in our 10" Dob, all three stars appear a sparkling blue-white.

33

Ophiuchus, The Serpent Bearer

NAME: Ophiuchus (oh-FEE-uh-kuss)

SEASON: Early summer

CULMINATION: midnight, 11 June

ABBREVIATION: Oph

GENITIVE: Ophiuchi (oh-FEE-uh-kee)

NEIGHBORS: Aql, Her, Lib, Sco, Ser, Sgr

BINOCULAR OBJECTS: NGC 6171 (M 107), NGC 6218 (M 12), NGC 6254 (M 10), NGC 6266 (M 62), NGC 6273 (M 19), NGC 6333 (M 9), NGC 6402 (M 14), IC 4665, NGC 6633

URBAN OBJECTS: NGC 6218 (M 12), NGC 6254 (M 10), NGC 6266 (M 62), IC 4665, NGC 663

Ophiuchus is a large equatorial constellation that ranks 11th in size among the 88 constellations. Ophiuchus covers 948 square degrees of the celestial sphere, or about 2.3%. Ophiuchus is a moderately bright constellation, containing two second-magnitude and seven third-magnitude stars. Despite that, Ophiuchus is often overlooked, probably because it lies so close to the prominent "teapot" asterism of Sagittarius to the southeast and the scorpion-like outline of Scorpius to the south. Ophiuchus, the Serpent Bearer, lies between Serpens Caput (the head of the snake) to the west and Serpens Cauda (the tail of the snake) to the east, dividing the constellation Serpens into two parts.

Ophiuchus, formerly known by its Latin name Serpentarius, is an ancient constellation, one of Ptolemy's original 44. In the earliest Greek myths, Ophiuchus represents the god Apollo struggling with the gigantic snake that guarded the Oracle at Delphi. In later myths, Ophiuchus represents the Trojan Laocoön, who warned the Trojans about the Trojan Horse, and was later slain by a pair of sea serpents sent by the gods to punish him, or Asclepius,

TABLE 33-1.

Featured star clusters, nebulae, and galaxies in Ophiuchus

Object	Type	Mv	Size	RA	Dec	M	B	U	D	R	Notes
NGC 6171	GC	7.8	13.0	16 32.5	-13 03	◉	◉				M 107; Class X
NGC 6218	GC	6.1	16.0	16 47.2	-01 57	◉	◉	◉			M 12; Class IX
NGC 6254	GC	6.6	20.0	16 57.1	-04 06	◉	◉	◉			M 10; Class VII
NGC 6273	GC	6.8	17.0	17 02.6	-26 16	◉	◉				M 19; Class VIII
NGC 6266	GC	6.4	15.0	17 01.2	-30 07	◉	◉	◉			M 62; Class IV
NGC 6369	PN	12.9	38"	17 29.3	-23 46					◉	Little Ghost Nebula; Class 4+2
NGC 6333	GC	7.8	12.0	17 19.2	-18 31	◉	◉				M 9; Class VIII
NGC 6402	GC	7.6	11.0	17 37.6	-03 15	◉	◉				M 14; Class VIII
IC 4665	OC	4.2	40.0	17 46.2	+05 43		◉	◉			Cr 349; Mel 179; Class III 2 m
NGC 6572	PN	9.0	11.0"	18 12.1	+06 51					◉	Class 2a
NGC 6633	OC	4.6	27.0	18 27.7	+06 34		◉	◉		◉	Cr 380; Mel 201; Class III 2 m

TABLE 33-2.

Featured multiple stars in Ophiuchus

Object	Pair	M1	M2	Sep	PA	Year	RA	Dec	UO	DS	Notes
36*	SHJ 243AB	5.1	5.1	4.9	144	2006	17 15.3	-26 36		◉	
39-omicron	H 25	5.2	6.6	10.0	355	2002	17 18.0	-24 17		◉	
70*	STF 2272AB	4.2	6.2	5.1	137	2006	18 05.5	+02 30		◉	

CHART 33-1.

The constellation Ophiuchus (field width 60°)

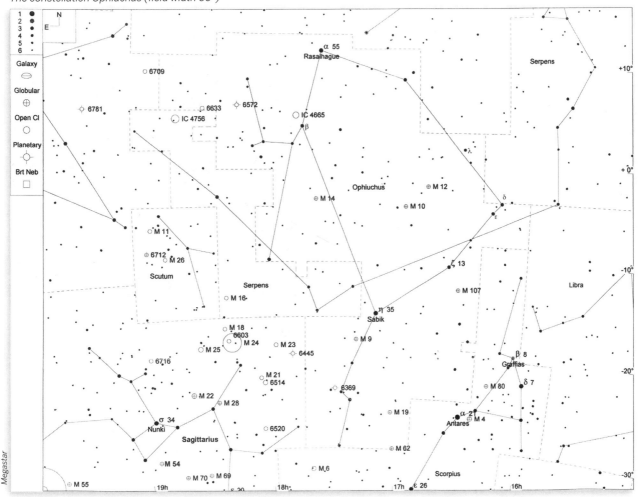

who learned the secret of immortality by watching snakes heal themselves and was slain by Zeus with a lightning bolt to prevent him from revealing the secret of immortality.

When we look at Ophiuchus, we are looking toward the galactic center and slightly above the galactic plane. Accordingly, Ophichus is rich in the objects associated with the galactic plane, namely globular clusters. Ophiuchus is home to no less than seven very bright Messier globular clusters, as well as a dozen other globs that are visible in amateur instruments. Surprisingly, given its location, Ophiuchus also contains several nice open clusters, most along its eastern edge and within the Milky Way.

Ophiuchus culminates at midnight on 11 June, and, for observers at mid-northern latitudes, is best placed for evening viewing from late spring through late summer.

Although we list all seven of the Messier globular clusters in this chapter as binocular objects, they differ significantly in difficulty. The Astronomical League Binocular Messier Club rates object difficulty as Easy, Tougher, or Challenge for both 50mm and 80mm binoculars. With a 50mm binocular, AL rates M 9 as a Challenge object, M 10 and M 12 as Easy objects, M 14, M 19, and M 62 as Tougher objects, and M 107 as not visible in a 50mm binocular. With an 80mm binocular, AL rates M 9 as a Tougher object, M 107 as a Challenge object, and all of the other Messier globulars as Easy objects.

Clusters, Nebulae, and Galaxies

M 107 (NGC 6171)	★★	❀❀❀❀	GC	MBUDR
Chart 33-2	Figure 33-1	m7.8, 13.0'	16h 32.5m	-13° 03'

NGC 6171, better known as Messier 107 or M 107, was one of the final objects to be added to the Messier catalog, in 1947, and one that Messier may never have observed himself. Pierre Méchain discovered this object in April, 1782, but there is no evidence that he reported this discovery to Messier or that Messier observed it.

M 107 is easy to locate, lying 2.7° SSW (south southwest) of the bright naked-eye star m2.6 13-ζ (zeta) Ophiuchi. Place 13-ζ on the NNE (north northeast) edge of your finder or binocular field and look for a 25' triangle of m7 stars about 2.5° to the SSW. M 107 lies about half a degree SSW of that triangle.

At 125X in our 10" reflector, M 107 shows a bright 3' core surrounded by a very loose 4.5' halo that is truncated at the S edge.

FIGURE 33-1.

NGC 6171 (M 107) (60' field width)

Image reproduced from Digitized Sky Survey courtesy Anglo-Australian Observatory and Space Telescope Science Institute

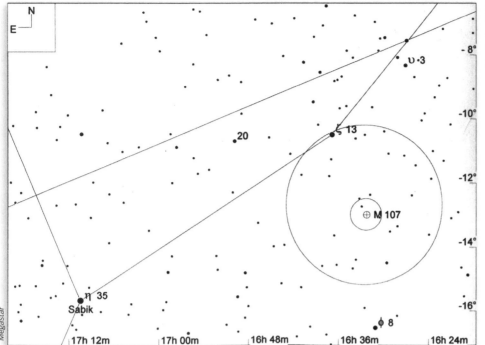

M 12 (NGC 6218)	★★★	☺☺☺	GC	MBUDR
Chart 33-3	Figure 33-2	m6.1, 16.0'	16h 47.2m	-01° 57'

NGC 6218, better known as Messier 12 or M 12, is a large, bright globular cluster. Charles Messier discovered M 12 on the night of 30 May 1764, the night after he discovered M 10. Messier's original log entry describes M 12 as "a nebula without stars."

M 12 is relatively easy to find. Begin by identifying the bright, wide naked-eye pair 1-δ (delta) Ophiuchi (Yed Prior) and 2-ε (epsilon) Ophiuchi (Yed Posterior). Look 7° NE for the m3.9 star 10-λ (lambda) Ophiuchi (Marfik). Place 10-λ on the NW edge of your finder or binocular field, and pan SE until M 12 comes into view, about 5.6° SE of Marfik.

With our 50mm binoculars, M 12 is a moderately large, very bright patch of nebulosity with no stars resolved. At 125X in our 10" Dob, M 12 is a beautiful glob, with hundreds of stars visible from m11 down to m13, embedded in a bright background haze. The loose 3' core has a granular appearance and is surrounded by a loose, rich halo that extends out to 10'.

FIGURE 33-2.

NGC 6218 (M 12) (60' field width)

Image reproduced from Digitized Sky Survey courtesy Anglo-Australian Observatory and Space Telescope Science Institute

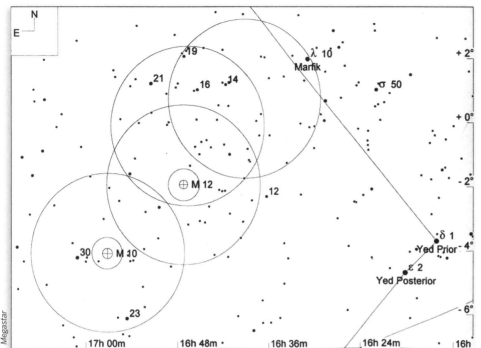

CHART 33-3.

*NGC 6218 (M 12) and NGC 6254
(M 10) (15° field width; 5° finder
circles; 1° eyepiece circles;
LM 9.0)*

M 10 (NGC 6254)	★★★	✎✎✎	GC	MBUDR
Chart 33-3	Figure 33-3	m6.6, 20.0'	16h 57.1m	-04° 06'

NGC 6254, better known as Messier 10 or M 10, is another large, bright globular cluster. Charles Messier discovered M 10 on the night of 29 May 1764, the night before he discovered M 12.

M 10 is relatively easy to locate, particularly if you have just observed M 12. M 10 lies just 3.3° SE of M 12, with both visible in the same 50mm finder field. Visually, M 10 is a near-twin of M 12, with M 10 just a bit dimmer. With our 50mm binoculars, M 10 is a bright, moderately large patch of nebulosity with no stars resolved. At 125X in our 10" Dob, M 10 shows a bright 4' core with hundreds of stars resolvable all the way to the core. The core is surrounded by a rich, evenly-distributed halo that extends out to 10'.

FIGURE 33-3.

NGC 6254 (M 10) (60' field width)

Image reproduced from Digitized Sky Survey courtesy Anglo-Australian Observatory and Space Telescope Science Institute

M 19 (NGC 6273)	★★★	◑◑◑	GC	MBUDR
Chart 33-4	Figure 33-4	m6.8, 17.0'	17h 02.6m	-26° 16'

NGC 6273, better known as Messier 19 or M 19, is another fine, bright globular cluster. Charles Messier was on a roll when he discovered M 19 on the night of 5 June 1764. A month prior, on 3 May, Messier had discovered M 2, followed by M 4 (8 May), M 5 (23 May), M 9 (28 May), M 10 (29 May), M 12 (30 May), M 13 (logged 1 June), M 14 (3 June), and M 15 (1 June). Although globular clusters are the rarest type of DSO, Messier by the night he discovered M 19 must have thought globular clusters were as common as dirt, particularly since he discovered yet another globular cluster, M 22, that same night.

M 19 is relatively easy to find. Start from Antares in Scorpius and look 12° E to identify m3.3 42-θ (theta), which is prominent to the naked eye, and by far the brightest star in the near vicinity. Place 42-θ on the E edge of your finder or binocular field, and look 4.5° WSW for M 19, which lies about 40' SSW of the m7 pair 28- and 31-Ophiuchi.

In our 50mm binoculars, M 19 is visible with averted vision as a moderate bright fuzzy star. At 125X in our 10" Dob, M 19 is notable for its striking oval appearance. Unlike most other globulars, which appear nearly perfectly spherical, both the core and halo of M 19 are distinct ovals, extending NNW-SSE (from north northwest to south southeast). The 3' x 4' core has a patchy, granular appearance. Averted vision reveals many, many faint stars flickering in and out, just on the edge of visibility. The 5' x 7' halo is partially resolved, with scores of m12 and dimmer stars unevenly distributed in clumps and chains.

FIGURE 33-4.

NGC 6273 (M 19) (60' field width)

Image reproduced from Digitized Sky Survey courtesy Anglo-Australian Observatory and Space Telescope Science Institute

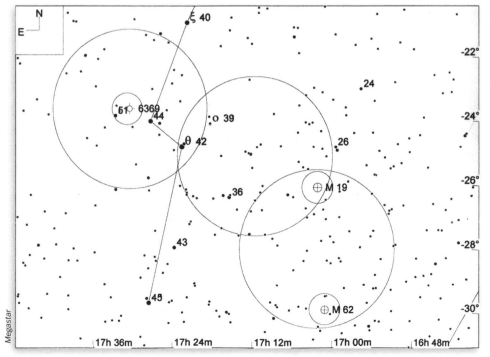

CHART 33-4.

NGC 6273 (M 19), NGC 6266 (M 62), and NGC 6369 (Little Ghost Nebula) (15° field width; 5° finder circles;

M 62 (NGC 6266)	★★★	�025	GC	MBUDR
Chart 33-4		m6.4, 15.0'	17h 01.2m	-30° 07'

NGC 6266, better known as Messier 62 or M 62, is another fine, bright globular cluster. Charles Messier discovered M 62 on the night of 7 June 1771, but did not determine its position accurately until nearly eight years later, on 4 June 1779, when he finally added it to his catalog. Had Messier cataloged M 62 on the night he discovered it, it would instead have been M 50.

M 62 is relatively easy to find by working from M 19. Place M 19 at the top edge of your binocular or finder field, and look for M 62 3.9° S, about 25' dead S of a close m8 pair that is easily visible in the finder.

In our 50mm binoculars, M 62 is visible with averted vision as a moderate bright fuzzy star. At 42X in our 10" Dob, the off-center core of M 62 gives it the appearance of a comet, with a bright head and a fainter tail extending NW. At 125X, M 62 appears very similar to its close neighbor M 19, but without the ovular appearance.

M 62 is notable for its striking oval appearance. Unlike most other globulars, which appear nearly perfectly spherical, both the core and halo of M 62 are distinct ovals, extending NNW-SSE. The 3' core has a patchy, granular appearance, with averted vision partially resolving many faint stars. The 7' halo is partially resolved, with scores of m12 and dimmer stars distributed in chains extending NW from the core.

NGC 6369	★★	�025	PN	MBUDR
Chart 33-4	Figure 33-5	m12.9, 38"	17h 29.3m	-23° 46'

NGC 6369, also called the Little Ghost Nebula, is a fairly ordinary planetary nebula. To locate NGC 6369, start from Antares and look 12° E to identify m3.3 42-θ (theta), which is prominent to the naked eye, and by far the brightest star in the near vicinity. From 42-θ, look 1.3° NE for m4.2 44-Ophiuchi. Extend the line from 42- to 44-Oph half its distance to the NE, and center your finder cross hairs there. NGC 6369 should be nearly centered in your low-power eyepiece.

Despite its relatively dim magnitude of 12.9, NGC 6369 has relatively high surface brightness. At 42X in our 10" reflector, NGC 6369 appears with direct vision as just another dim star. Averted vision reveals that it's slightly fuzzy. At 180X, NGC 6369 shows an almost featureless disc. With averted vision, some slight darkening toward the center is visible, but the object remains disc-like, in contrast to ring-like planetaries such as M 57. Using our Ultrablock narrowband filter does not increase the extent of the object, but makes the central darkening more evident and reveals some slight brightening along the N edge of the annulus.

FIGURE 33-5.

NGC 6369 (Little Ghost Nebula) (60' field width)

Image reproduced from Digitized Sky Survey courtesy Anglo-Australian Observatory and Space Telescope Science Institute

ILLUSTRATED GUIDE TO ASTRONOMICAL WONDERS

M 9 (NGC 6333)	★★	⬦⬦⬦	GC	MBUDR
Chart 33-5	Figure 33-6	m7.8, 12.0'	17h 19.2m	-18° 31'

NGC 6333, better known as Messier 9 or M 9, is another globular cluster that is considerably smaller and dimmer than the other Messier globulars in Ophiuchus with the exception of M 107. Charles Messier discovered and cataloged M 9 on the night of 28 May 1764, as the first of his globular clusters in Ophiuchus.

M 9 is relatively easy to find, lying just 3.5° SSE of the bright naked-eye star m2.4 35-η (eta) Ophiuchi, also known as Sabik. To locate M 9, place Sabik on the NW edge of your finder or binocular field, and look 3.5° SSE. M 9 lies about 45' S of an m6 star that appears prominently in the finder, and about the same distance NNE of another m6 star. Those two m6 stars form a 1.5° equilateral triangle with a third m6 star lying 1.3° dead E of M 9. Alternatively, once you have Sabik on the NW edge of your finder, you can drift the finder field 5.9° SSE until m4.4 40-ξ (xi) comes into view. Placing 40-ξ on the S edge of your finder field should nearly center M 9 in the finder.

We have never managed to see M 9 in our 50mm binoculars or finder, although many observers have reported seeing M 9 with 10X50 binoculars. At 125X in our 10" reflector, M 9 has a bright circular 2' core embedded in a 3.5' halo. The core is unresolved, but scattered strings and chains of m11 to m13 stars lie on the periphery.

FIGURE 33-6.

NGC 6333 (M 9) (60' field width)

Image reproduced from Digitized Sky Survey courtesy Anglo-Australian Observatory and Space Telescope Science Institute

CHART 33-5.

NGC 6333 (M 9) (10° field width; 5° finder circles; 1° eyepiece circle; LM 9.0)

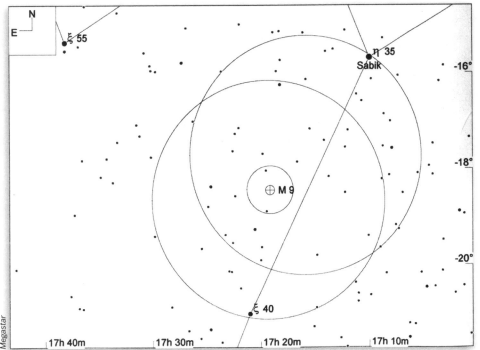

M 14 (NGC 6402)	★★	◎◎	GC	MBUDR
Chart 33-6	Figure 33-7	m7.6, 11.0'	17h 37.6m	-03° 15'

NGC 6402, better known as Messier 14 or M 14, is the seventh Messier globular cluster in Ophiuchus. Charles Messier discovered M 14 on the night of 1 June 1764. M 14 must have been a bit disappointing after the first globular cluster Messier cataloged that night, M 13 in Hercules. M 14 is one of the dimmest of the Messier globular clusters in Ophiuchus and is by far the hardest to find.

To locate M 14, begin by identifying m2.4 Sabik and m2.8 60-β (beta) Ophiuchi, also known as Cebalrai, which lies about 22° NNE of Sabik. Look halfway along the line from Sabik to Cebalrai for an m4.5 star, which is by far the brightest in the vicinity. Place that star on the SW edge of your finder field and look near the NE edge of the finder field for a prominent m6.2 star. M 14 lies 1.3° SW of that star, and 23' S of an m7.4 star that is easily visible in the finder.

With our 50mm binoculars and averted vision, M 14 is just visible as a fuzzy star. At 42X in our 10" Dob, M 14 looks more like a small elliptical galaxy than a globular cluster, with a condensed core and much fainter halo, but no stars resolved. At 125X, M 14 shows a bright 2.5' unresolved core that shows a slightly granular or mottled appearance, surrounded by a 5' halo with a few quite faint stars resolvable.

FIGURE 33-7.

NGC 6402 (M 14) (60' field width)

Image reproduced from Digitized Sky Survey courtesy Anglo-Australian Observatory and Space Telescope Science Institute

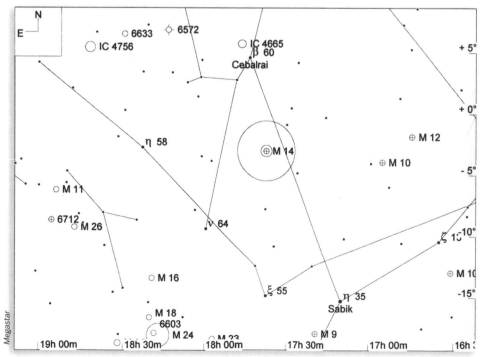

CHART 33-6.

NGC 6402 (M 14) (40° field width; 5° finder circle; LM 6.0)

IC 4665	★★	✦✦✦✦	OC	MBUDR
Chart 33-7	Figure 33-8	m4.2, 40.0'	17h 46.2m	+05° 43'

IC 4665 is a very loose, scattered, bright open cluster. IC 4665 is very easy to locate, lying just 1.3° NNE of the bright naked-eye star m2.8 β-Ophiuchi (Cebalrai). With our 50mm binoculars, IC 4665 comprises 25+ fairly bright stars distributed evenly over a 40' extent. As is so often true of large, scattered open clusters, IC 4665 appears better in a binocular than in a telescope. At 42X in our 10" Dob, IC 4665 shows three dozen bright stars distributed evenly over the field of view, which makes the object look more like a random collection of field stars than an open cluster.

FIGURE 33-8.

IC 4665 (60' field width)

Image reproduced from Digitized Sky Survey courtesy Palomar Observatory and Space Telescope Science Institute

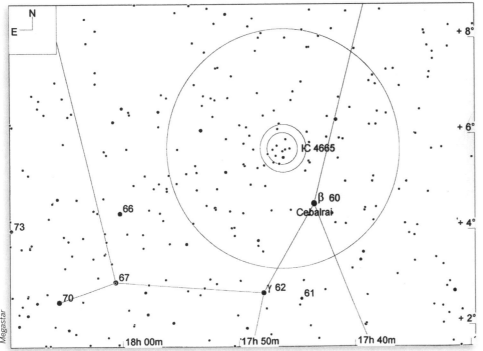

CHART 33-7.

IC 4665 (10° field width; 5° finder circle; 1° eyepiece circle; LM 9.0)

NGC 6572	★★	✧✧✧	PN	MBUDR
Chart 33-8	Figure 33-9	m9.0, 11.0"	18h 12.1m	+06° 51'

NGC 6572 is a tiny but very bright planetary nebula. To locate NGC 6572, begin by identifying the bright naked-eye stars m2.1 55-α (alpha) Ophiuchi (Rasalhague) and m2.8 60-β (beta) Ophiuchi (Cebalrai). These two stars form a prominent 8° triangle with m3.7 72-Oph as the E apex. Center 72-Oph in your finder and look for m4.6 71-Oph about one degree S. Place 71-Oph on the NW edge of your finder field to put NGC 6572 near the center of the field, where it should be within the field of your low-power eyepiece. NGC 6572 is visible in the finder, but appears stellar.

At 42X in our 10" Dob, NGC 6572 looks like a moderately bright star with a bluish tint, but no hint of fuzziness. At 125X that "star" becomes a tiny blob with no visible detail. At 250X, NGC 6572 shows a bright, small, bluish-green disc with no structural detail visible and no hint of central brightening. A narrowband filter increases the contrast but reveals no additional detail.

FIGURE 33-9.

NGC 6572 (60' field width)

Image reproduced from Digitized Sky Survey courtesy Palomar Observatory and Space Telescope Science Institute

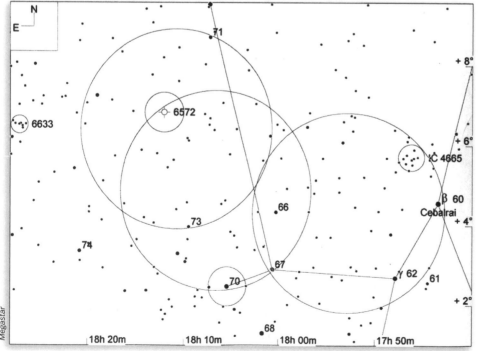

CHART 33-8.

NGC 6572 (12° field width; 5° finder circles; 1° eyepiece circles; LM 9.0)

NGC 6633	★★	⚹⚹⚹	OC	MBUDR
Chart 33-9	Figure 33-10	m4.6, 27.0'	18h 27.7m	+06° 34'

NGC 6633 is large, very bright, very loose open cluster. To locate NGC 6633, identify 71-Ophiuchi as described in the preceding section. Place 71-Oph on the NW edge of your finder or binocular field and look for NGC 6633 5.8° ESE (east southeast), where it appears prominently in the field of view. (IC 4756 in Serpens lies just 3.1° ESE of NGC 6633, and is visible in the same finder or binocular field.)

With our 50mm binocular, 15 or more m7 through m9 stars are visible, with another dozen or more down to m10 and beyond visible with averted vision. NGC 6633 lies in a very rich star field, and it's difficult to determine which stars are members of the cluster and which are field stars. The brightest stars form a distorted 15' triangular pattern, with additional chains and clumps of stars extending NE from the NE apex of the triangle. A very prominent m5.7 field star lies 20' SE of the center of the concentration. At 42X in our 10" Dob, dozens of additional stars are visible down to m13, and it's even more difficult to determine the boundaries of the cluster.

FIGURE 33-10.

NGC 6633 (60' field width)

Image reproduced from Digitized Sky Survey courtesy Palomar Observatory and Space Telescope Science Institute

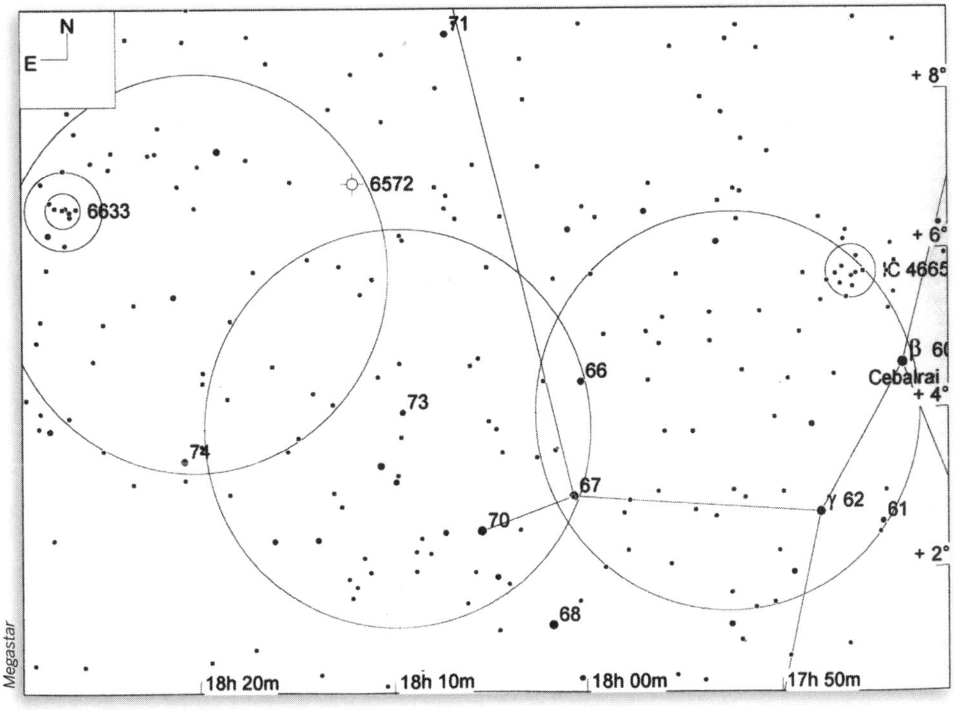

CHART 33-9.

NGC 6633 (12° field width; 5° finder circles; 1° eyepiece circle; LM 9.0)

Multiple Stars

36 (SHJ 243AB)	★★★	✹✹✹✹	MS	UD
Chart 33-4		m5.1/5.1, 4.9", PA 144° (2006)	17h 15.3m	-26° 36'

36-Ophiuchi is a very pretty double star, easily split even in small instruments. It's also easy to locate, lying 2.2° SW of the bright naked-eye star 42-θ (theta) Ophiuchi, and bright enough itself to be prominent in the finder. (See Chart 33-4.) At 100X in our 90mm refractor, the primary and companion are an evenly-matched pair with a striking orange coloration.

39-omicron (H 25)	★★★	✹✹✹✹	MS	UD
Chart 33-4		m5.2/6.6, 10.0", PA 355° (2002)	17h 18.0m	-24° 17'

39-o (omicron) Ophiuchi is another pretty double star that's easy to split with a small scope. 39-o lies just 1.2° NW of the bright naked-eye star 42-θ (theta) Ophiuchi, and is prominent in the finder. (See Chart 33-4.) At 100X in our 90mm refractor, the primary is yellow-orange and the noticeably fainter companion is lemon yellow.

70 (STF 2272AB)	★★★	✹✹✹	MS	UD
Chart 33-8		m4.2/6.2, 5.1", PA 137° (2006)	18h 05.5m	+02° 30'

70-Ophiuchi is a visual double, what we call a "fast mover". The primary and companion orbit each other in about 88 years, so the separation and position angle of this pair changes quickly. The values given are based on a 2006 observation. Fortunately, these two stars reached closest separation (1.5") in 1988, and the separation is currently increasing.

70-Ophiuchi is relatively easy to locate by using the star hop shown in Chart 33-8. At 100X in our 90mm refractor, the primary is a pretty yellow-orange color and the noticeably dimmer secondary a very nice reddish-orange.

34

Orion, The Hunter

NAME: Orion (oh-RYE-un)

SEASON: Early winter

CULMINATION: midnight, 13 December

ABBREVIATION: Ori

GENITIVE: Orionis (or-ee-OWN-is)

NEIGHBORS: Eri, Gem, Lep, Mon, Tau

BINOCULAR OBJECTS: NGC 1662, NGC 1981, NGC 1976 (M 42), NGC 2068 (M 78), NGC 2169

URBAN OBJECTS: NGC 1981, NGC 1976 (M 42), NGC 2169

Orion is a mid-size equatorial constellation that ranks 26th in size among the 88 constellations. Orion covers 594 square degrees of the celestial sphere, or about 1.4%. Orion is the most impressive constellation in the sky, and the one most likely to be familiar even to non-astronomers. The hourglass asterism of Orion is as familiar to most people as the Big Dipper or Plough asterism of Ursa Major. Orion includes two of the sky's ten brightest stars—Rigel and Betelgeuse—and, with Bellatrix, three of the top 25. Orion dominates the winter skies, even though it is surrounded by many of the sky's other brightest stars, including Sirius in Canis Major, Capella in Auriga, Procyon in Canis Minor, Aldebaran in Taurus, and Pollux in Gemini.

Orion is an ancient constellation that has been recognized as such since the earliest records made by the Assyrians, Sumerians, Egyptians, and Chinese. The ancient Chinese saw Orion as the center of the larger constellation Bai Hu (White Tiger). In Greek mythology, Orion is the Mighty Hunter. He stands with his two hunting dogs, Canis Major and Canis Minor on the shore of the river Eridanus. Their prey, Lepus the Hare, lies at the feet of Orion.

TABLE 34-1.

Featured star clusters, nebulae, and galaxies in Orion

Object	Type	Mv	Size	RA	Dec	M	B	U	D	R	Notes
NGC 1973/1975	EN/RN	99.9	29.0 x 20.0	05 35.4	-04 47					◉	
NGC 1981	OC	4.2	24.0	05 35.2	-04 26			◉	◉		Cr 73; Class III 3 p n
NGC 1976	EN/RN	3.0	60.0	05 35.0	-05 25	◉	◉	◉			M 42
NGC 1982	EN/RN	9.0	7.0 x 4.0	05 35.5	-05 17	◉					M 43
NGC 2024	EN	99.9	30.0 x 22.0	05 41.7	-01 48					◉	
NGC 2068	RN	8.3	8.4 x 7.8	05 46.8	+00 04	◉	◉				M 78
NGC 2022	PN	12.4	35.0"	05 42.1	+09 05					◉	Class 4+2
NGC 2169	OC	5.9	6.0	06 08.4	+13 58			◉	◉		Cr 83; Class III 3 m
NGC 2194	OC	8.5	10.0	06 13.8	+12 48					◉	Cr 87; Mel 43; Class II 2 r
NGC 1662	OC	6.4	20.0	04 48.5	+10 56				◉		Cr 55l Class II 3 m
NGC 1788	RN	99.9	5.5 x 3.0	05 06.9	-03 20					◉	

CHART 34-1.

The constellation Orion (field width 35°, north to the right)

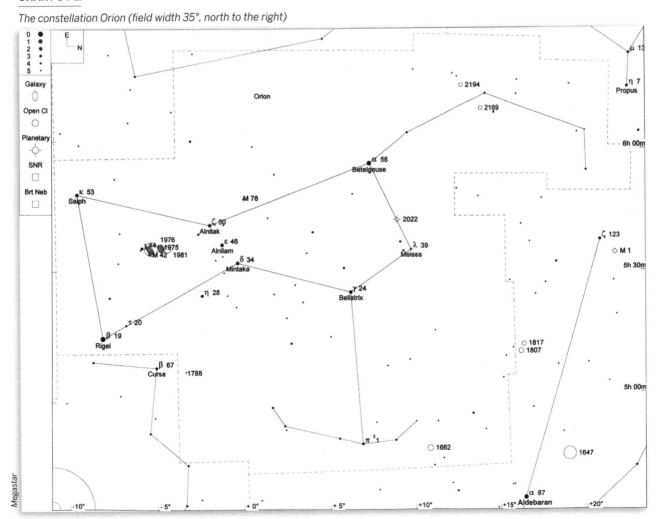

Megastar

Like many extraordinary humans in Greek mythology, Orion drew the wrath of the gods. Besotted with Orion because of his good looks and hunting prowess, Artemis (the Roman Diana), goddess of the hunt, fell deeply in love with Orion. Alas, Artemis was also goddess of the Moon, and she forgot her duty to light the night sky. Her twin brother, Apollo, was jealous of Artemis's affair with Orion and decided the only way to end it was to slay Orion. Apollo sent a giant scorpion to kill Orion. After an epic battle, Orion slew the scorpion, but during that battle the scorpion had stung him fatally. Artemis was bereft and appealed to Zeus, who cast the body of Orion into the night sky where he would live forever as the constellation that took his name. Apollo protested this favorable treatment to Zeus, who even-handedly granted the same favor to the scorpion, casting it into the night sky as the bright constellation Scorpius. Remembering the enmity between Orion and Scorpius, Zeus cast Scorpius as far away in the sky as possible from Orion, ensuring that the two constellations would never share the night sky.

Although it lacks galaxies, globular clusters, or bright planetary nebulae, Orion is home to an almost embarrassing richness of open clusters and emission/reflection nebulae. The area of Orion bounded by the belt stars and the sword is particularly rich in spectacular objects. Orion's wealth of deep-sky objects includes the most magnificent DSO of all, the Great Orion Nebula or M 42. Orion is also home to the famed Horsehead Nebula—a very difficult object in all but the largest amateur scopes—and to McNeil's Nebula, discovered by amateur Jay McNeil in 2003 to the embarrassment of professional astronomers, all of whom had missed it. The Trapezium is a collection of stars embedded in M 42—whether to consider it a multiple star system or an open cluster is still a matter of debate—that is actually a star nursery with baby stars being born as we watch. If you've never observed Orion, plan on spending an entire evening there, and that will only give you a start on exploring Orion's wonders. Most amateur astronomers return again and again to Orion. We suspect you'll do the same.

Orion culminates at midnight on 13 December, and, for observers at mid-northern latitudes, is best placed for evening viewing from mid-autumn through early spring.

TABLE 34-2.

Featured multiple stars in Orion

Object	Pair	M1	M2	Sep	PA	Year	RA	Dec	UO	DS	Notes
19-beta	STF 668A-BC	0.3	6.8	6.8	204	2004	05 14.5	-08 12		◉	Rigel
34-delta	STFA 14Aa-C	2.4	6.8	52.8	0	2003	05 32.0	-00 17		◉	Mintaka
44-iota	STF 752AB	2.9	7.0	11.3	141	2002	05 35.4	-05 55		◉	
SAO 132301	STF 747AB	4.7	5.5	36.0	224	2003	05 35.0	-06 00		◉	
39-lambda	STF 738AB	3.5	5.5	4.3	44	2003	05 35.1	+09 56		◉	Meissa
41-theta1	STF 748Aa-B	6.6	7.5	8.8	31	2004	05 35.3	-05 23		◉	Trapezium
41-theta1	STF 748Aa-C	6.6	5.1	12.7	132	2002	05 35.3	-05 23		◉	Trapezium
41-theta1	STF 748Aa-D	6.6	6.4	21.2	96	2002	05 35.3	-05 23		◉	Trapezium
41-theta2	STFA 16AB	5.0	6.2	52.2	93	2002	05 35.4	-05 24		◉	
48-sigma	STF 762AB-E	3.8	6.3	41.5	62	2003	05 38.7	-02 36		◉	
48-sigma	STF 762AB-D	3.8	6.6	12.7	84	2002	05 38.7	-02 36		◉	
50-zeta*	STF 774Aa-B	1.9	3.7	2.2	165	2006	05 40.7	-01 57		◉	
50-zeta	STF 774Aa-C	1.9	9.6	57.3	10	2003	05 40.8	-01 56		◉	

Clusters, Nebulae, and Galaxies

NGC 1973/1975	★★★	✺✺✺✺	EN/RN	MBUDR
Chart 34-2, 34-3	Figure 34-1	m99.9, 29.0' x 20.0'	05h 35.4m	-04° 47'

NGC 1973 and NGC 1975 (only NGC 1975 is labeled on the charts) are, with the much dimmer NGC 1977, individually numbered components of a single nebula complex that lies just S of NGC 1981 (visible at the top of Figure 34-1) and just N of NGC 1976 (M 42) and NGC 1982 (M 43). NGC 1975 is the smallest and most northerly of the three components. NGC 1973 is somewhat larger and brighter than NGC 1975, and lies about 5' SW of NGC 1975. NGC 1977 is much larger than either of the other two nebulae, and lies directly S of them, with 45- and 42-Orionis embedded in its nebulosity.

At 90X in our 10" reflector, NGC 1973/1975 are visible as small, moderately bright patches of nebulosity. NGC 1973 is a fairly prominent oval patch, about 4' x 2', elongated E-W (from east to west). NGC 1975 is a smaller and dimmer 1.5' circular patch. Seeing detail in either object is difficult because of the glare of the m5 pair 42/45-Orionis, which lie about 10' to the S.

Using our Ultrablock narrowband filter makes it obvious that these are combined emission/reflection nebulae. The emission components are greatly enhanced in contrast and detail, particularly because the narrowband filter dims the glare of 42/45-Orionis dramatically. The reflection components of the nebulae are dimmed to invisibility because they shine by reflected star light, which the narrowband filter blocks. With the narrowband

FIGURE 34-1.

NGC 1973, NGC 1975, and NGC 1981 (60' field width)

Image reproduced from Digitized Sky Survey courtesy Anglo-Australian Observatory and Space Telescope Science Institute

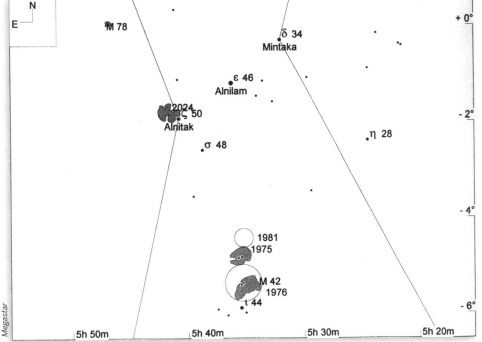

CHART 34-2.

The Belt and Sword regions of Orion (10° field width; LM 6.0)

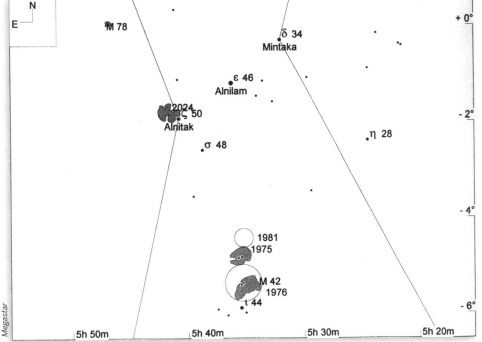

filter in place, both objects reveal a detailed wispy structure, and previously invisible tendrils of nebulosity are visible surrounding them. The narrowband filter also reveals structure and extent in NGC 1977, which is nearly invisible without the filter. With the narrowband filter, NGC 1977 is visible as a large, faint E-W nebulosity arcing S of and to both sides of 42-Orionis.

This group of objects is often overlooked, no doubt because of its close proximity to the magnificent M 42. If it were anywhere else in the sky, NGC 1973/1975/1977 would be considered a showpiece object.

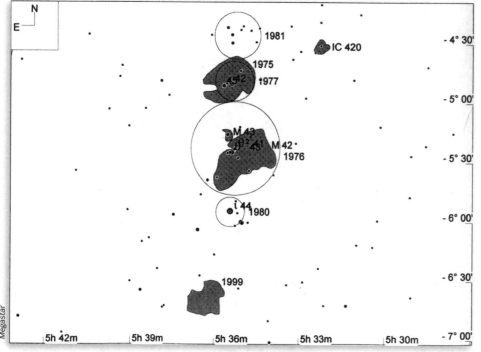

CHART 34-3.

The Sword of Orion (4° field width; LM 9.0)

NGC 1981	★★	✦✦✦✦	OC	MBUDR
Chart 34-2, 34-3	Figure 34-1	m4.2, 24.0'	05h 35.2m	-04° 26'

NGC 1981 is a sparse, loose, bright open cluster located 1° dead north of M 42, with the NGC 1973/1975/1977 complex intervening. With our 50mm binoculars, NGC 1981 is visible from a dark site as a dozen or so m6-9 stars embedded in a faint nebulosity of dimmer stars. At 90X in our 10" Dob, about 20 stars are visible down to m13. The most prominent feature is two S-NE arcs of stars, three m6/7 stars just E of center, and three m7/8 stars just W of center.

M 42 (NGC 1976)	★★★★★	✦✦✦✦	EN/RN	MBUDR
Chart 34-2, 34-3	Figure 34-2	m3.0, 60.0'	05h 35.0m	-05° 25'

M 43 (NGC 1982)	★★★★★	✦✦✦✦	EN/RN	MBUDR
Chart 34-2, 34-3	Figure 34-2	m9.0, 7.0' x 4.0'	05h 35.5m	-05° 17'

Yes, our rating system goes up only to four stars, but we decided that M 42 and M 43 deserve five stars anyway. (M 43 is actually a part of M 42, both physically and visually, but was assigned a separate identifier for historical reasons, about which more later.)

Other than Luna, there is no other object in the night sky that reveals as much detail as M 42. Most observers, including us, consider M 42, The Great Orion Nebula, to be the most magnificent object visible in the night sky. M 42 is visible to the naked eye as the center "star" in Orion's sword, and begins to reveal its majesty with even the slightest optical aid. With even a small telescope, you can spend hours observing M 42 without exhausting its wealth of detail.

M 42 ended up as a Messier object almost by accident. Charles Messier was a comet hunter. He had begun compiling his list of objects because he was frustrated at the time he'd wasted observing the object he later cataloged as Messier 1 (M 1). That object turned out to be fixed against the background stars, which established that it wasn't a comet. Unfortunately, it took Messier several nights to establish the non-cometary nature of M1. He decided to compile a list of objects that looked like comets, but weren't, in order to save others the aggravation of wasting time on non-comets.

Messier objects numbers 1 through 41 were all relatively dim objects that could easily be mistaken for comets. Messier patiently observed and logged these objects over a period of years, adding new objects to his list as he came across them. In early March of 1769, as the publication deadline for Messier's catalog drew near, Messier for some reason decided that 41 objects wasn't enough. Although no one knows for sure why Messier made this decision, the best guess is that Messier belatedly realized that the 1755 catalog compiled by Lacaille contained 42 objects. Presumably, Messier wanted his catalog to include more objects, and he must have decided that 45 was a nice, round number.

So, on the night of 4 March 1769, Messier set out to add four more items to his catalog. That presented a problem. The first 41 objects had taken Messier years to find, so he obviously had little hope of finding four new objects in one night. So Messier did the next best thing. He added four well-known objects to his catalog.

The first of these objects, M 42, had been known since antiquity as the fuzzy middle star in the sword of Orion, and was one of the first celestial objects to be examined in detail after telescopes were invented. (Although Galileo observed the area of Orion's sword and belt in 1609 and "added eighty other stars recently discovered in the vicinity," he somehow overlooked M 42. Because M 42 is clearly visible to the naked eye and shows as a hazy patch with binoculars, we suspect that Galileo noticed M 42 but assumed it was flaring caused by his primitive, uncoated optics.)

In the telescopes of Messier's time, M 42 and M 43 appeared to be distinctly separate objects, so Messier added M 43 as a separate entry in his catalog. Two down, two to go. Messier next turned his telescope to the constellation Cancer, where he found the bright open cluster known since antiquity as the Beehive Cluster or Praesepe. Messier logged the Beehive as M 44. One to go. Messier shifted his glance about 70° west and spotted the large, spectacularly bright open cluster known since ancient times as the Pleiades or The Seven Sisters. That one was really obvious, but time must have been growing short, so Messier must have gritted his teeth as he logged the Pleiades as the final object for that first edition of his catalog, M 45.

It's difficult for us to write a description of M 42, because doing it justice would require an entire chapter, if not an entire book. Figure 34-2 is a poor representation of M 42, but that's inevitable for any photograph of this object. Although a camera can show details too dim to be seen by the human eye, even using a very large telescope, a camera does a very poor job of representing objects that have very large differences in brightness. Exposing long enough to reveal the dim details grossly over-exposes the brighter areas, rendering them as the white blob visible in Figure 34-2. Conversely, using a short enough exposure time to show detail in the brighter areas grossly under-exposes the dimmer areas, preventing dim details from being recorded by the film or sensor.

Visually, it's a different story. The visible extent of M 42 is smaller than the photographic extent, but the level of detail visible at the eyepiece is orders of magnitude greater because the human eye easily accommodates much larger differences in brightness than film or CCD can manage. In the eyepiece, M 42 is a magnificent tapestry of knots, lumps, and tendrils of nebulosity. Many observers think that M 42 visually resembles a great bird, with M 43 as the head and the brighter parts of M 42 as the body and wings of the bird.

But, like the story of the three blind men and the elephant, what you see in M 42 depends on how you look at it. At low magnification, many scopes can fit all or nearly all of the visible extent of M 42 into one field of view, where it does indeed resemble a giant bird. Using higher magnification to zoom in on particular parts of M 42 will surprise you. Higher magnification reveals additional details that are invisible at low or medium magnification. If you have a narrowband filter and/or an O-III filter, use it (or them, but not both at the same time) at low, medium, and high magnification to observe the entire nebula and its individual parts in detail. You'll probably find new details every time you look, and, like us, you'll probably find yourself returning again and again to M 42 any time it's visible.

FIGURE 34-2.

NGC 1976 (M 42) and NGC 1982 (M 43) (60' field width)

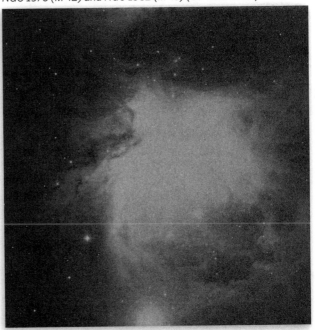

Image reproduced from Digitized Sky Survey courtesy Anglo-Australian Observatory and Space Telescope Science Institute

NGC 2024	★★	◦◦◦◦	EN	MBUDR
Chart 34-4	Figure 34-3	m99.9, 30.0' x 22.0'	05h 41.7m	-01° 48'

NGC 2024, sometimes called the Tank Track Nebula, is a bright emission nebula that lies just E of m1.9 Alnitak, the E star in the Belt of Orion. Unfortunately, the glare from Alnitak makes it very difficult to observe detail in NGC 2024. With a 50mm binocular, NGC 2024 is just visible as a faint NE elongation of the glow around Alnitak, but teasing out any detail requires a telescope of moderately large aperture and a narrowband or O-III filter, preferably the latter.

At 90X in our 10" reflector with a narrowband filter, NGC 2024 is visible as a 20' patch of nebulosity, elongated slightly E-W, and with a hint of the N-S dark lane dividing the nebula into about equal halves. At 125X, with an O-III filter and with Alnitak outside the field of view, the dark lane becomes wider and more pronounced, and the nebula is visible as three distinct segments. The N segment is the smallest and dimmest. The E segment is the largest of the three, and the W segment, nearest Alnitak, is slightly smaller and brighter. Averted vision reveals very faint tendrils linking the N segment to the two other segments.

FIGURE 34-3.

NGC 2024 (60' field width)

Image reproduced from Digitized Sky Survey courtesy Anglo-Australian Observatory and Space Telescope Science Institute

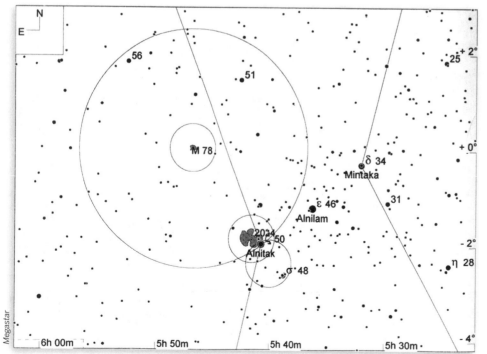

CHART 34-4.

NGC 2024 and NGC 2068 (M 78) (10° field width; 5° finder circle; 1° eyepiece circles; LM 9.0)

ILLUSTRATED GUIDE TO ASTRONOMICAL WONDERS

M 78 (NGC 2068)	★★★	๑๑๑	RN	MBUDR
Chart 34-4	Figure 34-4	m8.3, 8.4' x 7.8'	05h 46.8m	+00° 04'

NGC 2068, better known as Messier 78 or M 78, is a large, bright reflection nebula. Pierre Méchain discovered this object in early 1780. He reported his discovery to Charles Messier, who observed M 78 on the night of 17 December 1780 and added it to his catalog.

M 78 is easy to find, lying 2.5° NE of 50-ζ (zeta) Orionis (Alnitak). Place Alnitak on the SW edge of your binocular or finder field and look for M 78 near the center of the field, where it is visible with averted vision as a small, moderately faint patch of haze.

The Astronomical League Binocular Messier club list rates M 78 as an easy object with an 80mm binocular and a Tougher object (the middle category) with a 35mm or 50mm binocular. With our 50mm binoculars, M 78 appears as a moderately faint, small hazy patch. At 90X with our 10" reflector, M 78 is a bright patch shaped like a clamshell, brightest at the N edge and fading gradually to the S into diffuse nebulosity. Two m10 stars embedded in the body of the nebula illuminate it, and resemble the headlights of an oncoming car.

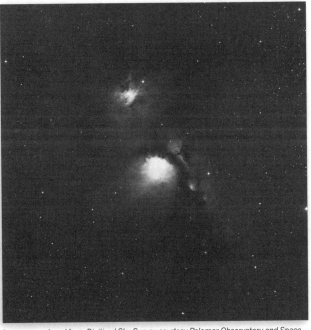

FIGURE 34-4.

NGC 2068 (M 78) (60' field width)

Image reproduced from Digitized Sky Survey courtesy Palomar Observatory and Space Telescope Science Institute

NGC 2022	★	๑๑	PN	MBUDR
Chart 34-5, 34-6	Figure 34-5	m12.4, 35.0"	05h 42.1m	+09° 05'

NGC 2022 is a tiny, very faint planetary nebula. It lies directly on the line from m0.6 58-α (alpha) Orionis (Betelgeuse) to m3.6 39-λ (lambda) Orionis (Meissa), and 1.9° ESE of Meissa, so it's very easy to find its general location. Verifying the identity of the object is another matter.

At low or medium magnification, NGC 2022 looks like just another dim star. It's usually possible to confirm planetary nebulae by "blinking" them with a narrowband or O-III filter, but this planetary is so faint that it disappears entirely with a narrowband filter at low or medium magnification. We locate NGC 2022 by placing our finder as shown in Chart 34-6 and looking for the two m8 stars visible in that chart in the NW quadrant of the eyepiece field. With those two relatively prominent stars, NGC 2022 forms an equally-spaced 22' ESE-WNW (from east southeast to west northwest) line, with NGC 2022 at the ESE end.

At 42X or 90X with our 10" Dob, NGC 2022 appears stellar. With averted vision at 180X, NGC 2022 is clearly an extended object. It appears as a tiny, very faint, grayish disc with no structure visible.

FIGURE 34-5.

NGC 2022 (60' field width)

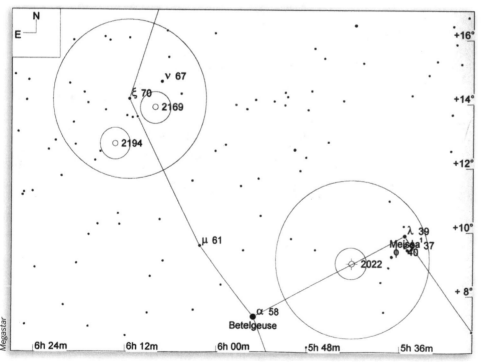

CHART 34-5.

NGC 2022, NGC 2169, and NGC 2194 (15° field width; 5° finder circles; 1° eyepiece circles; LM 7.0)

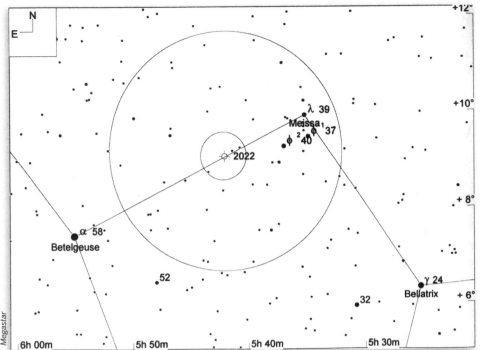

Megastar

NGC 2169	★★★	✿✿✿	OC	MBUDR
Chart 34-5, 34-7	Figure 34-6	m5.9, 6.0'	06h 08.4m	+13° 58'

NGC 2169 is a small, sparse, bright open cluster that is sometimes called the 37 Cluster. (To see why, look at Figure 34-6 and tilt your head slightly to the left. The SE group of bright stars forms the 3 and the NW group the 7.)

NGC 2169 is easy to find. To locate it, put Betelgeuse in your finder or binocular field and look for m4.1 61-μ (mu) Orionis 3° NE. Put 61-μ on the SW edge of the finder or binocular field and move the field slightly N until the bright pair m4.4 67-ν (nu) and m4.5 70-ξ (xi) come into view, about 5° NNE (north northeast) of 61-μ. NGC 2169 forms the S apex of an equilateral triangle with 67-ν and 70-ξ, and is visible with direct vision in a 50mm binocular or finder.

With our 50mm binocular, NGC 2169 is a loose scattering of stars over a 5' extent. Seven m7 through m9 stars are visible with direct vision, with another 3 or 4 stars visible with averted vision. No nebulosity is evident. At 90X in our 10" reflector, the 37 pattern is prominent, although inverted. Six m7 through m11 stars make up the "7" on the NW side of the cluster and eight m7 through m11 stars make up the "3" on the SE side. Another half dozen m10 and dimmer stars are scattered throughout the cluster.

FIGURE 34-6.

NGC 2169 (60' field width)

Image reproduced from Digitized Sky Survey courtesy Palomar Observatory and Space Telescope Science Institute

CHART 34-7.

*NGC 2169 and NGC 2194 (10°
field width; 5° finder circle; 1°
eyepiece circles; LM 9.0)*

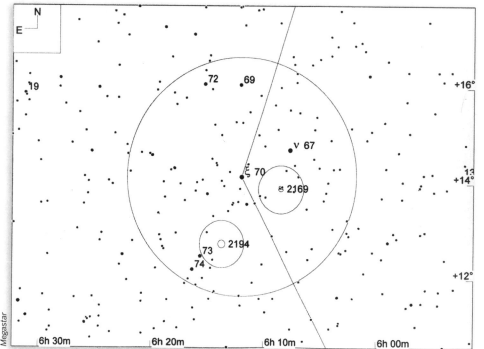

NGC 2194	★	♦♦♦	OC	MBUDR
Chart 34-5, 34-7	Figure 34-7	m8.5, 10.0'	06h 13.8m	+12° 48'

NGC 2194 is another open cluster that lies just 1.7° SE of NGC 2169. NGC 2194 is a bit larger than NGC 2169 and much richer and more concentrated, but it is also much dimmer.

To locate NGC 2194, center 70-ξ (xi) in your finder and look 2° SSE (south southeast) for the prominent m5 pair 73- and 74-Orionis. NGC 2194 lies 33' WNW of 73-Ori (the dimmer of that pair of stars). NGC 2194 is invisible in a 50mm finder, but easily visible with a low-power eyepiece. At 42X in our 10" Dob, NGC 2194 is a scattering of m10 and dimmer stars embedded in a very faint 5' patch of nebulosity. At 125X, about two dozen stars are visible from m10 down to m14.

FIGURE 34-7.

NGC 2194 (60' field width)

Image reproduced from Digitized Sky Survey courtesy Palomar Observatory and Space Telescope Science Institute

NGC 1662	★	✶✶✶	OC	MBUDR
Chart 34-8	Figure 34-8	m6.4, 20.0'	04h 48.5m	+10° 56'

NGC 1662 is a large, bright, but very sparse open cluster that is one of the objects required to complete the Astronomical League Deep-Sky Binocular club list. To locate NGC 1662, look 9° W of Bellatrix for the center of the 9° N-S chain of m3 through m5 stars π^1-Orionis through π^6-Orionis. By far the brightest of that chain is the prominent naked-eye star m3.2 1-π^3 (pi^3), which lies 8.8° dead W of Bellatrix. Place π^3 on the S edge of your binocular or finder field, and look for the bright stars m4.4 2-π^2 and m4.7 7-π^1 near the center of the field. NGC 1662 is the NW apex of the isosceles triangle it forms with 2-π^2 and 7-π^1.

With our 50mm binoculars, NGC 1662 is a moderately bright hazy patch with four or five m8/9 stars embedded. At 90X with our 10" Dob, about two dozen stars from m8 through m13 are visible, mostly scattered throughout the cluster, but with a 3' clump of half a dozen of the brightest stars clustered near the center.

FIGURE 34-8.

NGC 1662 (60' field width)

Image reproduced from Digitized Sky Survey courtesy Palomar Observatory and Space Telescope Science Institute

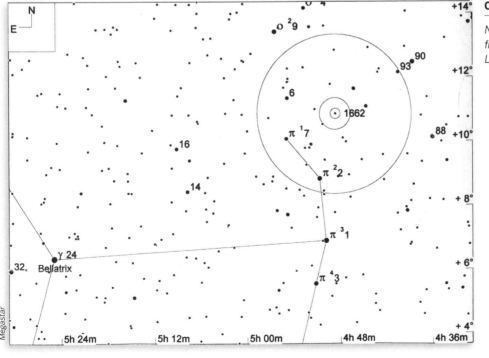

CHART 34-8.

NGC 1662 (15° field width; 5° finder circle; 1° eyepiece circle; LM 9.0)

NGC 1788	★★	✺✺✺	RN	MBUDR
Chart 34-9	Figure 34-9	m99.9, 5.5' x 3.0'	05h 06.9m	-03° 20'

NGC 1788 is a small, moderately bright reflection nebula. NGC 1788 is relatively easy to find. Look 3.5° NNW of Rigel to identify the bright naked-eye star m2.8 67-β (beta) Eridani (Cursa). Place Cursa near the S edge of your finder field, and look for a distorted 1° rectangle of four m6/m7 stars that appears very prominent near the center of the field. NGC 1788 lies just SE of center of that rectangle, where it is visible with averted vision in a low-power eyepiece as a small, moderately bright nebulous patch.

At 125X in our 10" reflector, NGC 1788 is a moderately bright 2' nebulous patch, elongated slightly E-W, and lying centered within a distorted box of three m9 and one m10 stars, with an m7 star lying SW of the SW corner of the box. The N edge of the nebula is the brightest, illuminated by a E-W pair of m10 stars that are embedded in the nebulosity. Averted vision reveals additional nebulosity extending NE about halfway to the NE m10 star in the box.

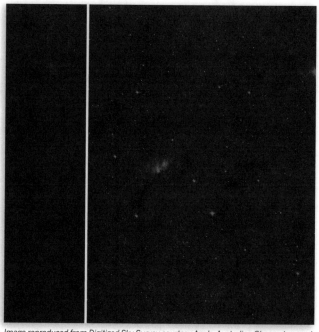

FIGURE 34-9.

NGC 1788 (60' field width)

Image reproduced from Digitized Sky Survey courtesy Anglo-Australian Observatory and Space Telescope Science Institute

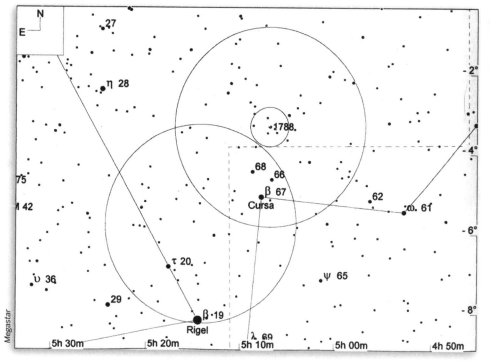

CHART 34-9.

NGC 1788 (12° field width; 5° finder circles; 1° eyepiece circle; LM 9.0)

Multiple Stars

19-beta (STF 668A-BC)	★★★	❄❄❄❄		MS	UD
Chart 34-1		m0.3/6.8, 6.8", PA 204° (2004)		05h 14.5m	-08° 12'

19-β (beta) Orionis (Rigel) is a pretty double star. At 125X in our 10" Dob, Rigel appears a blazing cool white with a slight bluish tinge. Its much dimmer companion is a pale, pretty blue color.

34-delta (STFA 14Aa-C)	★★★	❄❄❄❄		MS	UD
Chart 34-1, 34-2		m2.4/6.8, 52.8", PA 0° (2003)		05h 32.0m	-00° 17'

34-δ (delta) Orionis (Mintaka), is the NW star in Orion's belt. At 125X in our 10" Dob, Mintaka appears blue-white with a pale blue companion.

44-iota (STF 752AB)	★★★	❄❄❄❄		MS	UD
Chart 34-2, 34-3		m2.9/7.0, 11.3", PA 141° (2002)		05h 35.4m	-05° 55'

SAO 132301 (STF 747AB)	★★★	❄❄❄❄		MS	UD
Chart 34-2, 34-3 (not shown)		m4.7/5.5, 36", PA 224° (2003)		05h 35.0m	-06° 00'

44-ι (iota) Orionis is the S star in Orion's sword. At 125X in our 10" reflector, the primary is a sparkling pure white, and the much dimmer companion a subtle blue-white.

Just 8' SW of 44-ι and visible in the same medium or high-magnification eyepiece field is STF 747, which is actually a triple star. The pair required for the Astronomical League Double Star club is STF 747AB, which comprises the m4.7 star SAO 132301 and the m5.5 star SAO 132298 36" to the SW. A third m6.4 component lies about a third of the way between the primary and secondary. At 125X in our 10" reflector, all three stars are blue-white.

39-lambda (STF 738AB)	★★	❄❄❄❄		MS	UD
Chart 34-1		m3.5/5.5, 4.3", PA 44° (2003)		05h 35.1m	+09° 56'

39-λ (lambda) Orionis, also known as Meissa, is the head of Orion. As double stars go, it's pretty pedestrian. At 125X in our 10" reflector, the primary and its close companion are both pure white.

41-theta1 (STF 748Aa-B)	★★★★	✹✹✹✹		MS	UD
Chart 34-3		m6.6/7.5, 8.8", PA 31° (2004)		05h 35.3m	-05° 23'

41-theta1 (STF 748Aa-C)	★★★★	✹✹✹✹		MS	UD
Chart 34-3		m6.6/5.1, 12.7", PA 132° (2002)		05h 35.3m	-05° 23'

41-theta1 (STF 748Aa-D)	★★★★	✹✹✹✹		MS	UD
Chart 34-3		m6.6/6.4, 21.2", PA 96° (2002)		05h 35.3m	-05° 23'

41-theta2 (STFA 16AB)	★★★★	✹✹✹✹		MS	UD
Chart 34-3		m5.0/6.2, 52.2", PA 93° (2002)		05h 35.4m	-05° 24'

41-θ^1 (theta^1) Orionis, better known as the Trapezium, is the most spectacular multiple star system visible to us. The Trapezium is unique among multiple star systems in having its components in order of right ascension rather than in order of decreasing magnitude. Component A is m6.6, B 7.5, C 5.1, and D 6.4. These four brightest components form a trapezoid pattern, whence the name. At 125X in our 10" reflector, in decreasing order of brightness, the C star is white, the D star and A star are warm white, and the B star cool white.

41-θ^2 (theta^2) Orionis lies just 2' SE of the Trapezium, and is visible in the same high-power eyepiece field. The primary is cool white and the secondary white.

48-sigma (STF 762AB-E)	★★★	✹✹✹✹		MS	UD
Chart 34-2, 34-4		m3.8/6.3, 41.5", PA 62° (2003)		05h 38.7m	-02° 36'

48-sigma (STF 762AB-D)	★★★	✹✹✹✹		MS	UD
Chart 34-2, 34-4		m3.8/6.6, 12.7", PA 84° (2002)		05h 38.7m	-02° 36'

48-σ (sigma) Orionis is easily located 50' SW of Alnitak, the E star in Orion's belt. At 125X in our 10" reflector, the primary is a brilliant pure white and both companions are blue-white.

50-zeta (STF 774Aa-B)	★★★	ჿჿჿჿ		MS	UD
Chart 34-2, 34-4		m1.9/3.7, 2.2", PA 165° (2006)		05h 40.7m	-01° 57'

50-zeta (STF 774Aa-C)	★★★	ჿჿჿჿ		MS	UD
Chart 34-2, 34-4		m1.9/9.6, 57.3", PA 10° (2003)		05h 40.8m	-01° 56'

50-ζ (zeta) Orionis (Alnitak) is the E star in Orion's belt. At At 125X in our 10" reflector, the primary is a brilliant pure white. The m3.7 secondary, which is easy to lose in the glare of Alnitak, appears cool white by comparison, as does the much dimmer tertiary.

35

Pegasus, The Winged Horse

NAME: Pegasus (PEG-uh-sus)

SEASON: Summer

CULMINATION: midnight, 1 September

ABBREVIATION: Peg

GENITIVE: Pegasi (PEG-uh-see)

NEIGHBORS: And, Aqr, Cyg, Del, Equ, Lac, Psc, Vul

BINOCULAR OBJECTS: NGC 7078 (M 15)

URBAN OBJECTS: NGC 7078 (M 15)

Pegasus is a large north equatorial constellation that ranks 7th in size among the 88 constellations. Pegasus covers 1,121 square degrees of the celestial sphere, or about 2.7%. Although Pegasus is not particularly bright—it has only three second magnitude stars and three of the third magnitude—it is nonetheless prominent. Three of the four brightest stars in Pegasus, along with second magnitude Alpheratz in Andromeda, form a large, very prominent keystone-shaped asterism called the Great Square of Pegasus.

Pegasus is an ancient constellation, one of Ptolemy's original 44. In Greek mythology, Pegasus is the Winged Horse, and is the basis of the Mares of Diomedes, one of the Twelve Labors of Herakles.

Despite its size and prominence, Pegasus is relatively devoid of objects of interest to amateur astronomers. It can claim M 15, a very nice Messier globular cluster, and the bright galaxy NGC 7331. Otherwise, there's not much more in Pegasus to hold an amateur astronomer's interest.

Pegasus culminates at midnight on 1 September, and, for observers at mid-northern latitudes, is best placed for evening viewing from mid-summer through late autumn.

TABLE 35-1.

Featured star clusters, nebulae, and galaxies in Pegasus

Object	Type	Mv	Size	RA	Dec	M	B	U	D	R	Notes
NGC 7078	GC	6.3	18.0	21 30.0	+12 10	◉	◉	◉			M 15; Class IV
NGC 7331	Gx	9.4	14.5 x 3.7	22 37.1	+34 25					◉	Class SA(s)b

TABLE 35-2.

Featured multiple stars in Pegasus

Object	Pair	M1	M2	Sep	PA	Year	RA	Dec	UO	DS	Notes
8-epsilon	S 798AC	2.5	8.7	144.0	318	2000	21 44.2	+09 52		◉	Enif

CHART 35-1.

The constellation Pegasus (field width 50°)

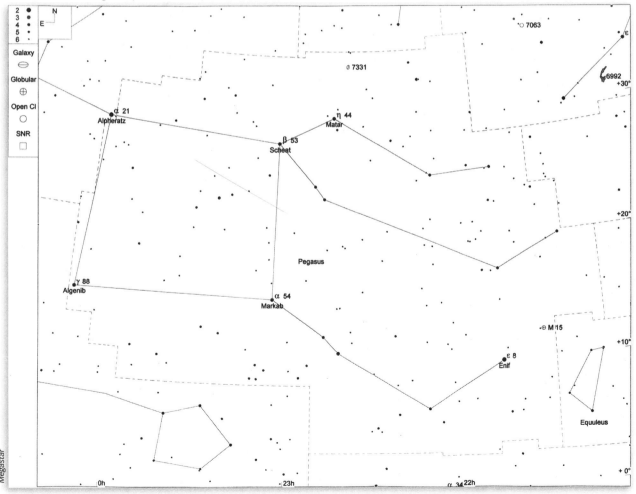

Clusters, Nebulae, and Galaxies

M 15 (NGC 7078)	★★★	✦✦✦✦	GC	MBUDR
Chart 35-2	Figure 35-1	m6.3, 18.0'	21h 30.0m	+12° 10'

NGC 7078, better known as Messier 15 or M 15, is a big, bright, beautiful globular cluster. Giovanni-Dominique Maraldi discovered this object on 7 September 1746 while he was scanning this area of the sky to locate a previously discovered comet. Charles Messier observed and cataloged this object on the night of 3 June 1764.

M 15 is very easy to find, lying 4.2° NW of m2.4 Enif, the brightest star in Pegasus. To locate M 15, put Enif on the SE edge of your binocular or finder field and look near the NE edge of the field. M 15 appears as a prominent patch of haze in even a 30mm binocular or finder.

With our 50mm binoculars, M 15 is a bright 10' patch of haze with no stars resolved. An m6 star lies 15' dead E, an m7 star the same distance SW, and an m7 star 8' NNE (north northeast). At 125X in our 10" reflector, M 15 is a beautiful globular. The dense, bright 5' core is unresolved, but is surrounded by a large halo of dim stars arranged in numerous streams, arcs, and chains. At 180X the core takes on a more granular appearance and allows a few mostly m13 stars to be resolved with difficulty using averted vision.

FIGURE 35-1.

NGC 7078 (M 15) (60' field width)

CHART 35-2.

NGC 7078 (M 15) (10° field width; 5° finder circle; 1° eyepiece circle; LM 9.0)

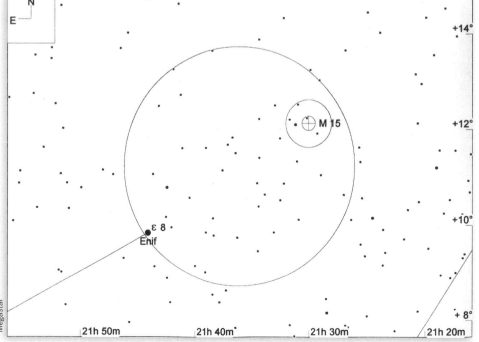

NGC 7331	★★	✺✺✺	GX	MBUDR
Chart 35-3	Figure 35-2	m9.4, 14.5' x 3.7'	22h 37.1m	+34° 25'

NGC 7331 is a medium-size bright galaxy that lies near the boundary of Pegasus with Lacerta. To locate NGC 7331, start at m2.5 Scheat (the NW corner of the Great Square). Look 5° WNW (west northwest) from Scheat for m2.9 44-η (eta) Pegasi (Matar). Put Matar on the SSE edge of the finder field and look near the E edge of the field for m5.6 38-Pegasi, which appears prominent in the finder. Put 38-Pegasi on the SW edge of the finder field, and NGC 7331 should be nearly centered in the finder, where it's readily visible with a low-power eyepiece.

At 125X in our 10" reflector, NGC 7331 shows a bright, condensed core elongated 0.75' x 3' NNW-SSE (from north northwest to south southeast) with a stellar nucleus in a moderately bright 2' x 6' halo.

FIGURE 35-2.

NGC 7331 (60' field width)

Image reproduced from Digitized Sky Survey courtesy Palomar Observatory and Space Telescope Science Institute

CHART 35-3.

NGC 7331 (15° field width; 5° finder circles; 1° eyepiece circle; LM 9.0)

Multiple Stars

8-epsilon (S 798AC)	★	✦✦✦✦	MS	UD
Chart 35-1		m2.5/8.7, 144.0", PA 318° (2000)	21h 44.2m	+09° 52'

About the best we can say about the double star 8-ε (epsilon) Pegasi (Enif) is that it's easy to find. From Markab (the SW corner of the Great Square), look 20° WSW for a prominent naked-eye star. That's m2.5 Enif, the brightest star in Pegasus, and it's a remarkably unimpressive double star. At 125X in our 10" reflector, the primary is a blazing warm white, with the much dimmer m9 white companion lying about 2.5' NW.

36

Perseus, The Hero

NAME: Perseus (PERS-ee-us)

SEASON: Autumn

CULMINATION: midnight, 7 November

ABBREVIATION: Per

GENITIVE: Persei (PERS-ee-ee)

NEIGHBORS: Aql, Her, Lib, Sco, Ser, Sgr

BINOCULAR OBJECTS: And, Ari, Aur, Cam, Cas, Tau, Tri

URBAN OBJECTS: NGC 869/884 (Double Cluster), Tr 2, NGC 1039 (M 34), Mel 20, NGC 1342

Perseus is a mid-size northerly constellation that ranks 24th in size among the 88 constellations. Perseus covers 615 square degrees of the celestial sphere, or about 1.5%. Perseus is a moderately bright constellation, containing eight stars brighter than fourth magnitude. Perseus is easily located, lying among the prominent constellations Auriga to the east, Andromeda to the west, Cassiopeia to the northwest, and Taurus to the south. Perseus is home to many fine DSOs, including several fine open clusters—the magnificent Double Cluster and Messier 34 among them—as well as the beautiful planetary nebula Messier 76 and, surprisingly, a bright galaxy and an emission nebula.

In Greek mythology, the Hero Perseus is best known for slaying the Gorgon Medusa, whose face was so ugly that merely looking upon it turned her victims to stone. Equipped with a pair of magic flying shoes provided by the god Hermes and a sword and mirror-like shield provided by the goddess Athena, Perseus set out to slay

TABLE 36-1.

Featured star clusters, nebulae, and galaxies in Perseus

Object	Type	Mv	Size	RA	Dec	M	B	U	D	R	Notes
NGC 650	PN	12.2	167.0"	01 42.3	+51 35	◉					M 76; Class 3+6
NGC 869	OC	5.3	29.0	02 19.0	+57 08			◉	◉	◉	Double Cluster; Cr 24; Mel 13; Class I 3 r
NGC 884	OC	6.1	29.0	02 22.3	+57 08			◉	◉	◉	Double Cluster; Cr 25; Mel 14; Class I 3 r
Cr 29	OC	5.9	20.0	02 36.8	+55 55			◉	◉		Tr 2; Class II 2 p
NGC 1023	Gx	10.4	8.7 x 2.3	02 40.4	+39 04					◉	Class SB(rs)0-; SB ???
NGC 1039	OC	5.2	35.0	02 42.1	+42 45	◉	◉				M 34; Cr 31; Class II 3 r
Cr 39	OC	2.3	184.0	03 24.3	+49 52			◉	◉		Alpha Perseii Association; Mel 20; Class III 3 m
NGC 1342	OC	6.7	14.0	03 31.6	+37 23			◉	◉		Cr 40; Mel 21; Class III 2 m
NGC 1491	EN	99.9	21.0	04 03.6	+51 18				◉		
NGC 1528	OC	6.4	23.0	04 15.3	+51 13				◉		Cr 47; Mel 23; Class II 2 m
NGC 1582	OC	7.0	37.0	04 31.7	+43 45				◉		Cr 51; Class IV 2 p

TABLE 36-2.

Featured multiple stars in Perseus

Object	Pair	M1	M2	Sep	PA	Year	RA	Dec	UO	DS	Notes
26-beta	n/a	n/a	n/a	n/a	n/a	n/a	03 08.2	+40 57	◉		Algol – variable, not multiple
15-eta	STF 307AB	3.8	8.5	28.5	301	2002	02 50.7	+55 53		◉	
SAO 23763	STF 331	5.2	6.2	11.9	85	2002	03 00.9	+52 21		◉	

CHART 36-1.

The constellation Perseus (field width 40°)

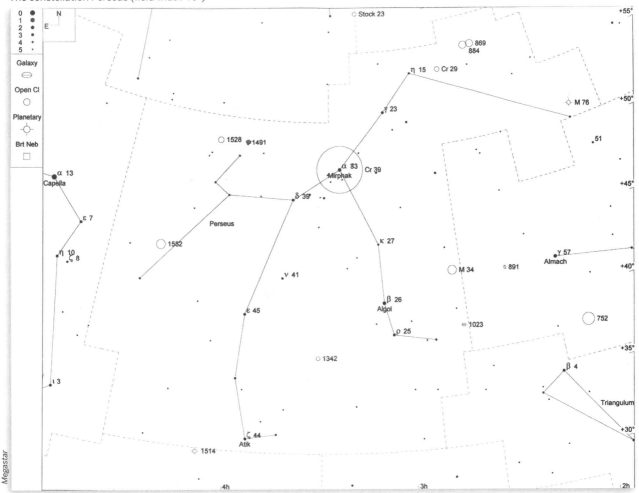

Medusa. Turnabout is fair play, so Perseus used his shield to force Medusa to look upon her own image, thereby turning herself to stone. After cutting off Medusa's head, Perseus set sail for home.

On his journey, Perseus happened to sail past the kingdom of Aethiopia, ruled by King Cepheus and Queen Cassiopeia. The vain Cassiopeia's boasts about her own beauty angered the Nereids, who appealed to Poseidon to punish Cassiopeia. Poseidon sent the sea monster Cetus to wreak havoc upon the kingdom. Cepheus consulted the Oracle at Ammon, who told him that the only way to save his country and his people was to sacrifice his daughter Andromeda to Cetus. Cepheus chained Andromeda to a giant rock near the shore, where she awaited her doom as Cetus approached.

As Perseus sailed past, he noticed Andromeda chained screaming to the rock, with Cetus approaching fast. Perseus whipped out the head of Medusa and showed it to Cetus, who was promptly petrified. Perseus freed Andromeda, and the two of them sailed off together and lived happily ever after.

Perseus culminates at midnight on 7 November, and, for observers at mid-northern latitudes, is best placed for evening viewing from late summer through late winter.

Clusters, Nebulae, and Galaxies

M 76 (NGC 650)	★★★	☺☺☺	PN	MBUDR
Chart 36-2, 36-3	Figure 36-1	m12.2, 167.0"	01h 42.3m	+51° 35'

NGC 650, better known as Messier 76 or M 76, is a large, bright planetary nebula, one of the finest planetaries in the sky. Pierre Méchain discovered M 76 on 5 September 1780, and reported his discovery to Charles Messier. Messier observed M 76 on the night of 21 October 1780 and added it to his catalog.

Although it's possible to locate M 76 by using a long star hop, it's much easier to locate it by starting with the bright stars m2.2 Almach in Andromeda and m2.3 Shedir in Cassiopeia, as shown in Chart 36-2. Draw an imaginary 19.5° line between those two stars and look along that line about 40% of the way from Almach to Shedir. You'll see the m3.6 star 51-Andromedae, which is by far the brightest star in the vicinity. Put 51-Andromedae on the S edge of your finder field and look for the m4.0 star φ (phi) Perseii, which is by far the brightest star in the finder field. About 50' dead N of φ you'll see an m6.7 star which is also quite prominent. M 76 lies 12' W of that star (which is visible to the left of M 76 in Figure 36-1), and is easily visible in a low-power eyepiece.

M 76 is often called the Little Dumbbell Nebula for its reputed resemblance to the Dumbbell Nebula, M 27 in Vulpecula. To us, though, it looks more like an elongated rectangle at low magnification and like a peanut or bowtie or figure-eight at medium and high magnification. At 125X in our 10" Dob, M 76 shows two distinct nodes. The SW node is noticeably brighter and has a sharply demarcated border. The dimmer NE node gradually fades into extremely faint tendrils of nebulosity. A narrowband or O-III filter noticeably improves both the contrast and the visible extent of this object.

FIGURE 36-1.

NGC 650 (M 76) (60' field width)

Image reproduced from Digitized Sky Survey courtesy Palomar Observatory and Space Telescope Science Institute

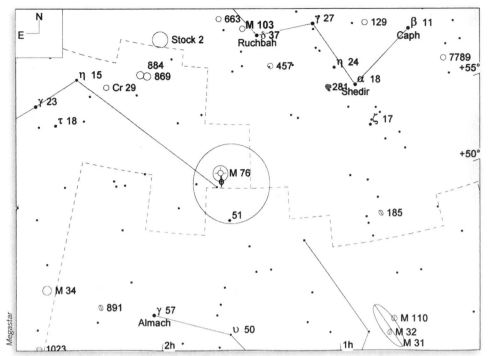

CHART 36-2.

NGC 650 (M 76) (30° field width; 5° finder circle; 1° eyepiece circle; LM 6.0)

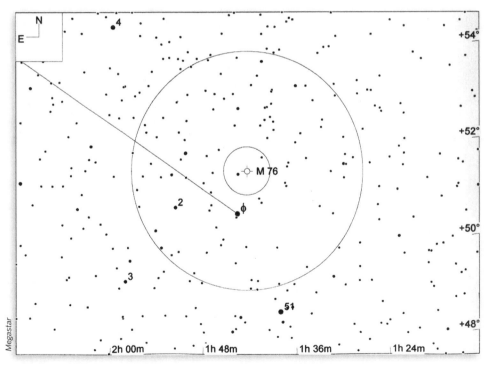

CHART 36-3.

NGC 650 (M 76) detail (10° field width; 5° finder circle; 1° eyepiece circle; LM 9.0)

NGC 869	★★★	◐◐◐	OC	MBUDR
Chart 36-4	Figure 36-2	m5.3, 29.0'	02h 19.0m	+57° 08'

NGC 884	★★★	◐◐◐	OC	MBUDR
Chart 36-4	Figure 36-2	m6.1, 29.0'	02h 22.3m	+57° 08'

NGC 869 and NGC 884 are two large, rich, bright open clusters whose centers lie only about half a degree apart. Because it is visible to the naked eye under good conditions, the Double Cluster has been known since at least Babylonian times. No one knows why Charles Messier didn't add the Double Cluster, individually or together, to his catalog. As we recounted in the Orion chapter, Messier found himself facing a publication deadline with only 41 objects in his catalog. He wanted more, so he added the naked-eye objects M 42/43, M 44, and M 45 to his catalog. Perhaps he was embarrassed at padding out his catalog with these easy objects, or perhaps he was just in such a hurry that he forgot to add the Double Cluster. Either way, the Double Cluster didn't make it into Messier's catalog.

The Double Cluster is easy to find, lying about 4.3° WNW of the naked-eye star m3.8 15-η (eta) Perseii on the N edge of the constellation. Put 15-η on the E edge of your finder or binocular field, and you'll see the Double Cluster near the WNW edge of the field.

With our 50mm binoculars, the Double Cluster is a beautiful sight. Dozens of stars from m6 down to m10 are visible in both clusters, with scores of dimmer stars in each cluster providing a nebulous background glow. At 42X in our 10" Dob (1.6° true field), 100+ stars are visible in each cluster. NGC 869 appears tighter, with a considerable concentration of bright stars near the center and many other stars scattered in knots and chains to the N and S of the central core. The bright stars in NGC 884 are more scattered, with an elongated 3' x 8' WNW-ESE (from west northwest to east southeast) central concentration and a second elongated 4' x 12' E-W concentration lying to the N.

FIGURE 36-2.

NGC 869 (right) and NGC 884 (Double Cluster) (60' field width)

Image reproduced from Digitized Sky Survey courtesy Palomar Observatory and Space Telescope Science Institute

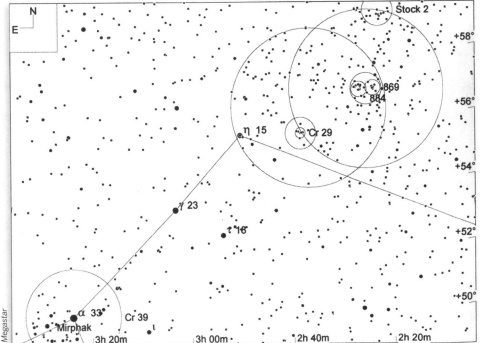

CHART 36-4.

NGC 869 and NGC 884 (The Double Cluster) and Collinder 29 (Cr 29) (15° field width; 5° finder circles; 1° eyepiece circle; LM 9.0)

Cr 29	★★	✦✦✦	OC	MBUDR
Chart 36-4	Figure 36-3	m5.9, 20.0'	02h 36.8m	+55° 55'

Collinder 29 is a large, bright open cluster that suffers from its close proximity to the Double Cluster. It's so close, in fact, that all three of these clusters are visible in the same low-power eyepiece field of our 4.5" Orion Starblast richest-field telescope, let alone any binocular or finder. The Double Cluster is so compelling that we've found ourselves failing to notice Collinder 29 even when it's visible in the same field.

Collinder 29 is easy to find, lying just 1.9° dead W of the bright naked-eye star m3.8 15-η (eta) Perseii, about a third of the way along and just S of the line from 15-η to the Double Cluster. With our 50mm binoculars, Cr 29 shows half a dozen m7 through m9 stars embedded in a soft background nebulosity. At 90X in our 10" reflector, about two dozen stars from m7 through m12 are visible. The cluster presents a very loose, scattered, poor appearance.

FIGURE 36-3.

Collinder 29 (60' field width)

Image reproduced from Digitized Sky Survey courtesy Palomar Observatory and Space Telescope Science Institute

NGC 1023	★★	❀❀❀	GX	MBUDR
Chart 36-5	Figure 36-4	m10.4, 8.7' x 2.3'	02h 40.4m	+39° 04'

NGC 1023 is a relatively bright galaxy that lies near the border of Perseus with Andromeda. To locate NGC 1023, begin by identifying the bright star 26-β (beta) Perseii (Algol). Algol is a variable star that ranges in brightness from m2.1 at maximum to m3.4 at minimum, but it's always easily visible to the naked eye. With your finder on Algol, follow the chain of bright stars, 2.2° SSW to m3.4 25-ϱ (rho) and then 2.9° W to m4.2 16-Per. With 16-Per on the E edge of your viewfinder, look for the m4.9 star 12-Per 2.5° NW. NGC 1023 lies 1.2° SSW (south southwest) of 12-Per, where it is visible in a low-power eyepiece.

At 125X in our 10" Newtonian, NGC 1023 presents a bright 1.5' x 5' lenticular halo elongated E-W (from east to west) with an embedded concentrated oval core that surrounds a stellar nucleus.

FIGURE 36-4.

NGC 1023 (60' field width)

Image reproduced from Digitized Sky Survey courtesy Palomar Observatory and Space Telescope Science Institute

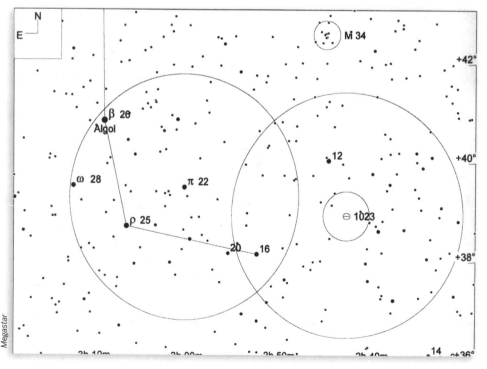

CHART 36-5.

NGC 1023 (10° field width; 5° finder circles; 1° eyepiece circle; LM 9.0)

M 34 (NGC 1039)	★★★	⊛⊛⊛	OC	MBUDR
Chart 36-6	Figure 36-5	m5.2, 35.0'	02h 42.1m	+42° 45'

NGC 1039, better known as Messier 34 or M 34, is a large, very bright open cluster. M 34 has probably been known since ancient times, because under good conditions it's visible to the naked eye as a faint nebulosity. Charles Messier independently re-discovered and cataloged M 34 on the night of 25 August 1764.

M 34 is very easy to find, lying about 40% of the way along and just N of a line from the bright naked-eye star 26-β (beta) Perseii (Algol) to the even brighter star 57-γ (gamma) Andromedae (Almach). With our 50mm binoculars, M 34 shows about 20 m8 and m9 stars embedded in a nebulous glow. At 90X in our 10" reflector, 50+ stars are visible from m8 to m12, scattered across an area the size of the full moon, with many of the brightest stars concentrated in the central half of the extent. Despite this moderate concentration, the cluster gives the impression of being very loose and scattered because many of its stars are clumped into small groups and chains.

FIGURE 36-5.

NGC 1039 (M 34) (60' field width)

Image reproduced from Digitized Sky Survey courtesy Palomar Observatory and Space Telescope Science Institute

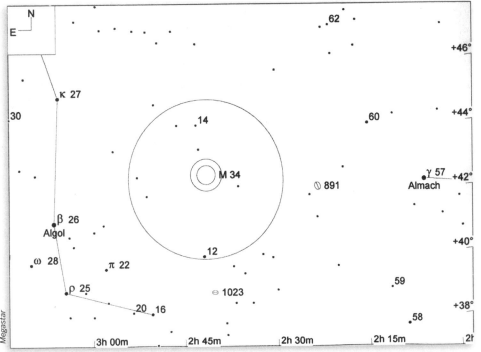

CHART 36-6.

NGC 1039 (M 34) (15° field width; 5° finder circle; 1° eyepiece circle; LM 7.0)

Cr 39	★★	⬦⬦⬦⬦	OC	MBUDR
Chart 36-1		m2.3, 184.0'	03h 24.3m	+49° 52'

Collinder 39, also known as Melotte 20 or the Alpha Perseii Association, is a gigantic, bright open cluster that centers on 33-α (alpha) Perseii (Mirphak). We didn't include an image of this cluster because we couldn't find one with a wide enough field of view to show the cluster as anything other than just a random collection of field stars. Even in a wide-field binocular, the gigantic size of this cluster—more than 3 full degrees—makes it difficult to take in as a discrete object. Note that the listed visual magnitude of 2.3 is very deceptive. Although you might think that an m2.3 object would be prominent to the naked eye, Cr 39 is anything but. That admittedly large amount of light (which does not include the light of m1.8 Mirphak) is distributed across such a large expanse of sky that the average surface brightness of this object is quite low.

With our 50mm binocular, Cr 39 is a scattering of 50+ stars from m5 down to m10 in small groups and chains across an extent that covers much of the field of view. The boundary of this cluster is not sharply delimited, particularly along the hemisphere of the border from E through S to W, but the density of stars gradually decreases. The border of the N hemisphere is more pronounced, with the star density falling off more rapidly beyond about 1.5° from the cluster's center at Mirphak. With our 4.5" Orion Starblast richest-field telescope (3.5° true field), it's very difficult to see this object as a cluster rather than as just a collection of hundreds of random field stars.

NGC 1342	★★	⬦⬦⬦	OC	MBUDR
Chart 36-7	Figure 36-6	m6.7, 14.0'	03h 31.6m	+37° 23'

NGC 1342 is a bright, medium size, loose, scattered open cluster. It lies far enough from bright stars that locating it by a star hop would take some time. Fortunately, it's very easy to locate NGC 1342 geometrically. To do so, draw an imaginary line between the bright stars 26-β (beta) Perseii (Algol) and m2.8 44-ζ (zeta) Perseii (Atik). NGC 1342 lies not quite halfway along and just N of that line.

With our 50mm binoculars, NGC 1342 is a relatively large, prominent nebulous patch with half a dozen m8/9 stars snaking along a 15' E-W line and another pair of m8/9 stars lying about 8' NE of center and just outside the nebulosity. At 90X in our 10" reflector, 50+ stars m8 and dimmer are scattered loosely across a 15' extent.

FIGURE 36-6.

NGC 1342 (60' field width)

Image reproduced from Digitized Sky Survey courtesy Palomar Observatory and Space Telescope Science Institute

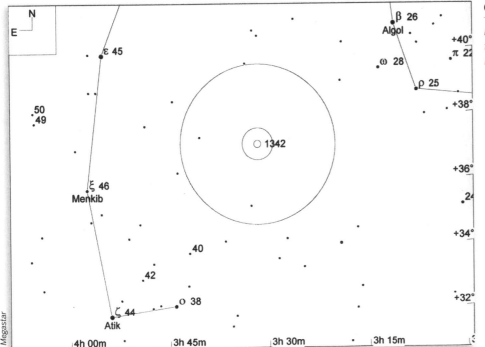

CHART 36-7.

NGC 1342 (15° field width; 5°
finder circle; 1° eyepiece circle;
LM 7.0)

NGC 1491	★★	✦✦✦	EN	MBUDR
Chart 36-8	Figure 36-7	m99.9, 21.0'	04h 03.6m	+51° 18'

NGC 1491 is a small, moderately bright emission nebula. Although its cataloged size is 21', that's a photographic extent; visually, NGC 1491 is much smaller.

It's relatively easy to find NGC 1491. Start by identifying the bright naked-eye star Mirphak. With Mirphak in your finder field, drift the finder about 7° dead E until the prominent group of three m4 stars—47-λ (lambda), 51-μ (mu), and 48-Perseii—come into view. NGC 1491 lies 1.2° NNW of 47-λ. You can judge its exact location by noting the triangle that NGC 1491 makes with 47-λ and m5.3 43-Perseii to the W.

At 125X in our 10" reflector, NGC 1491 is a relatively bright, diffuse 5' triangular nebulosity, elongated NNE-SSW. The SSW apex of the triangle is the brightest, with the nebulosity dimming and becoming more diffused to the NNE. Using our Orion Ultrablock narrowband filter greatly increases the contrast of the nebulosity, and increases its visible extent to 6' or so. With averted vision, the Ultrablock filter reveals extremely faint nebulosity extending 2' E from the SSW apex that is not visible without the filter.

FIGURE 36-7.

NGC 1491 (60' field width)

Image reproduced from Digitized Sky Survey courtesy Palomar Observatory and Space Telescope Science Institute

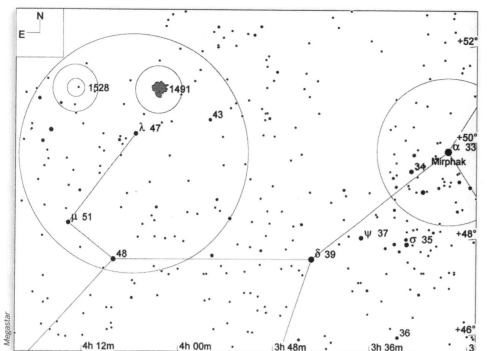

CHART 36-8.

*NGC 1491 and NGC 1528 (10°
field width; 5° finder circle; 1°
eyepiece circles; LM 9.0)*

NGC 1528	★★	✦✦✦	OC	MBUDR
Chart 36-8	Figure 36-8	m6.4, 23.0'	04h 15.3m	+51° 13'

NGC 1528 is a large, bright, relatively concentrated open cluster. It's easy to locate, lying 1.6° NE of the bright star 47-λ (lambda), which can be located as described in the preceding section.

With our 50mm binoculars, NGC 1528 is visible as a moderately bright, moderately large patch of haze without stars about half a degree NW of a prominent m6/7 field star. Averted vision reveals one or two very faint stars, with one or two others flickering in and out on the edge of visibility. At 90X in our 10" reflector, NGC 1528 shows 50+ stars over a 20' extent with considerable central concentration. Several close m10 pairs are visible in the E quadrant of the cluster, and two prominent chains of m9/10 stars extend from the central concentration, one straight W and the other to the SW.

FIGURE 36-8.

NGC 1528 (60' field width)

Image reproduced from Digitized Sky Survey courtesy Palomar Observatory and Space Telescope Science Institute

NGC 1582	★★	⚬⚬⚬	OC	MBUDR
Chart 36-9	Figure 36-9	m7.0, 37.0'	04h 31.7m	+43° 45'

NGC 1528 is a very large, moderately bright, loosely scattered open cluster. Although NGC 1582 lies in Perseus, we find it easiest to locate it by starting from Capella in Auriga. With Capella on the NE edge of your finder or binocular field, look SW for m3.0 7-ε (epsilon) and 10-η (eta). With those stars on the E edge of the field, move the field W until the prominent triangle of m6.2 57-, m4.3 58-, and m5.3 59-Perseii comes into view. NGC 1528 lies about 45' NNW of 57-Perseii.

With our 50mm binoculars, NGC 1582 is visible as a moderately large, fairly faint patch of nebulosity centering on a parallelogram of four m8/9 stars. At 90X in our 10" Dob, about two dozen stars down to m11 are visible, with a broad E-SW (from east to southwest) arc containing most of the brighter members. An arc of five m10/11 stars lies on the NW edge of the cluster, anchored by an m9 star at the W terminus. A E-W band of m9/11 stars extends across the S edge of the cluster.

FIGURE 36-9.

NGC 1582 (60' field width)

Image reproduced from Digitized Sky Survey courtesy Palomar Observatory and Space Telescope Science Institute

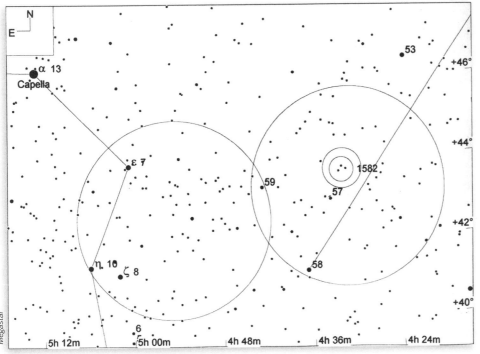

CHART 36-9.

NGC 1582 (12° field width; 5° finder circles; 1° eyepiece circle; LM 9.0)

Multiple Stars

26-beta	★★★	✧✧✧✧		variable star	UD
Chart 36-1		n/a		03h 08.2m	+40° 57'

26-β (beta) Perseii—better known as Algol, the Ghoul Star, or the Demon Star—is not a multiple star in the usual sense, despite the fact that it's included in the Astronomical League's Double Star list. Algol is a variable star, and was almost certainly the first variable star to be discovered. Algol made the AL Double Star list because it's the best-known example of an eclipsing binary variable star—one in which the variance in magnitudes is caused by an invisible companion that orbits Algol, periodically blocking some of its light.

Algol has been known to be variable since antiquity. Its period was first accurately measured and reported in 1782 by the English astronomer John Goodriche, who in an intuitive leap suggested that an invisible companion might be partially eclipsing Algol periodically. The orbital mechanics of stars were not well understood at that time, but as it turns out Goodriche nailed the proper explanation.

During minima, the brightness of Algol falls over a period of about five hours from its usual magnitude 2.1 to magnitude 3.4 as Algol is partially eclipsed by its companion. The eclipse occurs every 2.8674 days, or about 68 hours and 49 minutes. It's often possible to observe the transition from maximum to minimum or vice versa over the course of one observing session, and it's sometimes possible to observe the entire cycle maximum-minimum-maximum over one night. For detailed information about forthcoming times for Algol's cycle, check *Sky & Telescope* or *Astronomy* magazine or their web sites.

15-eta (STF 307AB)	★★★	✧✧✧	MS	UD
Chart 36-1		m3.8/8.5, 28.5", PA 301° (2002)	02h 50.7m	+55° 53'

15-η (eta) Perseii, sometimes called Miram, is one of the best color-contrast double stars other than Albireo. Miram is easy to locate with the naked eye. It lies 7.8° NW of Mirphak and just 3.1° NW of m2.9 23-γ (gamma) Perseii. At 100X in our 90mm refractor, 15-η is a beatiful pair. The primary is golden yellow, and the much dimmer companion is a pretty blue-white.

SAO 23763 (STF 331)	★★	✧✧✧	MS	UD
Chart 36-10		m5.2/6.2, 11.9", PA 85° (2002)	03h 00.9m	+52° 21'

STF 331 comprises the primary, SAO 23763, and its companion, SAO 23765. This double is easy to locate by noting the scalene triangle it forms with the prominent stars 23-γ (gamma) Perseii and 18-τ (tau) Perseii. At 100X in our 90mm refractor, STF 331 is a pretty pair. The secondary is noticeably dimmer than the primary, and both are distinctly blue-white.

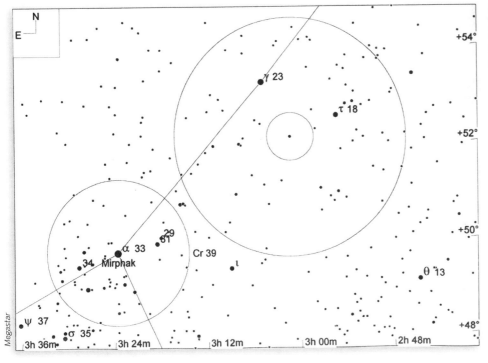

CHART 36-10.

SAO 23763 (STF 331) (10°
field width; 5° finder circle; 1°
eyepiece circle; LM 9.0)

N

E

+54°

γ 23

τ 18

+52°

+50°

29
67
α 33

Cr 39

34 Mirphak

ι

θ 13

+48°

Megastar

ψ 37

σ 35

3h 36m   3h 24m   3h 12m   3h 00m   2h 48m

37

Pisces, The Fishes

NAME: Pisces (PYE-seez)

SEASON: Autumn

CULMINATION: midnight, 27 September

ABBREVIATION: Psc

GENITIVE: Piscium (PYE-see-um)

NEIGHBORS: And, Aqr, Ari, Cet, Peg, Tri

BINOCULAR OBJECTS: none

URBAN OBJECTS: none

Pisces is a large, very dim north-equatorial constellation that ranks 14th in size among the 88 constellations. Pisces covers 889 square degrees of the celestial sphere, or about 2.2%. Despite the fact that its brightest star is only fourth magnitude, it's relatively easy to locate the large V outline of Pisces, assuming that you are fully dark-adapted and observing from a dark site.

Pisces is an ancient constellation, recognized as such since at least Babylonian times, and usually associated with fishes and water. In Greek mythology, Pisces usually represents the goddess Aphrodite and her son Eros, who transformed themselves into fishes to flee from the monster Typhon.

Despite its size, Pisces has few DSOs of interest to amateur astronomers. Pisces is home to numerous galaxies, but most of these are far too dim to be rewarding targets in any but the largest amateur instruments. The one exception is the relatively bright Messier galaxy M 74. That galaxy is known and dreaded by amateur astronomers who undertake the Messier Marathon—attempting to view and log all 110 Messier Objects in one night from dusk to dawn—because M 74 sets not long after the Sun on Marathon night and is therefore very difficult to locate and see in the lingering twilight.

Pisces culminates at midnight on 27 September, and, for observers at mid-northern latitudes, is best placed for evening viewing from early autumn through mid-winter.

TABLE 37-1.

Featured star clusters, nebulae, and galaxies in Pisces

Object	Type	Mv	Size	RA	Dec	M	B	U	D	R	Notes
NGC 628	Gx	10.0	10.5 x 9.5	01 36.7	+15 47	◉					M 74; Class SA(s)c; SB 13.7

TABLE 37-2.

Featured multiple stars in Pisces

Object	Pair	M1	M2	Sep	PA	Year	RA	Dec	UO	DS	Notes
65	STF 61	6.3	6.3	4.3	121	2004	00 49.9	+27 42		◉	
74-psi1	STF 88AB	5.3	5.5	30.0	161	2004	01 05.7	+21 28		◉	
86-zeta	STF 100AB	5.2	6.2	22.8	62	2004	01 13.7	+07 34		◉	
113-alpha*	STF 202AB	4.1	5.2	1.8	267	2006	02 02.0	+02 46		◉	Alrescha

CHART 37-1.

The constellation Pisces (field width 50°)

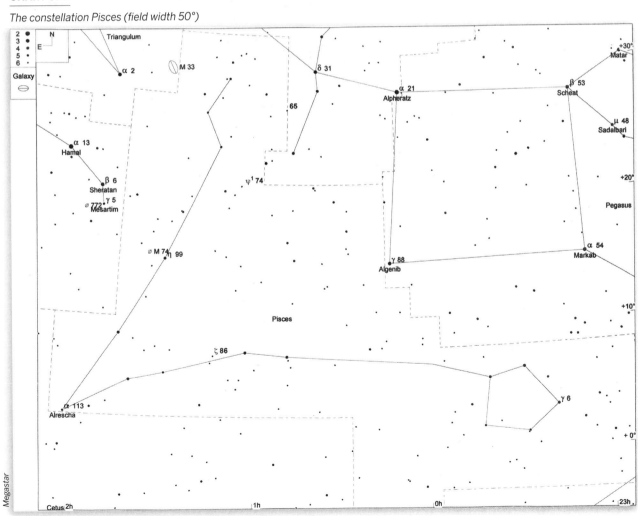

Megastar

Clusters, Nebulae, and Galaxies

M 74 (NGC 628)	★★	✦✦✦	GX	MBUDR
Chart 37-2	Figure 37-1	m10.0, 10.5' x 9.5'	01h 36.7m	+15° 47'

NGC 628, better known as Messier 74 or M 74, is a large, faint spiral galaxy that we see almost face-on. Pierre Méchain discovered M 74 in late September 1780 and reported his discovery to his friend and colleague Charles Messier. Messier observed this galaxy on the night of 18 October 1780 and added it to his catalog as object M 74.

Although M 74 lies in Pisces, it is easiest to locate it by starting from the prominent stars m2.0 13-α (alpha) Arietis (Hamal) and m2.7 6-β (beta) Arietis (Sheratan). Extend the 3.9° NE-SW line from Hamal to Sheratan another 7.7° SW to locate the m3.6 star 99-η (eta) Piscium, which is the brightest star in the vicinity (and indeed the brightest star in the constellation Pisces.) M 74 lies 1.3° ENE of 99-η. M 74 is invisible in a 50mm finder, but can be seen in a low-power eyepiece as what looks remarkably like a small, dim globular cluster.

At 42X in our 10" Newtonian, M 74 is visible with direct vision as a faint, fuzzy star lying about 18' dead W of a prominent m10 field star and about 12' dead S of another m10 field star. At 125X with averted vision, M 74 shows a moderately faint, ragged, nearly circular 7' halo with very faint mottling that brightens to a condensed core with a stellar nucleus.

FIGURE 37-1.

NGC 628 (M 74) (60' field width)

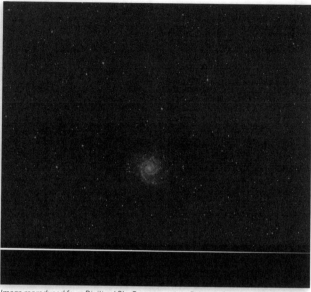

Image reproduced from Digitized Sky Survey courtesy Palomar Observatory and Space Telescope Science Institute

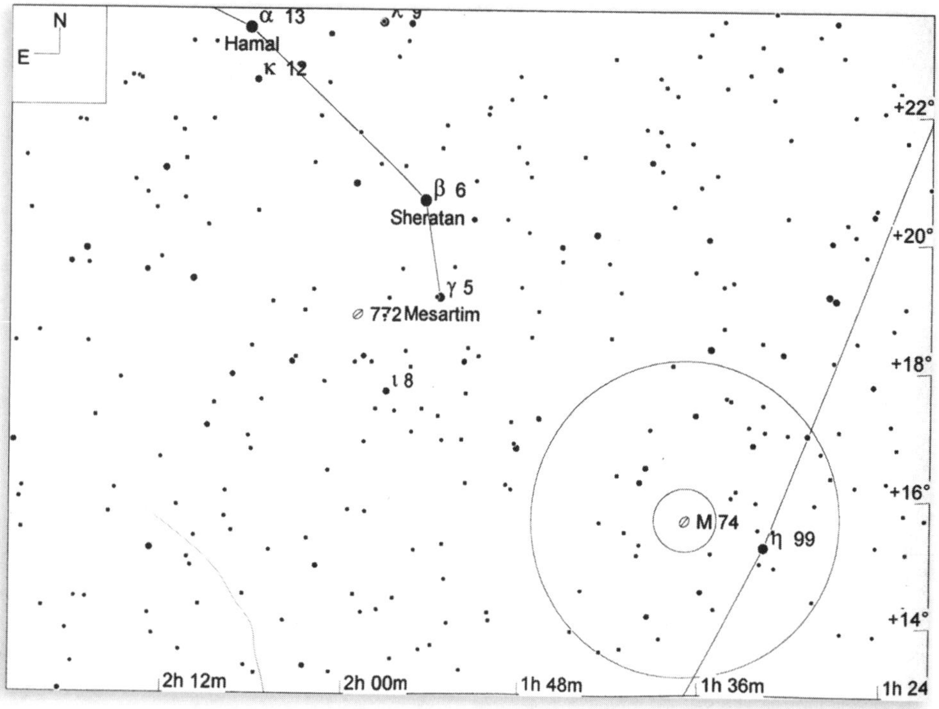

CHART 37-2.

NGC 628 (M 74) (10° field width; 5° finder circle; 1° eyepiece circle; LM 9.0)

Multiple Stars

65 (STF 61)	★★	✷✷✷	MS	UD
Chart 37-3		m6.3/6.3, 4.3", PA 121° (2004)	00h 49.9m	+27° 42'

Although 65-Piscium (STF 61) lies in Pisces, it's easiest to locate it by working from the bright naked-eye star m3.3 31-δ (delta) Andromedae. Once you locate 31-And, look in your finder for m4.4 30-ε (epsilon) Andromedae 1.6° to the S. Place 30-ε on the WNW edge of your finder field, as shown in Chart 37-3, and look for the prominent triangle formed by m5.4 68-Psc, m6.1 67-Psc, and m5.6 65-Psc. 65-Psc is the W apex of that triangle.

At 100X with our 90mm refractor, 65-Psc is a pretty double star, evenly matched in brightness and with both the primary and secondary a straw yellow shade.

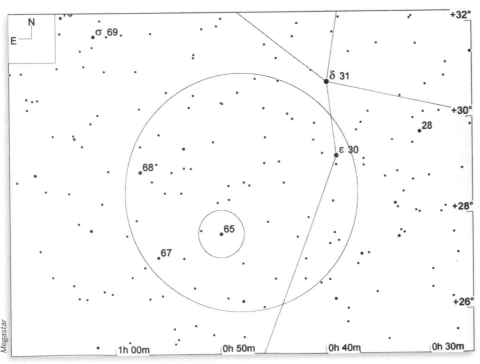

CHART 37-3.

65-Piscium (STF 61) (10° field width; 5° finder circle; 1° eyepiece circle; LM 9.0)

74-psi[1] (STF 88AB)	★★	✷✷✷	MS	UD
Chart 37-4		m5.3/5.5, 30.0", PA 161° (2004)	01h 05.7m	+21° 28'

74-ψ[1] (psi[1]) is also easiest to locate by working from 31-δ (delta) and 30-ε (epsilon) Andromedae. From those stars, move your finder SSE until m4.1 34-ζ (zeta) Andromedae comes into view. Once you identify 34-ζ, look 2.4° ESE for m4.4 38-η (eta), which is very prominent in the finder. Centering 38-η in the finder puts 74-ψ[1] at or just beyond the SE edge of the finder field.

At 100X with our 90mm refractor, 74-ψ[1] is a pretty double star, evenly matched in brightness and with both primary and companion the same blue-white color.

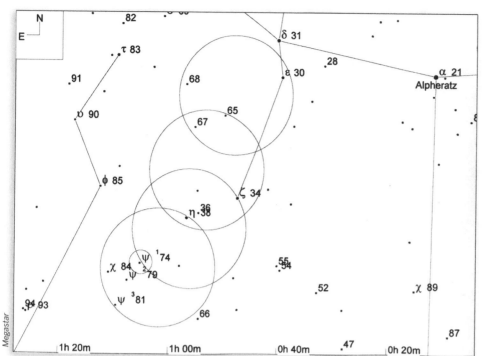

86-zeta (STF 100AB)	★★	◐◐	MS	UD
Chart 37-5		m5.2/6.2, 22.8", PA 62° (2004)	01h 13.7m	+07° 34'

86-ζ (zeta) is a rather ordinary double star that's a bit harder to find than most. Once you get there, you may wonder why you bothered. The easiest way we know to find 86-ζ (zeta) is to start at m3.8 113-α (alpha) Piscium, described in the next entry, which is the fairly dim naked-eye star that forms the point of the V in the constellation pattern. Use your finder to work your way W, as shown in Chart 37-5. 86-ζ is the m5 NE apex of a 4° equilateral triangle it forms with m5 98-μ (mu) and m5 89-Psc, all of which are very prominent in the finder.

At 100X with our 90mm refractor, 86-ζ is a rather ordinary double star, with a pure white primary and a noticeably dimmer yellow-white companion.

113-alpha (STF 202AB)	★★	◐◐◐◐	MS	UD
Chart 37-1		m4.1/5.2, 1.8", PA 267° (2006)	02h 02.0m	+02° 46'

113-α (alpha) Piscium, sometimes called Alrescha, forms the point of the V in the constellation pattern. This is an extremely tight double. We tried and failed to split it with our 90mm refractor. Even with our 10" Dob, this is a very difficult split that requires high magnification and excellent seeing. We finally got a (barely) clean split on a night of excellent seeing at 240X. Both components are cool white.

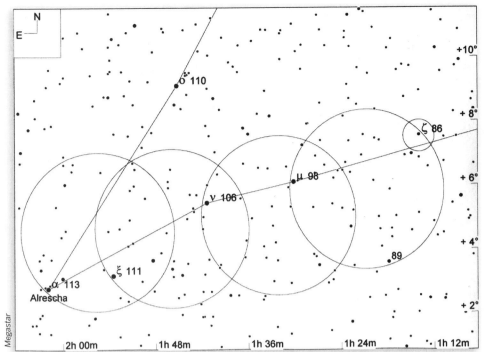

CHART 37-5.

86-zeta Piscium (STF 100AB)
(15° field width; 5° finder circles;
1° eyepiece circle; LM 9.0)

Megastar

38

Puppis, The Poop Deck

NAME: Puppis (PUPP-is)

SEASON: Winter

CULMINATION: midnight, 9 January

ABBREVIATION: Pup

GENITIVE: Puppis (PUPP-is)

NEIGHBORS: Car, CMa, Col, Hya, Mon, Pic, Pyx, Vel

BINOCULAR OBJECTS: NGC 2422 (M 47), NGC 2437 (M 46), NGC 2447 (M 93), NGC 2527, NGC 2539, NGC 2571

URBAN OBJECTS: NGC 2539

Puppis is a large, relatively dim south-equatorial constellation that ranks 20th in size among the 88 constellations. Puppis covers 673 square degrees of the celestial sphere, or about 1.6%. The most prominent stars in Puppis, ζ (zeta) Puppis (magnitude 2.2), π (pi) Puppis (magnitude 2.7), and 15-ϱ (rho) Puppis (magnitude 2.8), form a prominent 16° isosceles triangle that lies southeast of Sirius in Canis Major. 15-ϱ is the northern apex of that triangle, and is a convenient point of departure for locating most of the featured DSOs in Puppis.

Puppis is both a modern constellation and a part of an ancient constellation. The largest of Ptolemy's original 48 constellations was called Argo Navis, The Ship. Much of Argo Navis lay too far south in declination to be visible to northern European observers. Presumably for that reason, in the 1750s the French astronomer Abbé Nicolas Louis de Lacaille split Argo Navis into the modern constellations Carina (The Keel), Vela (The Sails), and Puppis (The Poop Deck).

When he split Argo Navis, de Lacaille left the original Bayer designations in place. Carina retains α (alpha), β (beta), ε (epsilon), η (eta), θ (theta), ι (iota), υ (upsilon), χ (chi), and ω (omega). Vela retains γ (gamma), δ (delta), ϰ (kappa), λ (lambda), μ (mu), o (omicron), φ (phi), and ψ (psi). Puppis retains ζ (zeta), ν (nu), ξ (xi), π (pi), ϱ (rho), σ (sigma), and τ (tau).

The modern constellation Pyxis (The Compass) was also a part of the ancient constellation Argo Navis, but de Lacaille treated Pyxis differently. When he split it off, he reassigned Bayer designations to its stars, so the Bayer designations in Pyxis duplicate those in Carina, Vela, and Puppis.

All of the featured objects in this chapter are in the northern section of Puppis, which lies square in the Milky Way. Like most Milky Way constellations, northern Puppis is rich in open clusters. Six of those, including three bright Messier open clusters, are featured in this chapter.

TABLE 38-1.

Featured star clusters, nebulae, and galaxies in Puppis

Object	Type	Mv	Size	RA	Dec	M	B	U	D	R	Notes
NGC 2422	OC	4.4	29.0	07 36.6	-14 29	◉	◉				M 47; Cr 152; Class I 3 m
NGC 2437	OC	6.1	27.0	07 41.8	-14 49	◉	◉				M 46; Cr 159; Class II 2 r
NGC 2440	PN	10.8	70"	07 41.9	-18 13					◉	Class 5+3
NGC 2539	OC	6.5	21.0	08 10.6	-12 49			◉	◉	◉	Cr 176; Mel 83; Class III 2 m
NGC 2447	OC	6.2	22.0	07 44.5	-23 51	◉	◉				M 93; Cr 160; Class I 3 r
NGC 2527	OC	6.5	15.0	08 04.9	-28 08				◉		Cr 174; Class II 2 m
NGC 2571	OC	7.0	13.0	08 19.0	-29 45				◉		Cr 181; Class II 3 m

TABLE 38-2.

Featured multiple stars in Puppis

Object	Pair	M1	M2	Sep	PA	Year	RA	Dec	UO	DS	Notes
kappa	ADS 6255	4.4	4.6	9.8	318	2002	07 38.8	-26 48	◉		H 27AB

The plethora of stars in the Milky Way also means that planetary nebulae are common, and Puppis has one fine example, which is also featured.

Puppis culminates at midnight on 9 January, and, for observers at mid-northern latitudes, is best placed for evening viewing during the winter months.

CHART 38-1.

The constellation Puppis (northern section) (field width 30°)

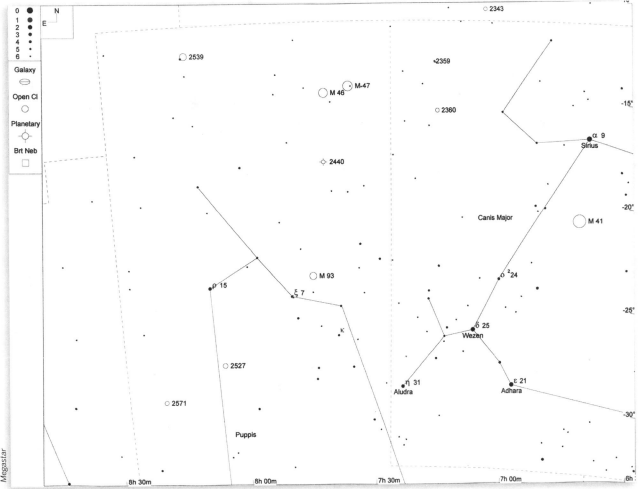

Clusters, Nebulae, and Galaxies

M 47 (NGC 2422)	★★★	✧✧✧	OC	MBUDR
Chart 38-1, 38-2	Figure 38-1	m4.4, 29.0'	07h 36.6m	-14° 29'

NGC 2422, better known as Messier 47 or M 47, is a large, bright open cluster. M 47 is so bright, in fact, that it can be glimpsed by the naked eye as a faint nebulosity if you are fully dark adapted and observing from a very dark site. Although, as a naked-eye object, M 47 had no doubt been observed by many people from ancient times on—including the Italian astronomer Giovanni Batista Hodierna, who logged it before 1654—Charles Messier independently re-discovered and cataloged M 47 on the night of 19 February 1771, and described it as brighter than nearby M 46. Unfortunately, Messier recorded the location of M 47 incorrectly, and it remained one of the "Missing Messier Objects" until 1959, when the error was finally corrected.

You can locate M 47 by a long star hop, but it's much easier to locate it geometrically. To do so, identify Sirius in Canis Major and m1.8 Wezen 11° to the SE. Those two stars form the baseline of a 13° isosceles triangle of which M 47 (and its companion cluster, M 46) form the NW apex. Simply point your binocular or finder to about where M 47 ought to be, and you'll see it and M 46 appear prominently in the eyepiece.

The Astronomical League Binocular Messier Club rates M 47 as an Easy object with any 35mm or larger binocular. With our 50mm binoculars, M 47 is visible as a large, bright hazy patch with 15+ stars visible, the brightest m6/7. One m7 star lies centered in the nebulosity, with an m6 star on the W edge of the nebulosity and an m7 star lying about 5' E. At 90X in our 10" reflector, 70+ stars are visible. The brightest stars, m6 and m7, are all blue-white or cool white. The cluster has an irregular appearance, with the stars gathered in clusters, clumps, and chains separated by lanes that are devoid of bright stars. Two of the brightest stars are doubles. The brightest star, near the W edge of the cluster, is a wide double (STF 1120) with a much dimmer companion. The bright star near the center of the cluster is a much closer double, STF 1121, with primary and companion of nearly identical brightness.

FIGURE 38-1.

NGC 2422 (M 47) (60' field width)

Image reproduced from Digitized Sky Survey courtesy Anglo-Australian Observatory and Space Telescope Science Institute

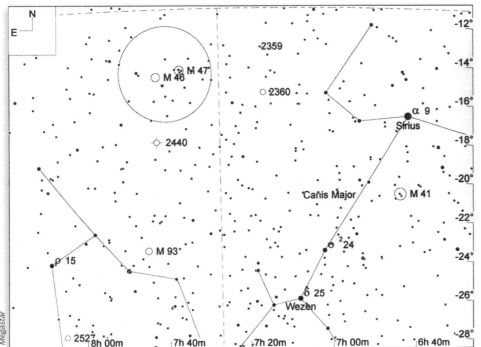

Megastar

M 46 (NGC 2437)	★★★	◍◍◍	OC	MBUDR
Chart 38-1, 38-2	Figure 38-2	m6.1, 27.0'	07h 41.8m	-14° 49'

NGC 2437, better known as Messier 46 or M 46, is a large, bright open cluster. Charles Messier discovered and cataloged M 46 on the night of 19 February 1771, the same night that he re-discovered M 47. M 46 was the first object Messier discovered after the publication of the first edition of his catalog, which contained 45 objects.

You can locate M 46 by a long star hop, but it's much easier to locate it geometrically, as described in the preceding section for M 47. Both are visible in the same binocular or finder field. M 46 is the smaller, dimmer nebulosity lying about 1.3° ESE (east southeast) of the more prominent M 47.

The Astronomical League Binocular Messier Club rates M 46 as an Easy object with any 35mm or larger binocular. With our 50mm binoculars, M 46 is visible as a large, bright hazy patch, although it is smaller and dimmer than M 47 in the same field. With averted vision, one m9 star is visible on the W edge of the nebulous patch and two m10 stars are faintly visible slightly separated to the S and E of the nebulosity.

At 42X in our 10" reflector, M 46 and M 47 fit (barely) into the same 1.7° true field of view. Although from their size and brightness you might expect the M 46 and M 47 pair to resemble the Double Cluster (NGC 869 and NGC 884) in Perseus, they do not. The Double Cluster pair are near twins, each with many bright stars. M 47 appears similar to the components of the Double Cluster, but M 46 is different. M 46 reminds us more of M 37 in Auriga, with no

FIGURE 38-2.

NGC 2437 (M 46) (60' field width)

Image reproduced from Digitized Sky Survey courtesy Anglo-Australian Observatory and Space Telescope Science Institute

very bright stars but an incredibly rich, even scattering of m10 and dimmer stars.

At 90X, 75+ stars are visible, and a mystery is solved. At lower magnification, one of the stars on the N edge of the cluster appears fuzzy. At 90X, that fuzzy star resolves as the disc of the planetary nebula NGC 2438. Although NGC 2438 appears to be embedded in M 46, it's really a foreground object that lies at about half the distance of the open cluster.

NGC 2440	★★	◐◐	PN	MBUDR
Chart 38-3	Figure 38-3	m10.8, 70"	07h 41.9m	-18° 13'

NGC 2440 is a small planetary nebula with high surface brightness. You can locate NGC 2440 by starting from M 47, as shown in Chart 38-3. Place M 47 near the NW edge of your finder field, and look for 6-Puppis, an m5 star that appears very prominently near the SE edge of the field. NGC 2440 lies 2.1° WSW (west southwest) of 6-Puppis and about 37' ESE of an m7.5 star that's visible in the finder. Center your finder cross hairs at the approximate location of NGC 2440 and use a narrowband or O-III filter with your low-power eyepiece to "blink" the nebula and confirm its identity.

At 180X in our 10" with an Orion Ultrablock narrowband filter and averted vision, NGC 2440 shows a distinct 15" disc elongated slightly ENE-WSW (from east northeast to west southwest). Some brightening is visible toward the center.

Image reproduced from Digitized Sky Survey courtesy Anglo-Australian Observatory and Space Telescope Science Institute

FIGURE 38-3.

NGC 2440 (60' field width)

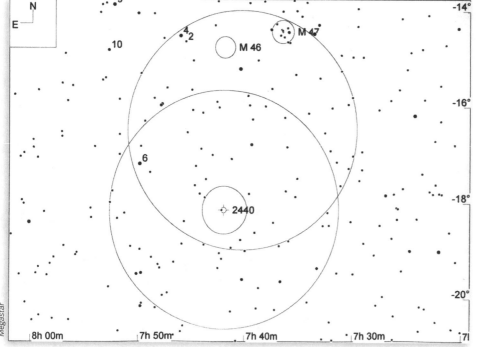

CHART 38-3.

NGC 2440 (10° field width; 5° finder circles; 1° eyepiece circle; LM 9.0)

Megastar

NGC 2539	★★★	◐◐◐	OC	MBUDR
Chart 38-4	Figure 38-4	m6.5, 21.0'	08h 10.6m	-12° 49'

NGC 2539 is a large, bright open cluster, and another of the objects that we're surprised Messier missed. We suspect he missed NGC 2539 because it lies about 7.5° ENE of the M 46/47 pair, and the telescopes of Messier's time had very narrow fields of view. Had Messier had access to a modern 7X50 binocular, he almost certainly would have logged NGC 2539 on the same night that he logged M 46 and M 47.

Although you can locate NGC 2539 by star hopping, the easiest way to find it is to put your finder or binocular on M 46 and M 47 and then sweep E and slightly N until NGC 2539 comes into view. NGC 2539 lies one degree N of the m5.5 star 18-Puppis, which appears very prominent in a binocular or finder, and just a few minutes NW of m4.7 19-Puppis.

With our 50mm binocular, NGC 2539 appears as a moderately large, bright patch of haze, with 19-Puppis glaring brightly at its SE edge. No stars are resolved within the nebulosity. At 90X in our 10" Dob, NGC 2539 is a 25' E-W oval, with 100+ m10 to m13 stars visible in clumps, chains, and arcs separated by dark, star-poor regions to the N and SE.

FIGURE 38-4.

NGC 2539 (60' field width)

Image reproduced from Digitized Sky Survey courtesy Anglo-Australian Observatory and Space Telescope Science Institute

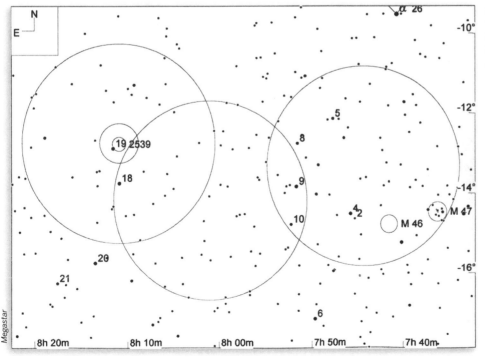

CHART 38-4.

NGC 2539 (12° field width; 5° finder circles; 1° eyepiece circle; LM 9.0)

M 93 (NGC 2447)	★★★	✿✿✿	OC	MBUDR
Chart 38-5	Figure 38-5	m6.2, 22.0'	07h 44.5m	-23° 51'

NGC 2447, better known as Messier 93 or M 93, is a moderately large, bright open cluster, and one of the final objects that Messier discovered himself. Messier discovered this object on the night of 20 March 1781 and cataloged it as M 93.

The easiest way to locate M 93 is to work from the m1.8 star Wezen in Canis Major. With Wezen in your finder or binocular, just pan 8.6° slightly N of E until M 93 comes into the field of view.

The Astronomical League Binocular Messier Club rates M 93 as an Easy object in a 35mm or larger binocular. With our 50mm binoculars, M 93 is a large, bright patch of haze 1.5° NW of the yellow-white m3.3 7-ξ (xi), which appears glaringly bright. Half a dozen m9/10 stars are visible with averted vision embedded in the nebulosity. At 90X in our 10" reflector, 75+ stars are visible from m9 to m13, forming a triangular shape with its apex pointing WSW. The two brightest stars in the cluster form the WSW point of the triangle, and both are wide doubles.

FIGURE 38-5.

NGC 2447 (M 93) (60' field width)

Image reproduced from Digitized Sky Survey courtesy Anglo-Australian Observatory and Space Telescope Science Institute

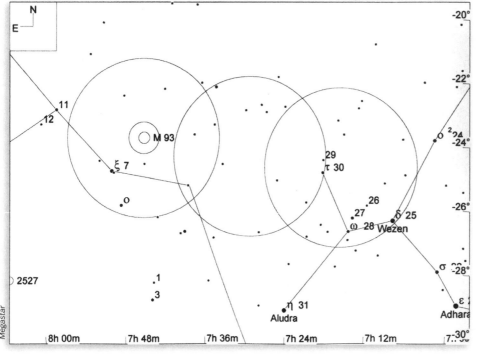

CHART 38-5.

NGC 2447 (M 93) (15° field width; 5° finder circles; 1° eyepiece circle; LM 7.0)

NGC 2527	★★	✧✧✧	OC	MBUDr
Chart 38-6		m6.5, 15.0'	08h 04.9m	-28° 08'

NGC 2527 is a moderately large, bright open cluster. It's relatively easy to locate from m3.3 7-ξ (xi), which you can locate by following the instructions in the preceding section. Once you've identified 7-ξ, look 4.2° E for the bright naked-eye star m2.8 15-ϱ (rho) Puppis. Put 15-ϱ on the N edge of your finder or binocular field, and look 3.8° S for NGC 2527.

With our 50mm binocular, NGC 2527 is visible as a moderately large, bright patch of haze. No embedded stars are visible with direct vision, but averted vision shows half a dozen very faint stars flickering in and out on the edge of visibility. At 90X in our 10" reflector, this cluster presents a very loose, scattered appearance. About three dozen stars are visible from m9 down to m13. Most of those lie within a 10' to 15' central area, but some likely outliers can be seen up to the edges of a 25' circle. It's very difficult to identify cluster members, because the cluster lies within a very rich Milky Way star field and is not well detached.

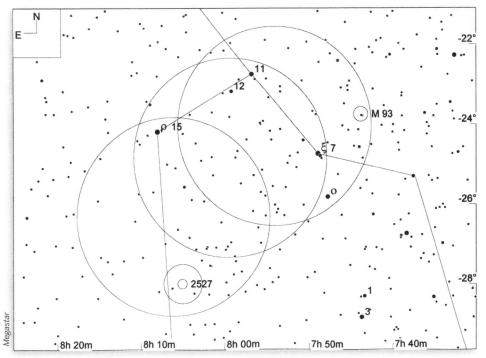

CHART 38-6.

NGC 2527 (12° field width; 5° finder circles; 1° eyepiece circle; LM 9.0)

NGC 2571	★★	❀❀❀	OC	MBUDr
Chart 38-7		m7.0, 13.0'	08h 19.0m	-29° 45'

NGC 2571 is a medium size, moderately bright open cluster. It's relatively easy to locate, lying 3.5° ESE of NGC 2527 and 6° SSE of m2.8 15-ϱ (rho) Puppis.

With our 50mm binocular, NGC 2571 is visible as a moderately large, fairly bright nebulous patch. Two embedded stars are faintly visible with direct vision, and one or two more with averted vision. At 90X in our 10" reflector, this cluster is a large, loose group of 20+ stars scattered across a 12' oval elongated NW-SE (from northwest to southeast). A wide m9 pair lies at the center of the cluster, with the other cluster members, mostly m12/13 extending NW and SE.

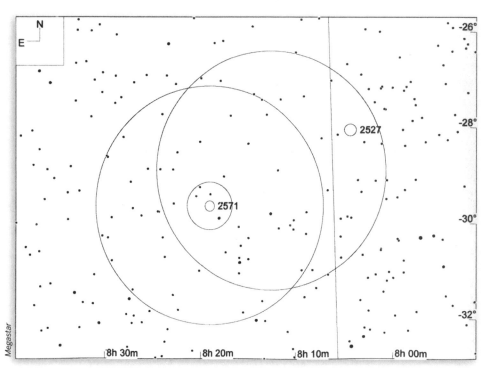

CHART 38-7.

NGC 2571 (10° field width; 5° finder circles; 1° eyepiece circle; LM 9.0)

Multiple Stars

"Kappa Puppis" (ADS 6255)	★★	✹✹✹		MS	UD
Chart 38-8		m4.4/4.6, 9.8", PA 318° (2002)		07h 38.8m	-26° 48'

The Astronomical League Double Star Club list includes "Kappa Puppis" as a required object, but, as we noted in the introduction to this chapter, Puppis has no star designated kappa. The real ϰ lies 32.9° SSW (south southwest) in the constellation Vela. So what happened? Whoever compiled that list mistook a lower-case Roman k for the Greek ϰ (kappa). The correct designation of the star in question is k-Puppis rather than ϰ-Puppis. It's an easy mistake to make.

k-Puppis is relatively easy to find. As shown in Chart 38-8, you can start from either m2.5 31-η (eta) CMa (Aludra) or m2.8 15-ϱ (rho) Puppis. In fact, k-Puppis lies just a bit more than halfway on the line from Aludra to 15-ϱ, so you can probably eyeball it.

At 90X in our 10" Dob, k-Puppis is a matched pair of cool white stars.

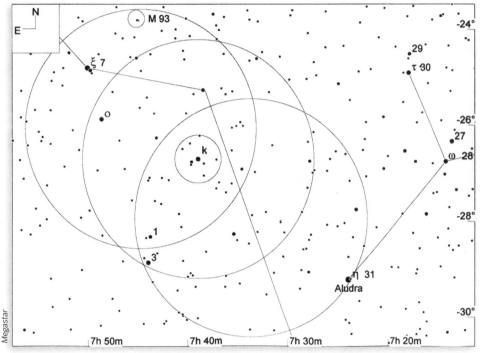

CHART 38-8.

Kappa Puppis (ADS 6255) (10° field width; 5° finder circles; 1° eyepiece circle; LM 9.0)

39

Sagitta, The Arrow

NAME: Sagitta (SADG-ih-taw)

SEASON: Summer

CULMINATION: midnight, 17 July

ABBREVIATION: Sge

GENITIVE: Sagittae (SADG-ih-tye)

NEIGHBORS: Aql, Del, Her, Vul

BINOCULAR OBJECTS: NGC 6838 (M 71)

URBAN OBJECTS: none

Sagitta is a tiny, faint, north equatorial constellation that ranks 86th in size among the 88 constellations. Sagitta covers only 80 square degrees of the celestial sphere, or about 0.2%. Of the constellations visible to observers at mid-northern latitudes, only Equuleus is smaller. The brightest stars in Sagitta are fourth magnitude, and it has only five of those.

Despite its tiny size and lack of prominent stars, Sagitta is an ancient constellation, one of Ptolemy's original 48. Although the names are similar, Sagitta (The Arrow) is not always associated in mythology with Sagittarius (The Archer). In what is probably the oldest of several Greek myths related to Sagitta, Sagitta represents the arrow shot by Herakles in defense of Prometheus. Punished by Zeus for bringing fire to mortals, Prometheus was chained to a rock and tormented by Aquila the Eagle. Herakles came to his rescue, slaying Aquila with the arrow Sagitta from his mighty bow. Another Greek myth relates the story of Herakles as one of his Twelve Labors using Sagitta to slay the Stymphalian Birds. Still other myths associate Sagitta with the arrow shot by Sagittarius at Scorpius (although Sagittarius is facing the wrong direction for this to be likely) or with the arrow shot by Centaurus the Centaur at Aquila.

Sagitta has few interesting DSOs or multiple stars. The one exception is the fine Messier globular cluster M 71, which makes Sagitta worth a visit.

Sagitta culminates at midnight on 17 July, and, for observers at mid-northern latitudes, is best placed for evening viewing from late spring through early autumn.

TABLE 39-1.

Featured star clusters, nebulae, and galaxies in Sagitta

Object	Type	Mv	Size	RA	Dec	M	B	U	D	R	Notes
NGC 6838	GC	8.4	7.2	19 53.8	+18 47	◉	◉				M 71; Class ???

CHART 39-1.

The constellation Sagitta (field width 35°)

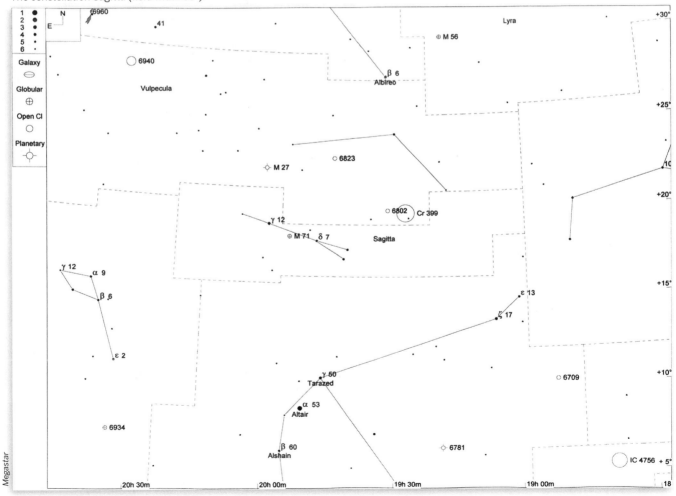

Clusters, Nebulae, and Galaxies

M 71 (NGC 6838)	★★★	ஓஓ	GC	MBUDR
Chart 39-2	Figure 39-1	m8.4, 7.2'	19h 53.8m	+18° 47'

NGC 6838, better known as Messier 71 or M 71, is a fine, bright, very loose globular cluster. Philippe Loys de Chéseaux discovered M 71 in 1745 or 1746. Charles Messier observed this object and added it to his catalog on the night of 4 October 1780. M 71 is a very unusual globular. Until 50 years ago astronomers weren't even sure that M 71 was a globular cluster. Many believed that M 71 was a very tight open cluster like M 11 in Scutum. Detailed photometric measurements finally confirmed that M 71 is in fact a glob, but in amateur instruments it's still impossible to differentiate it from a very tight open cluster.

M 71 is relatively easy to find, lying about halfway along and just S of the 2.8° ENE-WSW (from east northeast to west southwest) line from m3.8 7-δ (delta) Sagittae to m3.5 12-γ (gamma) Sagittae. Both of those stars lie SE of the brightest stars in Vulpecula, and are visible to the naked eye. The Astronomical League Binocular Messier Club rates M 71 as a Challenge object (the most difficult category) with a 35mm or 50mm binocular. That may be a bit optimistic, as we have never seen M 71 with our 50mm binocular or finder. However, the m6.3 star 9-Sagittae is prominent in the finder, and M 71 lies just 21' ENE of that star.

At 125X in our 10" reflector, M 71 shows a ragged 4' halo with scores of stars resolvable all the way to a poorly concentrated core.

FIGURE 39-1.

NGC 6838 (M 71) (60' field width)

Image reproduced from Digitized Sky Survey courtesy Palomar Observatory and Space Telescope Science Institute

CHART 39-2.

NGC 6838 (M 71) (10° field width; 5° finder circle; 1° eyepiece circle; LM 9.0)

Multiple Stars

There are no featured multiple stars in Sagitta.

40
Sagittarius, The Archer

NAME: Sagittarius (SADJ-ih-TAIR-ee-us)

SEASON: Summer

CULMINATION: midnight, 5 July

ABBREVIATION: Sgr

GENITIVE: Sagittarii (SADJ-ih-TAIR-ee-ih)

NEIGHBORS: Aql, Cap, CrA, Mic, Oph, Sco, Sct, Ser, Tel

BINOCULAR OBJECTS: NGC 6494 (M 23), NGC 6514 (M 20), NGC 6523 (M 8), NGC 6520, NGC 6603 (M 24), NGC 6613 (M 18), NGC 6618 (M 17), NGC 6626 (M 28), IC 4725 (M 25), NGC 6637 (M 69), NGC 6656 (M 22), NGC 6681 (M 70), NGC 6716, NGC 6715 (M 54), NGC 6809 (M 55), NGC 6864 (M 75)

URBAN OBJECTS: NGC 6523 (M 8), NGC 6520, NGC 6618 (M 17), NGC 6656 (M 22), NGC 6818

Sagittarius is a large, prominent south-equatorial constellation that ranks 15th in size among the 88 constellations. Sagittarius covers 867 square degrees of the celestial sphere, or about 2.1%. Although Sagittarius has only two second-magnitude stars, its eight third-magnitude stars fill out the constellation outline. Other than perhaps Orion and that portion of Ursa Major made up by the Big Dipper (or Plough) asterism, Sagittarius is the most familiar of constellations even to non-astronomers.

The famous "teapot" asterism in Sagittarius is a familiar sight to anyone who has scanned the southern horizon during the summer months, made even more striking by the whisps of "steam" (the bright Milky Way star field) coming out of the "spout." In fact, many people who are completely unfamiliar with astronomy recognize the Teapot, even if they don't know the actual name of the constellation.

Sagittarius is an ancient constellation, one of Ptolemy's original 48, and has been recognized as a constellation at least since the earliest written records. The ancients, unfamiliar with teapots, saw Sagittarius as an archer, often mounted on a horse, or as a Centaur or similar mythological beast.

When we look toward Sagittarius, we are looking almost directly toward the center of our own galaxy, the Milky Way. Accordingly, Sagittarius is extraordinarily rich in bright, interesting DSOs, including no less than fifteen Messier objects. (More than one eighth of all Messier objects lie in this one constellation.) Sagittarius is so rich, as many of the DSS images in this chapter illustrate, that it's often difficult to tell where one object stops and the next one starts. Many cataloged objects in Sagittarius are very close to or even embedded in other cataloged objects.

In addition to being a rich hunting ground for even small telescopes, Sagittarius is full of binocular objects. Of the 15 Messier objects, 14 are on the Astronomical League Binocular Messier Club lists. Eight of those—M 8, 17, 18, 22, 23, 24, 25, and 55—are rated Easy with a 35mm or larger binocular. M 28 is a Tougher object (the middle category) with a 35mm or 50mm binocular, and Easy with an 80mm binocular. M 54 and M 75 are rated Challenge objects (the most difficult category) with a 35mm or 50mm binocular, and as Tougher objects with an 80mm binocular. M 20, M 69, and M 70 are not considered possible with a 35mm or 50mm binocular, and are Challenge objects with an 80mm binocular. M 21 is not considered possible with even an 80mm binocular. In addition to the Binocular Messier objects, two open clusters, NGC 6520 and NGC 6716, appear on the Deep Sky Binocular Club list, and are relatively easy objects with a 50mm binocular.

Sagittarius culminates at midnight on 5 July, and, for observers at mid-northern latitudes, is best placed for evening viewing from late spring through early autumn.

TABLE 40-1.

Featured star clusters, nebulae, and galaxies in Sagittarius

Object	Type	Mv	Size	RA	Dec	M	B	U	D	R	Notes
NGC 6637	GC	7.7	9.8	18 31.4	-32 21	◉	◉				M 69; Class V
NGC 6681	GC	7.8	8.0	18 43.2	-32 18	◉	◉				M 70; Class V
NGC 6715	GC	7.7	12.0	18 55.1	-30 29	◉	◉				M 54; Class III
NGC 6656	GC	5.2	32.0	18 36.4	-23 54	◉	◉	◉			M 22; Class VII
NGC 6626	GC	6.9	13.8	18 24.5	-24 52	◉	◉				M 28; Class IV
NGC 6523	EN	5.0	50.0 x 40.0	18 04.1	-24 18	◉	◉	◉			M 8; IC 1271
NGC 6514	EN/RN + OC	9.0 + 6.3	17.0 x 12.0 + 30.0	18 02.4	-22 59	◉	◉				M 20; Class n
NGC 6531	OC	5.9	13.0	18 04.2	-22 30	◉					M 21; Cr 363; Class I 3 r
NGC 6603 + IC 4715 (M 24)	OC + OC	11.5	4.0 + 120.0	18 17.0	-18 36	◉	◉				M 24; Class: Star cloud
NGC 6613	OC	6.9	9.0	18 20.0	-17 06	◉	◉				M 18; Class II 3 p n
NGC 6618	EN	6.9	11.0 x 6.0	18 20.8	-16 10	◉	◉	◉			M 17
IC 4725	OC	4.9	40.0	18 28.8	-19 17	◉	◉				M 25; Cr 382; Class I 3 m
NGC 6494	OC	5.5	27.0	17 56.9	-19 01	◉	◉				M 23; Class II 2 r
NGC 6445	PN	13.2	44.0" x 30.0"	17 49.2	-20 01					◉	Class 3b+3
NGC 6520	OC	7.6	6.0	18 03.4	-27 53			◉	◉	◉	Cr 361; Mel 187; Class I 2 r n
NGC 6716	OC	7.5	6.0	18 54.6	-19 52				◉		Cr 393; Class IV 1 p
NGC 6818	PN	9.9	48.0"	19 44.0	-14 09			◉		◉	Little Gem Nebula; Class 4
NGC 6809	GC	6.3	19.0	19 40.0	-30 58	◉	◉				M 55; Class XI
NGC 6864	GC	8.6	6.8	20 06.1	-21 55	◉	◉				M 75; Class I

CHART 40-1.

The constellation Sagittarius (field size width 40°)

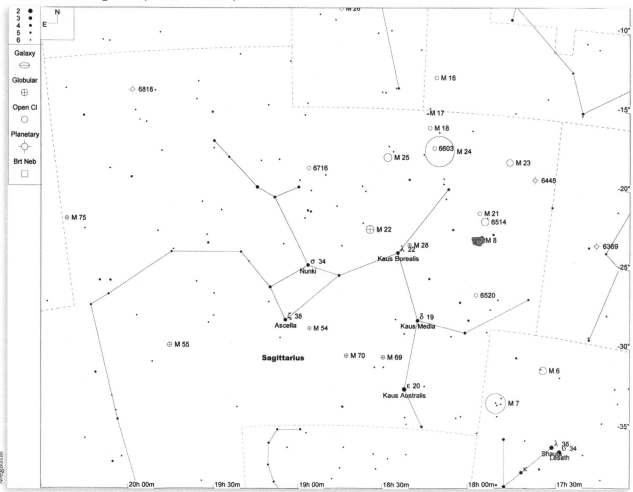

Clusters, Nebulae, and Galaxies

NGC 6637 (M 69)	★★	⊛⊛⊛⊛	GC	MBUDR
Chart 40-2	Figure 40-1	m7.7, 9.8'	18h 31.4m	-32° 21'

NGC 6637, better known as Messier 69 or M 69, is one of the smaller, dimmer globular clusters in Messier's catalog. Charles Messier discovered and logged M 69 on the night of 31 August 1780 while he was searching for an object previously reported by Nicholas Louis de la Caille. Messier initially believed that M 69 was the object that had been reported by de la Caille, although the position of M 69 differed by more than a full degree from the position reported by de la Caille, but it was later established that the two objects were unrelated and that M 69 was therefore an original Messier discovery.

M 69 is quite easy to find. It lies 2.5° NE of the bright naked-eye star m1.8 20-ε (epsilon) Sagitarii (Kaus Australis), which is the SW star in the pentagon of the teapot asterism and the brightest star in Sagittarius. M 69 is visible in a 50mm binocular or finder as a dim, fuzzy star. At 90X in our 10" reflector, M 69 shows a bright 2.5' halo with a concentrated center that brightens to a very bright core that appears nearly stellar. No stars are resolved in the core.

FIGURE 40-1.

NGC 6637 (M 69) (60' field width)

Image reproduced from Digitized Sky Survey courtesy Anglo-Australian Observatory and Space Telescope Science Institute

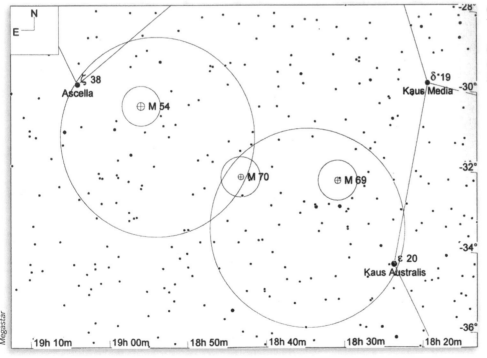

CHART 40-2.

NGC 6637 (M 69), NGC 6681 (M 70), and NGC 6715 (M 54) (12° field width; 5° finder circles; 1° eyepiece circles; LM 9.0)

M 70 (NGC 6681)	★★	✧✧✧✧	GC	MBUDR
Chart 40-2	Figure 40-2	m7.8, 8.0'	18h 43.2m	-32° 18'

NGC 6681, better known as Messier 70 or M 70, is a globular cluster that is slightly smaller and slightly dimmer than M 69, but otherwise a near twin. Charles Messier discovered and cataloged M 70 on the night of 31 August 1780, the same night he discovered M 69. M 70 played a role in the discovery of one of the great comets of the 20th century. In 1995, astronomers Alan Hale and Thomas Bopp were observing M 70 when they noticed a faint fuzzy nearby that shouldn't have been there. That faint fuzzy turned out to be the Great Hale-Bopp Comet.

M 70 is very easy to find. It lies almost exactly halfway along a line from m1.8 20-ε (epsilon) Sagitarii (Kaus Australis), the SW star in the teapot, to m2.6 38-ζ (zeta) Sagitarii (Ascella), the SE star in the teapot. With a 50mm binocular or finder, M 70 looks just like M 69, a dim, fuzzy star. At 90X in our 10" reflector, M 70 shows a bright 2' halo with a well concentrated core that brightens to the center. No stars are resolved in the core.

FIGURE 40-2.

NGC 6681 (M 70) (60' field width)

Image reproduced from Digitized Sky Survey courtesy Anglo-Australian Observatory and Space Telescope Science Institute

M 54 (NGC 6715)	★★	✧✧✧✧	GC	MBUDR
Chart 40-2	Figure 40-3	m7.7, 12.0'	18h 55.1m	-30° 29'

NGC 6715, better known as Messier 54 or M 54, is the brightest of the three globular clusters at the base of the teapot. Charles Messier discovered and cataloged M 54 on the night of 24 July 1778. Although M 54 is a very prominent globular for mid-northerly observers, Messier described it as a "very faint" nebula, no doubt because the southerly declination of M 54 meant that it was "down in the muck" when observed from Messier's Paris observatory at about 48° N latitude.

M 54 is very easy to find. It lies 1.7° WSW (west southwest) of m2.6 38-ζ (zeta) Sagitarii (Ascella), the SE star in the teapot. With a 50mm binocular or finder, M 70 appears almost stellar, but close examination shows the characteristic fuzziness of a globular cluster at low magnification. At 90X in our 10" reflector, M 70 shows a bright 2' halo with a dense, very bright 1' core. No stars are resolved in the core, although averted vision shows a few very faint stars in the surrounding halo.

FIGURE 40-3.

NGC 6715 (M 54) (60' field width)

Image reproduced from Digitized Sky Survey courtesy Anglo-Australian Observatory and Space Telescope Science Institute

M 22 (NGC 6656)	★★★★	✧✧✧✧	GC	MBUDR
Chart 40-3	Figure 40-4	m5.2, 32.0'	18h 36.4m	-23° 54'

NGC 6656, better known as Messier 22 or M 22, is a very large, very bright globular cluster. In fact, M 22 is larger and brighter than the Great Hercules Cluster, M 13, which is much better known to observers at mid-northerly latitudes. Among globular clusters, M 22 is outshone only by two southern clusters, Omega Centauri (NGC 5139) at declination -47° 29' and 47 Tucanae (NGC 104) at declination -72° 05'. Omega Centauri never rises above the southern horizon for observers north of latitude 42° 31' N, and 47 Tucanae never rises for observers north of latitude 17° 55' N, so most observers in the northern hemisphere have never seen either of these objects.

There is some uncertainty about the discoverer of M 22. This object has been known as a star since antiquity, but was probably first identified as a non-stellar object by the German amateur astronomer Johann Abraham Ihle in 1655. M 22 was later observed and logged by numerous astronomers, including Halley, De Chéseaux, Le Gentil, de la Caille, Bevis, and others. Charles Messier acknowledged M 22 as a well-known object when he observed and cataloged it on the night of 5 June 1764.

M 22 is very easy to find. It lies 2.4° NE of the bright naked-eye star m2.8 22-λ (lambda) Sagitarii (Kaus Borealis), the top of the teapot. With a 50mm binocular or finder, M 22 appears as a large, bright hazy patch with no stars resolved. At 90X in our 10" reflector, M 22 is spectacular. The halo extends out to about 25', roughly the size of the full moon, and is slightly ovular, extending ENE-WSW (from east northeast to west southwest). Hundreds of m9 and dimmer stars extend outward from a concentrated core, many arranged in chains, loops, and knots extending in every direction except W and NE, where there are darker, relatively star-poor voids. At 125X, the core is partially resolved, with a granular appearance and a few individual stars resolved.

FIGURE 40-4.

NGC 6656 (M 22) (60' field width)

Image reproduced from Digitized Sky Survey courtesy Anglo-Australian Observatory and Space Telescope Science Institute

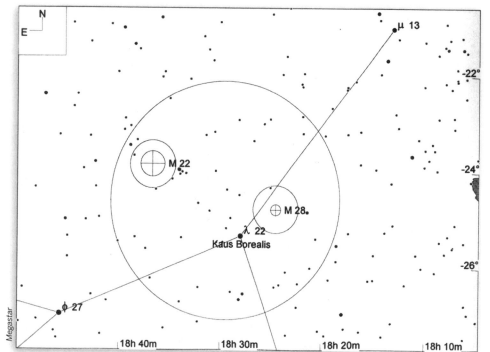

M 28 (NGC 6626)	★★★	⦾⦾⦾⦾	GC	MBUDR
Chart 40-3	Figure 40-5	m6.9, 13.8'	18h 24.5m	-24° 52'

NGC 6626, better known as Messier 28 or M 28, is a large, bright globular cluster. Although it suffers by comparison with nearby M 22, M 28 is a very impressive globular cluster in its own right. Charles Messier discovered and cataloged M 22 on the night of 27 July 1764.

M 28 is very easy to find. It lies 57' NW of the bright naked-eye star m2.8 22-λ (lambda) Sagitarii (Kaus Borealis), the top of the teapot. With a 50mm binocular or finder, M 22 appears as a moderately large, moderately bright hazy patch with no stars resolved. At 90X in our 10" reflector, M 28 shows a well-resolved 4' halo surrounding a dense, concentrated 2' core.

FIGURE 40-5.

NGC 6626 (M 28) (60' field width)

Image reproduced from Digitized Sky Survey courtesy Anglo-Australian Observatory and Space Telescope Science Institute

M 8 (NGC 6523)	★★★★	✸✸✸	EN	MBUDR
Chart 40-4	Figure 40-6	m5.0, 50.0' x 40.0'	18h 04.1m	-24° 18'

NGC 6523, better known as Messier 8, M 8, or the Lagoon Nebula, is a huge, bright, spectacular emission nebula. M 8 surrounds and is illuminated by the open cluster NGC 6530. The English astronomer John Flamsteed discovered and reported NGC 6530 in about 1680, but apparently his telescope was insufficient to reveal the surrounding nebulosity of NGC 6523. Abbe Philippe Loys de Chéseaux again observed and cataloged the open cluster in 1746, but it was not until 1747 that Le Gentil finally reported observing the nebula that was later to be cataloged as NGC 6523 and M 8.

Charles Messier observed and cataloged this object on the night of 23 May 1764, but his detailed description and position report make it clear that Messier intended to catalog the open cluster rather than the nebula as M 8. Messier did report nebulosity surrounding the bright star 9-Sgr, but mentioned this only in passing. Despite that, most modern astronomers consider M 8 to be the nebula NGC 6523 rather than the open cluster NGC 6530.

M 8 is easy to find. It forms the SW apex of a 5.5° isosceles triangle, with the bright naked-eye star m2.8 22-λ (lambda) Sagitarii (Kaus Borealis), the top of the teapot, as the SE apex and m3.8 13-μ (mu) Sagittarii as the N apex. With a 50mm binocular or finder, M 8 appears as a large, moderately bright nebulosity surrounding m7.9 9-Sgr. More than a dozen m7 through m9 stars of NGC 6530 are visible embedded in the nebulosity to the E of 9-Sgr.

At 90X in our 10" reflector, M 8 is big, bright, and magnificent, filling the 45' field of view. The brightest parts of this nebula are so bright that, like M 42 in Orion, they have a gray-green color even

FIGURE 40-6.

NGC 6523 (M 8) (60' field width)

Image reproduced from Digitized Sky Survey courtesy Anglo-Australian Observatory and Space Telescope Science Institute

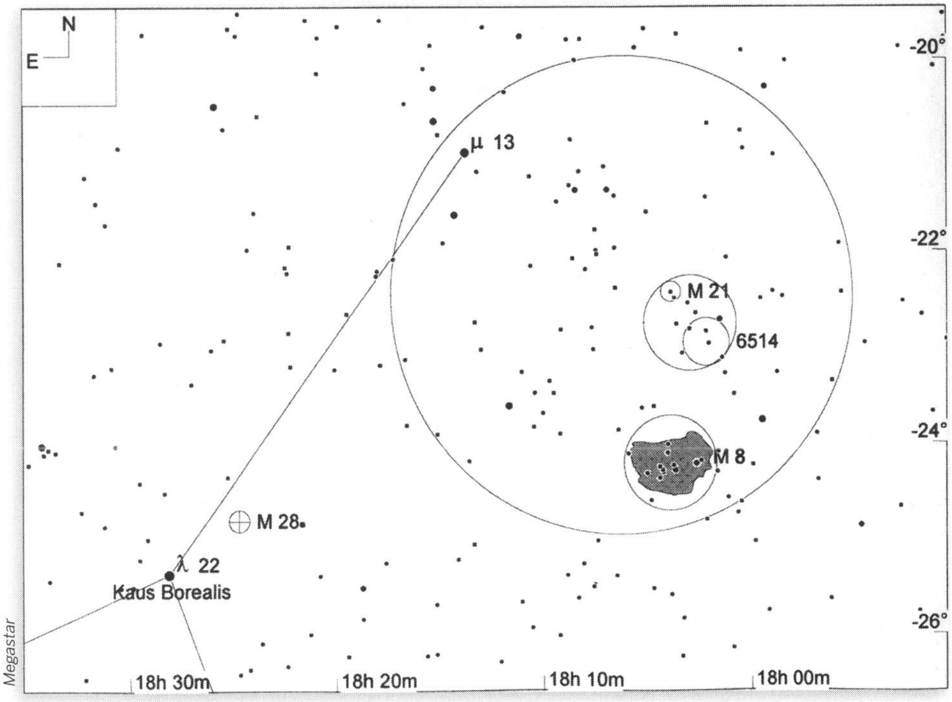

CHART 40-4.

NGC 6523 (M 8), NGC 6514 (M 20), and NGC 6531 (M 21) (10° field width; 5° finder circle; 1° eyepiece circles; LM 9.0)

with direct vision. The most prominent feature of M 8 is a wedge-shaped dark lane separating the two brightest segments of the nebula. The largest, brightest segment of the nebula lies W of the dark lane. 9-Sgr lies at the NE tip of the dark lane, with NGC 6530 to its E. The second-brightest segment of the nebula surrounds NGC 6530 to the E.

Although M 8 is a spectacular object without filtration, seeing the maximum possible extent and amount of detail requires using a narrowband filter (or, ideally, an O-III filter). Using either type of filter dims down 9-Sgr and the bright stars of NGC 6530, allowing the nebulosity to shine through. Nebulosity that is visible unfiltered only with averted vision becomes visible with direct vision. With filtration and averted vision, fine tendrils, loops, and whorls that are otherwise invisible become strikingly visible against the pure black filtered background.

M 20 (NGC 6514)	★★★★	ଚ୍ଚ		EN/RN + OC	MBUDR
Chart 40-4	Figure 40-7	m9.0 + 6.3, 17.0' x 12.0' + 30.0'		18h 02.4m	-22° 59'

NGC 6514, better known as Messier 20, M 20, or the Trifid Nebula, is another spectacular nebula that is associated with an open cluster. NGC 6514 refers both to the emission/reflection nebula and the open cluster that is embedded in and surrounds it. Le Gentil cataloged the open cluster in or before 1750, but apparently did not see the nebulosity associated with it. Charles Messier discovered and cataloged M 20 on the night of 5 June 1764. Although he did not note any nebulosity in his original description of M 20, in his notes for the nearby M 21, which he discovered the same night, Messier commented about the nebulosity surrounding both M 21 and M 20.

M 20 is easy to find, particularly if you have just observed M 8. M 20 lies about 1.3° NNW (north northwest) of M 8, and is visible in the same 50mm binocular or finder field. With our 50mm binoculars, M 20 is a moderately bright N-S oval nebulosity that surrounds an m6 star, with an m8 star lying 8' NNE. At 90X in our 10" reflector, M 20 is moderately large and very bright. The most prominent feature is the triple radial dark lane that separates the emission nebula portion of M 20 into three unequal segments. The E dark lane is the longest and most prominent, and is of medium width. The S lane is short and narrow, and the W lane is short and thick. The lanes converge into a 2' central core that is noticeably brighter than the dark lanes, but much dimmer than the surrounding emission nebulosity. Averted vision shows distinct mottling or granularity in this central area.

The dark lanes divide the emission nebula into three distinct segments. The N segment is the largest and brightest. The SW segment is the smallest and dimmest. The SE segment is a bit smaller and dimmer than the N segment, but noticeably larger and brighter than the SW segment. In addition the the emission nebula components that surround the m6 central star, averted vision shows very faint nebulosity from the reflection nebula that surrounds the m8 star that lies 8' NNE.

Although M 20 is a fine object without filtration, you really need a narrowband or O-III filter to see all of the detail available. When we use our Orion Ultrablock narrowband filter, the reflection nebulosity surrounding the N m8 star disappears entirely, but the Trifid itself simply pops. The extent of the object increases from about 12' to 15' or so, and very faint nebulosity that is invisible without a filter becomes easily visible with averted vision.

FIGURE 40-7.

NGC 6514 (M 20) (60' field width)

Image reproduced from Digitized Sky Survey courtesy Anglo-Australian Observatory and Space Telescope Science Institute

M 21 (NGC 6531)	★★	✦✦✦	OC	MBUDR
Chart 40-4	Figure 40-8	m5.9, 13.0'	18h 04.2m	-22° 30'

NGC 6531, better known as Messier 21 or M 21, is moderately large, moderately bright open cluster that is in the running for the least impressive Messier object in Sagittarius. Charles Messier discovered and cataloged M 21 on the night of 5 June 1764, the same night that he observed and cataloged M 20.

M 21 is easy to find. It lies about 40' NE of M 20, where it is visible in the same low-power eyepiece field. Despite its relatively bright visual magnitude, M 21 is invisible in a 50mm finder and binocular other than its two or three brightest stars, which appear as just field stars. At 90X in our 10" reflector, M 21 is a small, tight cluster of fairly bright stars. The center of the cluster is marked by a prominent m8 star with an m9 companion. A dozen m8 through m10 stars form the tight central concentration of the cluster, with another three dozen dimmer stars filling out the cluster. A chain of m9/10 stars extends E of center, and group of three prominent m10 stars lies about 3' N of center. The SW edge of the cluster is marked by a very prominent m8 star, and the W edge by a widely spaced pair of m10 stars.

FIGURE 40-8.

NGC 6531 (M 21) (60' field width)

Image reproduced from Digitized Sky Survey courtesy Anglo-Australian Observatory and Space Telescope Science Institute

NGC 6603 + M 24 (IC 4715)	★★ + ★★★★	✦✦✦ + ✦✦✦	OC + OC (star cloud)	MBUDR
Chart 40-5	Figure 40-9	m11.5, 4.0' + 120.0'	18h 17.0m	-18° 36'

M 24 has the distinction of being one of only three Messier objects (along with M 25 and M 40) that was not assigned an NGC number. The small open cluster NGC 6603 is embedded in the gigantic Milky Way star cloud that comprises M 24. That star cloud was later assigned the Index Catalog number IC 4715. Charles Messier observed and cataloged M 24 on the night of 20 June 1764. From his original description—"Cluster on the parallel of the preceding & near the end of the bow of Sagittarius, in the Milky Way: a large nebulosity in which there are many stars of different magnitudes: the light which is spread throughout this cluster is divided into several parts; it is the center of this cluster which has been determined. (diam. 1d 30')"—it is clear that Messier was referring to IC 4715 rather than just NGC 6603.

M 24 is easy to locate. It lies 7.2° NNW of the bright naked-eye star m2.8 22-λ (lambda) Sagittarii (Kaus Borealis), the top of the teapot, and 2.5° NNE of m3.8 13-μ (mu) Sgr. Because of its gigantic size, M 24 is best viewed with a binocular—the larger the better—

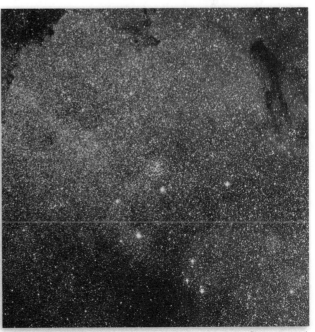

FIGURE 40-9.

NGC 6603 + IC 4715 (M 24) (60' field width)

Image reproduced from Digitized Sky Survey courtesy Anglo-Australian Observatory and Space Telescope Science Institute

or with a richest-field telescope at very low magnification. With our 50mm binoculars, M 24 is a magnificent sight. 100+ stars are visible from m6 down to m10, embedded in a bright nebulosity of stars too dim to be visible directly. At 42X in our 10" Dob (1.7° true field of view), M 24 is breathtaking. Zillions of stars are visible, and zillions more below the threshold of visibility provide a bright background nebulosity. Figure 40-9 gives an impression of the incredible richness of this star cloud, but no photograph can convey the beauty of this object. To do that, you have to see it with your own eyes.

NGC 6603 is visible as a small, tight open cluster in Figure 40-9, just above the central star in the E-W line of three m7 stars. At 90X in our 10" reflector, NGC 6603 is well detached from the surrounding star field. NGC 6603 is a 6' circular cluster with 50+ faint stars visible. The m8 field star that lies 5' SSW of NGC 6603 is a striking reddish-orange color.

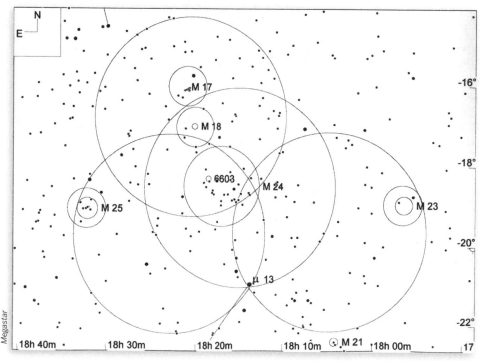

CHART 40-5.

NGC 6603 and IC 4715 (M 24), NGC 6613 (M 18), NGC 6618 (M 17), NGC 6494 (M 23), and IC 4725 (M 25) (12° field width; 5° finder circles; 1° eyepiece circles; LM 9.0)

M 18 (NGC 6613)	★★	◑◑◑	OC	MBUDR
Chart 40-5	Figure 40-10	m6.9, 9.0'	18h 20.0m	-17° 06'

NGC 6613, better known as Messier 18 or M 18, is a relatively large, relatively bright open cluster. Charles Messier discovered and cataloged M 18 on the night of 3 June 1764, just after he discovered and cataloged M 17.

M 18 is easy to find, lying about midway between M 22 and M 17. With our 50mm binoculars, M 18 is a small but prominent nebulosity lying between M 22 and M 17. With averted vision, a few faint stars are visible, surrounding and embedded in a small nebulous patch. At 90X in our 10" reflector, M 18 is very loose and scattered, with about two dozen stars visible from m9 down to m13. The dozen or so brightest stars are arranged in two parallel chains that run NNE-SSW.

FIGURE 40-10.

NGC 6613 (M 18) (60' field width)

Image reproduced from Digitized Sky Survey courtesy Anglo-Australian Observatory and Space Telescope Science Institute

M 17 (NGC 6618)	★★★★	◍◍◍	EN	MBUDR
Chart 40-5	Figure 40-11	m6.9, 11.0' x 6.0'	18h 20.8m	-16° 10'

FIGURE 40-11.

NGC 6618 (M 17) (60' field width)

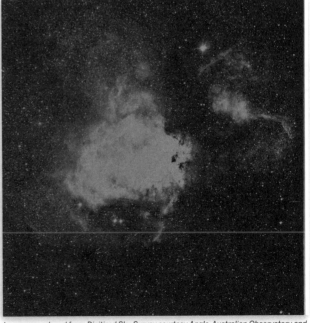

Image reproduced from Digitized Sky Survey courtesy Anglo-Australian Observatory and Space Telescope Science Institute

NGC 6618 probably has more common names than any other deep-sky object. It is known as Messier 17, M 17, the Swan Nebula, the Omega Nebula, the Horseshoe Nebula, the Lobster Nebula, the Checkmark Nebula, and probably others we haven't heard about. M 17 is a spectacular emission nebula. Although it is much smaller than M 8 or M 42 in Orion, it is very bright and rich in detail. Although M 17 is a naked-eye object under very good conditions and so may have been known since antiquity, the first recorded observation of M 17 was made by Philippe Loys de Chéseaux in 1746. This discovery was not widely reported, and Charles Messier independently rediscovered and cataloged M 17 on the night of 3 June 1764.

M 17 is easy to find simply by sweeping your binocular or finder 9.4° slightly W of N from the bright naked-eye star m2.8 22-λ (lambda) Sagittarii (Kaus Borealis). With our 50mm binoculars, M 17 is a relatively small but very bright patch of haze extending about 5' E-W. At 90X in our 10" reflector, M 17 is a very bright nebulosity extending 1.5' x 10' ESE-WNW, anchored on the SE corner by an m7 star, with a very prominent, closely-spaced m8 double star lying 15' SE. A small part of the much fainter nebulosity that extends N and E of the brighter nebulosity is visible with averted vision, surrounding several of the brightest members of the associated open cluster. The brightest part of the nebulosity reveals significant detail, with loops, whorls, knots, and tendrils visible throughout, particularly in the W portion near the m7 star.

As impressive as M 17 is without filtration, a narrowband or O-III filter improves it dramatically. Using our Orion Ultrablock narrowband filter dims down the bright stars nearest the nebula, revealing significant additional detail.

M 25 (IC 4725)	★★★	✿✿✿	OC	MBUDR
Chart 40-5	Figure 40-12	m4.9, 40.0'	18h 28.8m	-19° 17'

IC 4725, better known as Messier 25 or M 25, is a very large, very bright open cluster. M 25 is bright enough to be visible to the naked eye under excellent observing conditions as a faint nebulosity, so it has probably been known since antiquity. Philippe Loys de Chéseaux made the first recorded observation of this object in 1746, but Charles Messier, apparently unaware of that discovery, independently rediscovered, observed, and cataloged M 25 on the night of 20 June 1764. M 25 shares with M 24 and M 40 the distinction of being one of only three Messier objects that were not assigned NGC numbers.

M 25 is easy to find simply by sweeping your binocular or finder 6.4° N of the bright naked-eye star m2.8 22-λ (lambda) Sagittarii (Kaus Borelis), which is the top of the teapot. With our 50mm binoculars, M 25 is a large, moderately bright patch of haze with 15+ m7 through m9 stars embedded. At 90X in our 10" reflector, M 25 is an impressive cluster of 100+ stars m7 and dimmer. The center of the cluster includes five of the most prominent stars, which anchor two prominent E-W (from east to west) chains. Two more very prominent stars lie near the NE and NNW edges of the cluster, at the ends of another chain of stars that arcs N. Scores of dimmer stars fill out the cluster, arranged in loops, chains, arcs, and knots.

FIGURE 40-12.

IC 4725 (M 25) (60' field width)

Image reproduced from Digitized Sky Survey courtesy Anglo-Australian Observatory and Space Telescope Science Institute

M 23 (NGC 6494)	★★★	✿✿✿	OC	MBUDR
Chart 40-5	Figure 40-13	m5.5, 27.0'	17h 56.9m	-19° 01'

NGC 6494, better known as Messier 23 or M 23, is yet another large, very bright open cluster. Charles Messier discovered and cataloged M 23 on the night of 20 June 1764.

M 23 is easy to find, lying 4.5° WNW of m3.8 13-μ (mu) Sagittarii. To locate M 23, put 13-μ on the E edge of your binocular or finder field and look for M 23 on the NW edge of the field, where it appears very prominently. With our 50mm binoculars, M 23 is a bright patch of haze about the size of the full moon. One m7 star and one m8 star are visible with direct vision, and four or five much fainter stars are just at the limited of visibility with averted vision. At 90X in our 10" reflector, 100+ m9 and dimmer stars are visible in a 25' extent. The two stars readily visible with a binocular—the m8 star about 9' NE of the cluster center and the m7 star about 19' NW—are actually foreground field stars. The m9/10 and dimmer stars that are actually members of the cluster are arranged in numerous arcs, chains, and clumps with dark voids separating them.

FIGURE 40-13.

NGC 6494 (M 23) (60' field width)

Image reproduced from Digitized Sky Survey courtesy Anglo-Australian Observatory and Space Telescope Science Institute

NGC 6445	★★	◌◌◌	PN	MBUDR
Chart 40-6	Figure 40-14	m13.2, 44.0" x 30.0"	17h 49.2m	-20° 01'

NGC 6445 is a small planetary nebula with surface brightness much higher than its visual magnitude of 13.2 indicates. NGC 6445 is surprisingly easy to find, lying 2.1° WSW of M 23. To locate NGC 6445, center M 23 in your finder and look for a prominent 1° semicircle of five m7/8 stars, the northernmost of which lies 1.1° dead W of M 23. About 40' dead W of the bottom of this arc lies another m8 star that is prominent in the finder, with NGC 6445 5' dead W of that star and visible in a low-power eyepiece as a bright, fuzzy star, only slightly dimmer than the globular cluster NGC 6440 that lies 23' to its S. At 90X and 180X in our 10" reflector, NGC 6445 shows an obvious disc, but no detail.

FIGURE 40-14.

NGC 6445 (center) and NGC 6440 (bottom center) (60' field width)

Image reproduced from Digitized Sky Survey courtesy Anglo-Australian Observatory and Space Telescope Science Institute

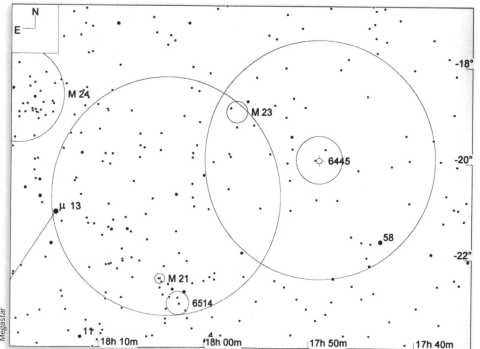

NGC 6520	★★	✧✧✧✧	OC	MBUDR
Chart 40-7	Figure 40-15	m7.6, 6.0'	18h 03.4m	-27° 53'

NGC 6520 is a bright, medium size open cluster that lies in the midst of a dense Milky Way star field. NGC 6520 is notable for lying just E of a dense patch of interstellar dust that blocks nearly all light from the background stars. This dark nebula, Barnard 86, is commonly known as the Ink Spot.

NGC 6520 is easy to locate. Put the bright naked-eye star m2.7 19-δ (delta) Sgr (Kaus Media) on the SE edge of your binocular or finder field and m3.6 10-γ (gamma) Sgr (Alnasi) on the SW edge. The entire field is filled with nebulosity from the thousands of dim stars that make up the Milky Way star field, but NGC 6520 is relatively prominent at the WNW edge of the field as a noticeably brighter patch that forms the NW apex of a right triangle with m4.7 $γ^1$ (gamma^1) Sgr as the S apex and m4.6 SAO 186328 as the NE apex.

At 90X in our 10" Dob, NGC 6520 is a relatively concentrated 5' cluster centered on an m8 star with about two dozen m9 and dimmer stars. An m9 star that appears prominently on the SE edge of the cluster appears to be a foreground star. The Ink Spot is clearly visible to the W of the cluster as a pitch black patch devoid of stars.

FIGURE 40-15.

NGC 6520 (60' field width)

Image reproduced from Digitized Sky Survey courtesy Anglo-Australian Observatory and Space Telescope Science Institute

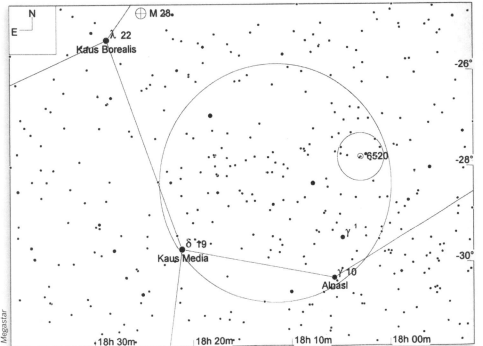

NGC 6716	★	✹✹✹	OC	MBUDR
Chart 40-8	Figure 40-16	m7.5, 6.0'	18h 54.6m	-19° 52'

NGC 6716 is a rather underwhelming open cluster that lies N of the handle of the teapot. To locate NGC 6716, identify the bright naked-eye star m2.8 41-π (pi) Sgr, which lies 6.2° NE of m2.1 34-σ (sigma) Sagittarii (Nunki), the N star in the handle of the teapot. From 41-π, look 1.4° WSW for m3.8 39-o (omicron) Sgr, and then 1.7° WNW for m3.5 37-ξ² (xi²) Sagittarii. NGC 6716 lies 1.4° NNW of 37-ξ² and is visible in a 50mm finder or binocular with averted vision as a relatively faint nebulosity with no stars visible. At 90X in our 10" reflector, NGC 6716 is a loose, scattered group of 15+ m10 and dimmer stars.

FIGURE 40-16.

NGC 6716 (60' field width)

Image reproduced from Digitized Sky Survey courtesy Anglo-Australian Observatory and Space Telescope Science Institute

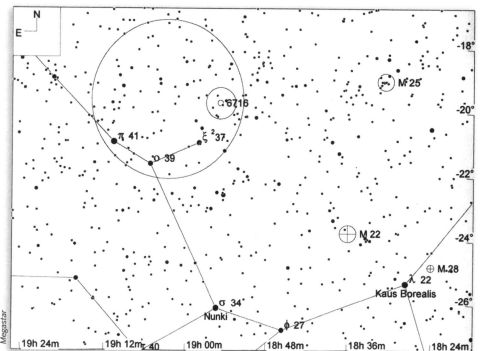

CHART 40-8.

NGC 6716 (15° field width; 5° finder circle; 1° eyepiece circle; LM 9.0)

NGC 6818	★★	◐◐	PN	MBUDR
Chart 40-9	Figure 40-17	m9.9, 48.0"	19h 44.0m	-14° 09'

NGC 6818, better known as the Little Gem Nebula, is a small but bright planetary nebula. To locate NGC 6818, start by placing the bright naked-eye star m2.8 41-π (pi) Sagittarii on the SW edge of your finder field. Look 4.2° NE near the NE edge of the finder for m3.9 44-ϱ^1 (rho^1) Sgr, which appears very prominent in the finder. Place 44-ϱ^1 and m4.6 46-υ (upsilon) Sgr 1.9° to its N on the W edge of the finder field and look for the prominent m5 pair 54- and 55-Sagittarii on or just beyond the E edge of the finder field. NGC 6818 lies about 2.1° NNE of that pair, forming the NE apex of a 1.5° right triangle with two m5.5 stars that are prominent in the finder.

Although NGC 6818 is invisible in our 50mm finder, it's relatively easy to point the scope geometrically close enough to put NGC 6818 near the center of the field of a low-power eyepiece, where it can be confirmed by "blinking" it with a narrowband or O-III filter. At 42X in our 10" reflector, NGC 6818 appears nearly stellar, with just a hint of fuzziness. At 250X, NGC 6818 is a tiny, bright, bluish-green disc with slight central darkening. A narrowband or O-III filter increases the contrast, but reveals no larger extent or additional detail.

FIGURE 40-17.

NGC 6818 (60' field width)

Image reproduced from Digitized Sky Survey courtesy Anglo-Australian Observatory and Space Telescope Science Institute

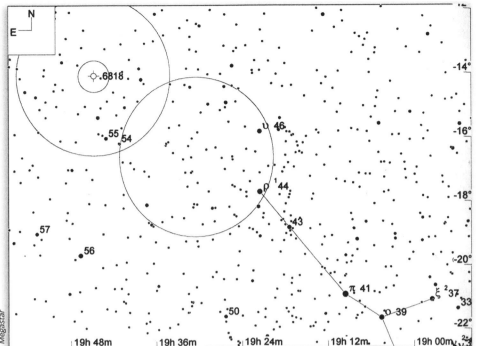

CHART 40-9.

NGC 6818 (Little Gem Nebula)
(15° field width; 5° finder circles;
1° eyepiece circle; LM 9.0)

NGC 6809 (M 55)	★★★	❂❂❂	GC	MBUDR
Chart 40-10	Figure 40-18	m6.3, 19.0'	19h 40.0m	-30° 58'

NGC 6809, better known as Messier 55 or M 55, is a big, bright, extremely loose globular cluster. Abbe Nicholas Louis de la Caille discovered and cataloged this object on 16 June 1752 during his trip to South Africa. Charles Messier attempted to locate it several times, beginning in 1764, but was frustrated in his attempts because the nearly 31° S declination of this object meant that it never rose more than about 11° above the S horizon at Messier's Paris observatory. Messier finally succeeded in observing the object on the night of 24 July 1778 and added it to his catalog.

M 55 is easy to find. Put the bright naked-eye stars that form the bottom of the teapot handle, m2.6 38-ζ (zeta) Sagittarii and m3.3 40-τ (tau) Sagittarii, on the W edge of your binocular or finder field, and sweep the field about 8° ESE until M 55 comes into view. With our 50mm binocular, M 55 is an impressive object. Unlike most globular clusters, which are simply round nebulous patches without stars, M 55 actually looks like a globular cluster, or perhaps even an extremely tight open cluster. Although of course it's not actually resolvable with a 50mm binocular, the large, bright halo appears almost granular and gives the impression of numerous stars right on the edge of being resolved.

At 90X in our 10" reflector, M 55 is a beautiful sight, one of our favorite globular clusters. Hundreds of stars are resolved in the halo and all the way to the core. Numerous chains and arcs of stars are visible, particularly to the N and S of the core.

FIGURE 40-18.

NGC 6809 (M 55) (60' field width)

Image reproduced from Digitized Sky Survey courtesy Anglo-Australian Observatory and Space Telescope Science Institute

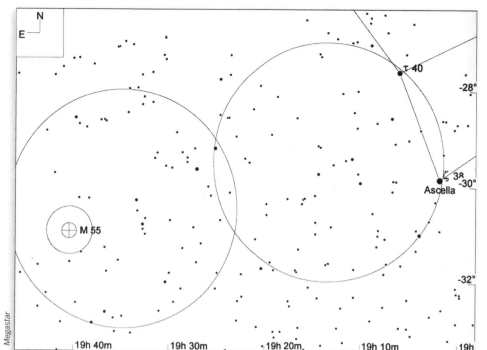

M 75 (NGC 6864)	★★	◐◐	GC	MBUDR
Chart 40-11		m8.6, 6.8'	20h 06.1m	-21° 55'

NGC 6864, better known as Messier 75 or M 75, is a relatively small, dim globular cluster. Pierre Méchain discovered this object on the night of 27 August 1780 and reported it to Charles Messier, who observed and cataloged it on the night of 18 October 1780.

The distance of M 75 is not well known, with estimates ranging as high as 100,000 light years, or about twice the distance from us to the galactic center. Whatever its true distance, M 75 is certainly one of the most distant Messier objects that lies within our own galaxy, and probably the most distant. Because M 75 is both very remote from us and an extremely tight globular cluster, it's very difficult to resolve individual stars in it.

Although M 75 lies within Sagittarius, it's easier to locate it by starting from the bright stars Dabih and Algedi in Capricornus, as shown in Chart 40-11. Although M 75 is invisible to us with our 50mm finder or binoculars, it lies 2.8° dead W of m5.9 4-Capricorni and just under one degree NE of a prominent pair of m6 stars that's easily visible in the finder, which makes it relatively easy to center in the finder using the geometric relationship of those stars.

At 125X in our 10" reflector, M 75 shows a moderately faint 2' featureless halo surrounding a slightly brighter core. Averted vision reveals a few faint stars embedded in the halo.

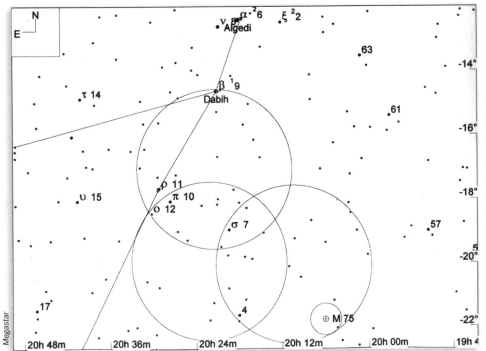

Multiple Stars

There are no featured multiple stars in Sagittarius.

41
Scorpius, The Scorpion

NAME: Scorpius (SKOR-pee-us)

SEASON: Early summer

CULMINATION: midnight, 3 June

ABBREVIATION: Sco

GENITIVE: Scorpii (SKOR-pee-ih)

NEIGHBORS: Ant, Cen, CMi, Cnc, Crt, Crv, Leo, Lib, Lup, Mon, Pup, Pyx, Sex, Vir

BINOCULAR OBJECTS: NGC 6093 (M 80), NGC 6121 (M 4), NGC 6405 (M 6), NGC 6475 (M 7)

URBAN OBJECTS: NGC 6121 (M 4), NGC 6405 (M 6), NGC 6475 (M 7)

Scorpius is a prominent, mid-size, south equatorial constellation that ranks 33rd in size among the 88 constellations. Scorpius covers 497 square degrees of the celestial sphere, or about 1.2%. First-magnitude Antares, the brightest star in Scorpius, keeps company with nearly a dozen second- and third-magnitude stars. Scorpius is unusual among the constellations because it actually resembles its namesake. Scorpius really does look like a scorpion, from its head and claws in the region NW of Antares to the bright star Shaula, which forms its stinger.

Scorpius is an ancient constellation, one of Ptolemy's original 48. Originally, Scorpius included parts of Libra as its claws, which made the constellation outline even more scorpion-like. In Greek mythology, Scorpius is the giant scorpion sent by the goddess Hera to attack Orion, the Mighty Hunter. Orion killed the scorpion, but during the battle was fatally stung. An angry Zeus immortalized Orion and Scorpius by casting them into the sky, acknowledging their enmity toward each other by placing them as far apart as possible. Accordingly, Scorpius never rises while Orion is visible in the night sky, and vice versa.

Scorpius is home to four featured DSOs, all of them Messier clusters. M 4 and M 80 are nice globular clusters, and M 6 and M 7 are bright open clusters. All four of these objects are visible with a binocular (although a telescope is needed to do them justice). The Astronomical League Binocular Messier Club rates M 4, M 6, and M 7, as Easy objects with a 35mm or 50mm binocular, and M 80 as a Tougher object (the middle category). We think AL is a being a bit conservative on M 80, which we find to be easy enough with our 50mm binoculars. AL rates all four Messier clusters as Easy objects with binoculars larger than 50mm.

Scorpius culminates at midnight on 3 June, and, for observers at mid-northern latitudes, is best placed for evening viewing from late spring through mid-summer.

TABLE 41-1.

Featured star clusters, nebulae, and galaxies in Scorpius

Object	Type	Mv	Size	RA	Dec	M	B	U	D	R	Notes
NGC 6093	GC	7.3	10.0	16 17.0	-22 59	◉	◉				M 80; Class II
NGC 6121	GC	5.4	30.0	16 23.6	-26 32	◉	◉	◉			M 4; Class IX
NGC 6405	OC	4.2	30.0	17 40.3	-32 16	◉	◉	◉			M 6; Cr 341; Class II 3 r
NGC 6475	OC	3.3	80.0	17 53.9	-34 47	◉	◉	◉			M 7; Cr 354; Class I 3 r

TABLE 41-2.

Featured multiple stars in Scorpius

Object	Pair	M1	M2	Sep	PA	Year	RA	Dec	UO	DS	Notes
8-beta	ADS 9913	2.6	4.5	13.6	20	2003	16 05.4	-19 48	◉	◉	Graffias; H 7AC
14-nu	ADS 9951	4.2	6.6	40.8	337	2003	16 12.0	-19 27		◉	H 6Aa-C
xi	STF 1998AC	4.9	7.3	7.5	48	2004	16 04.4	-11 22		◉	ADS 9909
SAO 159668	STF 1999AB	7.5	8.1	11.8	98	2003	16 04.4	-11 26		◉	

CHART 41-1.

The heart of the constellation Scorpius (field width 45°)

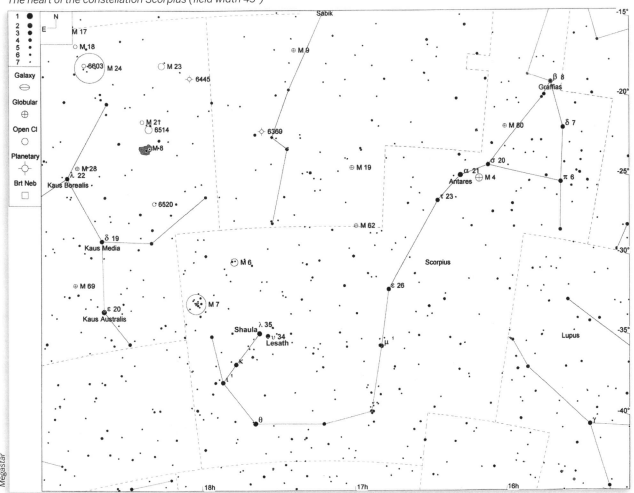

Clusters, Nebulae, and Galaxies

M 80 (NGC 6093)	★★★	✿✿✿✿	GC	MBUDR
Chart 41-2	Figure 41-1	m7.3, 10.0'	16h 17.0m	-22° 59'

NGC 6093, better known as Messier 80 or M 80, is a fine, bright globular cluster. Charles Messier discovered this globular on the night of 4 January 1781 and added it to his catalog as M 80. This was to be the last globular that Messier discovered himself. (The globulars M 92 and M 107 were added to his catalog later, but were not original discoveries by Messier.)

M 80 is very easy to find. It lies almost exactly halfway along the line between m1.1 21-α (alpha) Scorpii (Antares) and m2.6 8-β (beta) Scorpii (Graffias). With our 50mm binoculars, M 80 is a faint, small patch of nebulosity. At 125X in our 10" reflector, M 80 is a bright 5' ball. The core has a granular, mottled appearance with many stars just on the edge of resolution. Numerous outliers are visible, mostly in clumps and chains that extend primarily N and W from the core.

FIGURE 41-1.

NGC 6093 (M 80) (60' field width)

Image reproduced from Digitized Sky Survey courtesy Anglo-Australian Observatory and Space Telescope Science Institute

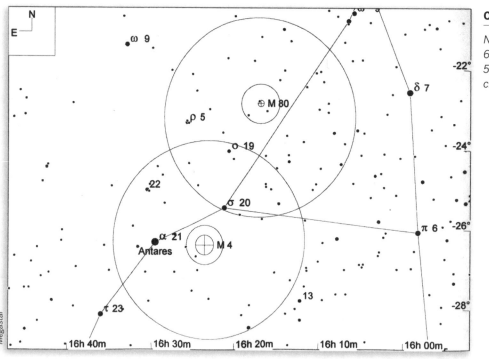

CHART 41-2.

NGC 6093 (M 80) and NGC 6121 (M 4) (12° field width; 5° finder circles; 1° eyepiece circles; LM 9.0)

M 4 (NGC 6121)	★★★	✧✧✧✧	GC	MBUDR
Chart 41-2	Figure 41-2	m5.4, 30.0'	16h 23.6m	-26° 32'

NGC 6121, better known as Messier 4 or M 4, is a large, very bright globular cluster. Philippe Loys de Chéseaux discovered and cataloged this cluster in 1745 or 1746. Charles Messier had discovered and cataloged the globular cluster M 3 on 3 May 1764. The next few nights must have been cloudy, because it wasn't until the night of 8 May 1764 that Messier turned his telescope to M 4, which he already knew about from reading the 1745 catalog of de Chéseaux. Presumably, Messier wanted to compare this earlier discovery with M 3, his own first original discovery.

With the telescopes available to him, Messier had been unable to resolve any stars in the supernova remnant M 1 or the globular cluster M 2, both of which had been discovered and cataloged earlier by other observers. Both of these objects appeared to Messier as nebulous patches of haze without stars, as did his first original discovery, M 3. Messier must therefore have been surprised when he first put M 4 in his eyepiece and saw more than just a hazy patch. Messier reported M 4 as "a cluster of very small stars" that were individually resolvable. M 4 became the first globular cluster to be resolved as anything more than nebulosity.

M 4 is very easy to find. In fact, under very dark conditions, M 4 is a naked-eye object. M 4 lies 1.3° dead W of Antares, where both are visible in the same low-power eyepiece field with our 10" Dob. With our 50mm binoculars, M 4 is a large, bright patch of hazy, easily visible with direct vision even with the glare of first-magnitude Antares in the same field. At 125X in our 10" reflector, M 4 is a very impressive sight. The 10' core is very bright, although less concentrated than most globulars. The cluster appears as a mass of stars of widely different brightness, which can be resolved all the way to the core. A prominent N-S (from north to south) chain of bright stars extends from the halo across the central core. The halo contains numerous chains and clumps of brighter stars embedded in the bright nebulosity of many dimmer stars. This is a beautiful globular.

FIGURE 41-2.

NGC 6121 (M 4) (60' field width)

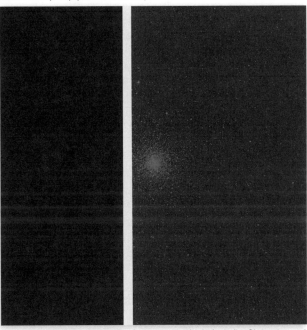

Image reproduced from Digitized Sky Survey courtesy Anglo-Australian Observatory and Space Telescope Science Institute

M 6 (NGC 6405)	★★★	✸✸✸✸	OC	MBUDR
Chart 41-3	Figure 41-3	m4.2, 30.0'	17h 40.3m	-32° 16'

NGC 6405 is a large, bright open cluster that's better known as Messier 6, M 6, or the Butterfly Cluster. As a naked-eye object, M 6 (and its near neighbor, M 7) has been known since ancient times. In the second century, Ptolemy reported M 6 and M 7 as small clouds near the scorpion's stinger, and it's almost certain that these two clusters were known well before that time. By Charles Messier's time, this object was well known to all astronomers, so Messier made no claim to discovering it. He observed M 6 and added it to his catalog on the night of 23 May 1764 as the first of the many open clusters that would eventually appear in his catalog.

M 6 is easy to find, lying 5° NNE (north northeast) of m1.6 Shaula, the stinger in the tail of Scorpius. With our 50mm binoculars, M 6 is a bright cluster of about three dozen mostly m8/9 stars. At 42X in our 10" reflector, 100+ stars are visible. The brightest, a warm-white m6 star, lies near the ENE edge of the cluster and forms the tip of one of the butterfly's wings. A 5' x 20' band of stars extends 20' WSW, forming the two wings. A clump of m8/9 stars extends NE from the center of this band, forming the tail. The two antennae are two chains of m10/11 stars that extend SSE of the cluster center.

FIGURE 41-3.

NGC 6405 (M 6) (60' field width)

Image reproduced from Digitized Sky Survey courtesy Anglo-Australian Observatory and Space Telescope Science Institute

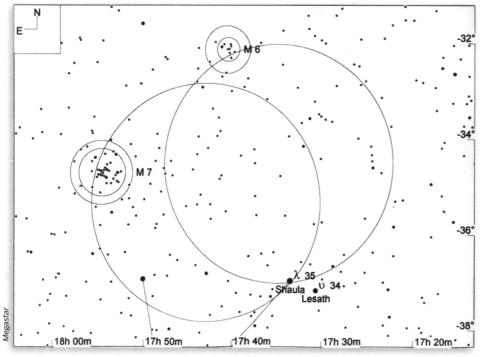

CHART 41-3.

NGC 6405 (M 6) and NGC 6475 (M 7) (10° field width; 5° finder circles; 1° eyepiece circles; LM 9.0)

M 7 (NGC 6475)	★★★	◔◔◔◔	OC	MBUDR
Chart 41-3		m3.3, 80.0'	17h 53.9m	-34° 47'

NGC 6475, better known as Messier 7, M 7, or Ptolemy's Cluster, is a huge, bright, poorly detached open cluster, even larger and brighter than M 6, 4° to its NW. Like M 6, M 7 has been known since antiquity, and was well known to all astronomers in Messier's time. Charles Messier observed M 7 and added it to his catalog on the same night, 23 May 1764, that he observed M 6. At declination -34° 47', M 7 is the most southerly Messier object. M 7 is so far south, in fact, that Messier did well to observe it at all. From Messier's Paris observatory at about 48° N latitude, M 7 never rises more than about 7° above the southern horizon, where atmospheric haze and air pollution (a real problem in Paris in 1764) would have made even an object as bright as M 7 difficult to see.

M 7 is easy to locate, even if you can't see it naked-eye. It lies just over halfway on the line from the bright star Kaus Media (the NW corner of the Sagittarius teapot) to m1.6 Shaula (the stinger in the tail of Scorpius). With our 50mm binoculars, M 7 is a bright, concentrated patch of haze embedded in Milky Way nebulosity. About two dozen stars are visible from m6 down to m9, with many more stars visible with averted vision. At 42X in our 10" reflector (1.7° true field), 100+ stars are visible down to m11 with (to use the technical term) zillions of dimmer background stars. Most of the brighter stars are blue-white or cool white, and many are doublets or triplets. Several chains extend outward from a box of bright stars near the center of the cluster, and a prominent 30' arc of m6/7 stars lies 25' N of center. Even with a 1.7° true field of view, it's not obvious where the edges of the cluster end and the Milky Way star field begins.

Multiple Stars

8-beta (ADS 9913)	★★	◔◔◔◔	MS	UD
Chart 41-1, 41-4		m2.6/4.5, 13.6", PA 20° (2003)	16h 05.4m	-19° 48'

8-β (beta) Scorpii (Graffias) is actually a triple star, but the B component is separated from the primary by only 0.5", much too close to achieve a clean split even with a large telescope on a night of superb seeing. (Few amateurs are lucky enough observe from locations where arcsecond seeing occurs even once a year.)

The Astronomical League Double Star Club list includes the Graffias A-C pair, which is separated by a comfortable 13.6" and can be split cleanly in even a 4" scope. At 90X in our 10" reflector, the primary is a blazing pure white, and the much dimmer companion a soft blue-white.

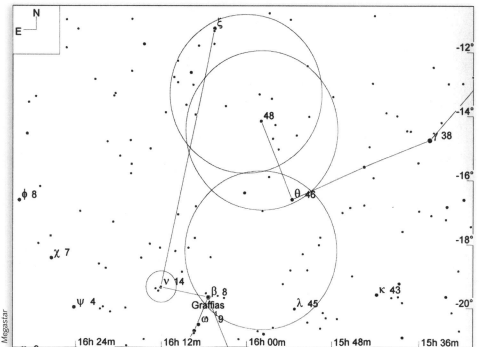

14-nu (ADS 9951)	★★	✦✦✦✦		MS	UD
Chart 41-1, 41-4		m4.2/6.6, 40.8", PA 337° (2003)		16h 12.0m	-19° 27'

14-ν (nu) Scorpii is easy to find, lying 1.6° slightly N of E of Graffias, and prominent in the finder. 14-ν is actually a quadruple system, reminiscent of the famous Double Double in Lyra. The A-B pair are separated by less than an arcsecond, which makes it impossible to split with anything but a large telescope on a night of superb seeing. At 2.3", the separation of the C-D pair is much wider, about the same as the close pairs of the Double Double. The Astronomical League Double Star Club doesn't require splitting either of these close pairs, but only the AB-CD pair, which is separated by more than 40" and can be split cleanly with even a 60mm refractor.

At 90X in our 10" reflector, the AB primary pair are pure white, and the CD companion pair are yellow white.

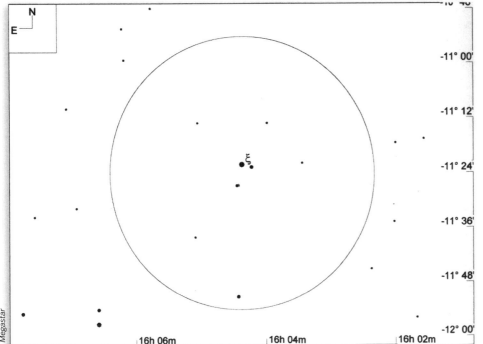

N
E

Megastar

-11° 00'

-11° 12'

-11° 24'

-11° 36'

-11° 48'

-12° 00'

16h 06m   16h 04m   16h 02m

xi (STF 1998AC)	★★	✦✦✦	MS	UD
Chart 41-4, 41-5		m4.9/7.3, 7.5", PA 48° (2004)	16h 04.4m	-11° 22'

SAO 159668 (STF 1999AB)	★★	✦✦✦	MS	UD
Chart 41-4, 41-5		m7.5/8.1, 11.8", PA 98° (2003)	16h 04.4m	-11° 26'

ξ (xi) Scorpii lies 8.4° dead N of Graffias. Although it's a dim naked-eye star, it's easiest to locate it by using the star hop shown in Chart 41-4, from Graffias to 46-θ (theta) Lib to 48-Lib and thence to ξ-Scorpii. Like Graffias, ξ-Scorpii is actually triple, but the AB components are separated by only 0.5" and so can be split only at high magnification on a night of superb seeing. (The best we've ever managed was to see the AB pair as an elongated shape at 240X on a very steady night.) The AB-C pair, which is the one required by the AL Double Star club, has a 7.5" separation and can be split even in a small scope. At 90X in our 10" reflector, the AB primary is yellow and the dimmer companion blue-white.

SAO 159668 (STF 1999AB) lies 4.7' S of ξ-Scorpii, and is visible in the same medium- or high-power eyepiece field. The primary and secondary are of similar brightness, and both are yellow.

42

Sculptor, The Sculptor

NAME: Sculptor (SKULP-tor)

SEASON: Early autumn

CULMINATION: midnight, 27 September

ABBREVIATION: Scl

GENITIVE: Sculptoris (SKULP-tor-is)

NEIGHBORS: Aqr, Cet, For, Gru, Phe, PsA

BINOCULAR OBJECTS: NGC 0253

URBAN OBJECTS: none

Sculptor is an extremely faint, mid-size, mid-southerly const-ellation that ranks 36th in size among the 88 constellations. Sculptor covers 475 square degrees of the celestial sphere, or about 1.2%. Sculptor's brightest stars are only fourth magnitude, and it has only three of those. Sculptor is a modern constellation, created in the 18th century by the French astronomer Abbé Nicolas Louis de Lacaille, and so has no mythology associated with it.

Sculptor has only one featured DSO, the bright galaxy NGC 253, and no featured multiple stars. Sculptor culminates at midnight on 27 September, and, for observers at mid-northern latitudes, is best placed for evening viewing from mid-autumn through early winter.

TABLE 42-1.

Featured star clusters, nebulae, and galaxies in Sculptor

Object	Type	Mv	Size	RA	Dec	M	B	U	D	R	Notes
NGC 253	Gx	8.0	27.7 x 6.7	00 47.5	-25 17				◉	◉	Class SAB(S)c; SB 12.8

CHART 42-1.

The constellation Sculptor (field width 40°)

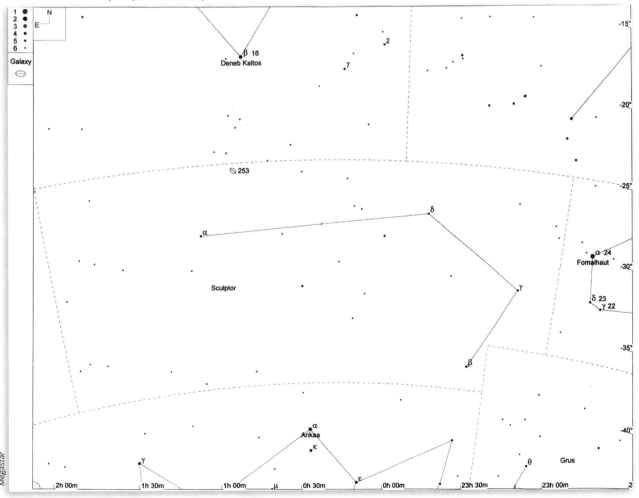

Clusters, Nebulae, and Galaxies

NGC 253	★★★	✸✸✸	GX	MBUDR
Chart 42-2	Figure 42-1	m8.0, 27.7' x 6.7'	00h 47.5m	-25° 17'

NGC 253 is a large, bright galaxy that's the brightest member of the Sculptor Galaxy Group, the galaxy group closest to our own Local Group. NGC 253 is bright enough to be visible in a 50mm binocular or finder, and is an impressive galaxy in a 10" or larger telescope.

NGC 253 is relatively easy to locate, lying 7.3° S of the bright naked-eye star m2.0 16-β (beta) Ceti (Deneb Kaitos). To find NGC 253, put Deneb Kaitos on the N edge of your finder or binocular field, and look in the SE quadrant for a prominent right-triangle formed by three m5/6 stars. Place that triangle at the N edge of the field and look S of center in the field for NGC 253, which is visible in a 50mm finder or binocular with averted vision as a faint but distinct streak of light.

At 90X in our 10" reflector, NGC 253 is a bright, elongated 4' x 20' halo extending NE-SW (from northeast to southwest). An extended, well-concentrated core brightens gradually to a non-stellar nucleus. Averted vision reveals very faint mottling in the halo, particularly to the SW.

FIGURE 42-1.

NGC 253 (60' field width)

Image reproduced from Digitized Sky Survey courtesy Anglo-Australian Observatory and Space Telescope Science Institute

CHART 42-2.

NGC 253 (15° field width; 5° finder circles; 1° eyepiece circle; LM 9.0)

Multiple Stars

There are no featured multiple stars in Sculptor.

43

Scutum, The Shield

NAME: Scutum (SKEW-tum)

SEASON: Summer

CULMINATION: midnight, 1 July

ABBREVIATION: Sct

GENITIVE: Scuti (SKEW-tee)

NEIGHBORS: Aql, Ser, Sgr

BINOCULAR OBJECTS: NGC 6694 (M 26), NGC 6705 (M 11)

URBAN OBJECTS: NGC 6705 (M 11)

Scutum is a tiny, faint, south equatorial constellation that ranks 84th in size among the 88 constellations. Scutum covers only 109 square degrees of the celestial sphere, or about 0.3%. Its brightest stars are only fourth magnitude. It has only two of those, and only nine of the fifth magnitude. The ancients considered the area of Scutum to be just a dark patch of sky, and included it in none of their constellations.

Scutum is a modern constellation, created in 1683 by Polish astronomer Johannes Hevelius to mark the victory of Polish forces under the command of Jan Sobieski in the battle of Vienna. Under its original name, Scutum Sobiescianum (The Shield of Sobieski), Scutum had the distinction of being the only constellation other than Coma Berenices to have been named for an historical personage. The latter part of the appellation was soon dropped, and Scutum became known under its present name, which is associated with the shields carried by the soldiers of ancient Rome.

Scutum has no bright stars and only three DSOs of interest to amateur astronomers, but two of those three are excellent Messier open clusters. The third DSO, NGC 6712, is a small, rather faint globular cluster.

Scutum culminates at midnight on 1 July, and, for observers at mid-northern latitudes, is best placed for evening viewing from late spring through early autumn.

TABLE 43-1.

Featured star clusters, nebulae, and galaxies in Scutum

Object	Type	Mv	Size	RA	Dec	M	B	U	D	R	Notes
NGC 6705	OC	5.8	13.0	18 51.1	-06 16	◉	◉	◉			M 11; Cr 391; Class I 2 r
NGC 6694	OC	8.0	14.0	18 45.2	-09 23	◉	◉				M 26; Cr 389; Class II 3 m
NGC 6712	GC	8.1	9.8	18 53.1	-08 42					◉	Class IX

CHART 43-1.

The constellation Scutum (field width 30°)

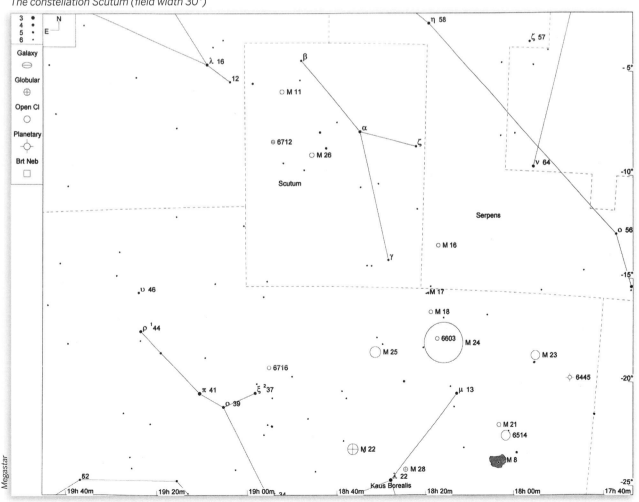

Clusters, Nebulae, and Galaxies

M 11 (NGC 6705)	★★★	✸✸✸	OC	MBUDr
Chart 43-2	Figure 43-1	m5.8, 13.0'	18h 51.1m	-06° 16'

NGC 6705, better known as Messier 11, M 11 or the Wild Duck Cluster, is a large, bright, very rich open cluster. M 11 was discovered in 1681 by the German astronomer Gottfried Kirch. Charles Messier observed and cataloged M 11 on the night of 30 May 1764.

The easiest way to locate M 11 is to start from the naked-eye star m3.4 16-λ (lambda) Aquilae. Place that star on the E edge of your binocular or finder field and look 4° WSW (west southwest), where M 11 appears prominently.

The Astronomical League Binocular Messier Club rates M 11 as an Easy object with a 35mm or larger binocular. With our 50mm binoculars, M 11 is visible with direct vision as a bright hazy patch of tiny faint stars with one m8 star lying near the center. At 90X in our 10" reflector, M 11 looks more like a very loose globular cluster than an open cluster. There are (we guesstimate) 150 to 200 stars from m11 down to m14. Other than the one m8 star near the center of the cluster, which appears to be a foreground field star, the brightest 100 or so stars are all of m11/12 and uniformly distributed across the 15' extent of the cluster.

Despite repeated efforts with everything from 50mm binoculars to a 17.5" Dob at various magnifications, we can't see any resemblance to a duck, wild or otherwise.

FIGURE 43-1.

NGC 6705 (M 11) (60' field width)

Image reproduced from Digitized Sky Survey courtesy Anglo-Australian Observatory and Space Telescope Science Institute

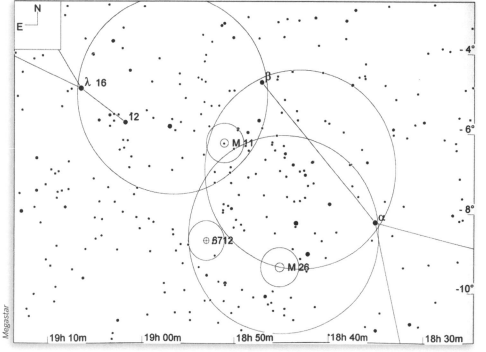

CHART 43-2.

NGC 6705 (M 11), NGC 6694 (M 26), and NGC 6712 (12° field width; 5° finder circles; 1° eyepiece circles; LM 9.0)

M 26 (NGC 6694)	★★★	✹✹✹	OC	MBUDR
Chart 43-2	Figure 43-2	m8.0, 14.0'	18h 45.2m	-09° 23'

NGC 6694, better known as Messier 26 or M 26, is a large, bright, open cluster. Charles Messier discovered and cataloged M 26 on the night of 20 June 1764. Although M 26 suffers by comparison to the nearby open cluster M 11—even Messier noted that he needed a "better instrument" to view it—it's a very nice cluster nonetheless.

M 26 is difficult to locate directly, but easy to locate if you work from M 11. To do so, put M 11 near the NE edge of your finder field and look for m4.2 β (beta) Scuti 1.8° NNW(north northwest) and m3.9 α (alpha) Scuti 4.4° WSW. Both of these stars are prominent in the finder. In the S quadrant of the finder field, you'll see two very prominent stars, m4.9 ε (epsilon) is the N star of that pair, with m4.7 δ (delta) about 50' to its SSW. M 26 lies about 50' ESE (east southeast) of δ Scuti, forming the SE apex of a right triangle with those two stars.

The Astronomical League Binocular Messier Club rates M 26 as a Challenge object (the most difficult category) with 35mm and 50mm binoculars, and an Easy object with an 80mm binocular. We agree with that rating, as we've never seen even a hint of M 26 with our 50mm finder or 50mm binoculars. At 90X with our 10" reflector, M 26 is a fine, bright, concentrated open cluster that's well detached from the surrounding star field. About three dozen stars, mostly m12/13 are visible in a 10' area, with the brighter stars concentrated in clumps in the S half. The N half contains about the same number of mostly dimmer stars, which are more uniformly distributed.

FIGURE 43-2.

NGC 6694 (M 26) (60' field width)

Image reproduced from Digitized Sky Survey courtesy Anglo-Australian Observatory and Space Telescope Science Institute

NGC 6712	★	✧✧	GC	MBUDR
Chart 43-2	Figure 43-3	m8.1, 9.8'	18h 53.1m	-08° 42'

NGC 6712 is a small, rather dim globular cluster, much smaller and dimmer than its cataloged size and magnitude suggest. To locate NGC 6712, begin by centering M 26 in your eyepiece (and therefore your finder). Look for a prominent 1.7° ENE-WSW (from east northeast to west southwest) line of three m6 stars, the W member of which lies about 50' SSE of M 26. NGC 6712 lies 52' dead N of the E star in that chain, near the ENE edge of the finder field, and is clearly visible in a low-power eyepiece.

At 125X in our 10" reflector, NGC 6712 is a moderately faint 3' circular nebulosity with some central brightening. No stars are resolved.

FIGURE 43-3.

NGC 6712 (60' field width)

Image reproduced from Digitized Sky Survey courtesy Anglo-Australian Observatory and Space Telescope Science Institute

Multiple Stars

There are no featured multiple stars in Scutum.

44

Serpens, The Serpent

NAME: Serpens (SUR-penz)	
SEASON: Late spring	
CULMINATION: midnight, 3 June	
ABBREVIATION: Ser	
GENITIVE: Serpentis (sur-PEN-tis)	
NEIGHBORS: Aql, Boo, CrB, Her, Lib, Oph, Sct, Sgr, Vir	
BINOCULAR OBJECTS: NGC 5904 (M 5), NGC 6611 (M 16), IC 4756	
URBAN OBJECTS: NGC 5904 (M 5), IC 4756	

Serpens is a large, relatively dim equatorial constellation that ranks 23rd in size among the 88 constellations. Serpens covers 637 square degrees of the celestial sphere, or about 1.5%. Its brightest stars are third magnitude, and it has only two of those. Serpens is unique among the constellations in not being contiguous. Serpens Caput (the Head of the Serpent) lies to the northwest and Serpens Cauda (the Tail of the Serpent) to the southeast of Ophiuchus (the Serpent Bearer), which divides the two parts of Serpens.

Serpens is an ancient constellation, one of Ptolemy's original 48, but has been recognized as a constellation since at least the dawn of recorded history. In early Greek mythology, Serpens and Ophiuchus were considered a single constellation, the Snake Bearer, and from that grew the myths that surround Ascelpius and the dawn of medicine.

Despite its size, Serpens has relatively few interesting objects. It is home to two bright Messier objects, the globular cluster M 5 and the open cluster M 16. as well as one of the largest, brightest open clusters that wasn't assigned an NGC number, IC 4756.

Serpens culminates at midnight on 3 June, and, for observers at mid-northern latitudes, is best placed for evening viewing from mid-spring through late summer.

TABLE 44-1.

Featured star clusters, nebulae, and galaxies in Serpens

Object	Type	Mv	Size	RA	Dec	M	B	U	D	R	Notes
NGC 5904	GC	5.7	23.0	15 18.6	+02 05	◉	◉	◉			M 5; Class V
NGC 6611	OC	6.0	6.0	18 18.7	-13 48	◉	◉				M 16; Cr 375; Class II 3 m n
IC 4756	OC	4.6	52.0	18 38.9	+05 26			◉	◉		Cr 386; Mel 210; Class II 3 r

TABLE 44-2.

Featured multiple stars in Serpens

Object	Pair	M1	M2	Sep	PA	Year	RA	Dec	UO	DS	Notes
13-delta*	STF 1954AB	4.2	5.2	4.0	173	2006	15 34.8	+10 32		◉	
63-theta1	STF 2417AB	4.6	4.9	22.3	104	2004	18 56.2	+04 12		◉	Alya

CHART 44-1.

The constellation Serpens Caput (field width 45°)

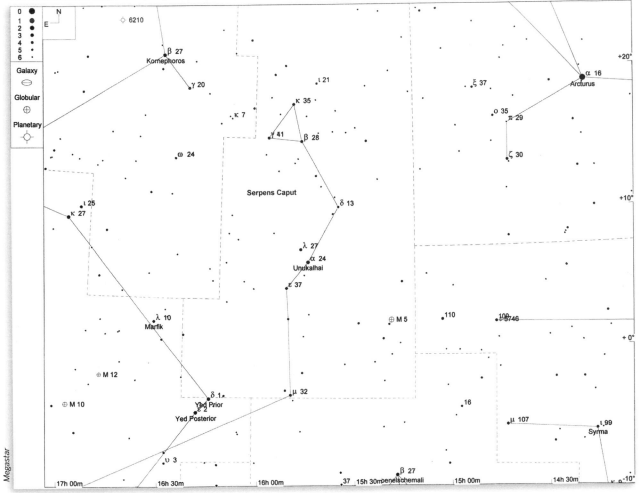

Megastar

CHART 44-2.

The constellation Serpens Cauda (field width 45°)

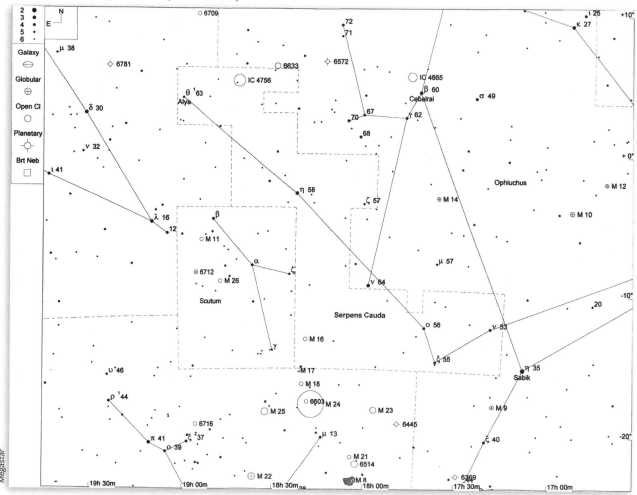

Clusters, Nebulae, and Galaxies

M 5 (NGC 5904)	★★★	◊◊◊	GC	MBUDR
Chart 44-3	Figure 44-1	m5.7, 23.0'	15h 18.6m	+02° 05'

NGC 5094, better known as Messier 5 or M 5, is a large, bright globular cluster. The German astronomers Gottfried and Maria Margarethe Kirch discovered M 5 on the night of 5 May 1702 as they were observing the comet of that year. They described the object as a "nebulous star" and apparently took no more interest in it. Charles Messier independently re-discovered and cataloged M 5 on the night of 23 May 1764.

Although M 5 lies in a region that's nearly devoid of bright stars, it's still relatively easy to find. Start by identifying the bright naked-eye star m2.6 24-α (alpha) Serpentis, which lies in the middle of Serpens Caput and is by far the most prominent star in the vicinity. M 5 lies 7.7° SW of 24-α, and can be found easily by sweeping your binocular or finder field SW from 24-α. M 5 is so bright that some observers have reported glimpsing it with the naked eye, and it is very prominent in any optical finder or binocular.

With our 50mm binocular, M 5 is a large, very bright nebulous patch lying about 20' NW of the bright star m5.1 5-Serpentis. No stars are visible within the nebulosity, although averted vision shows a 45' arc of m10 and dimmer stars surrounding the nebulosity to the N. At 125X in our 10" reflector, M 5 shows a bright core with a granular, mottled appearance embedded in a 12' halo with a slightly oval shape extending NE-SW (from northeast to southwest). Chains of mostly m12/13 stars extend from the NNW (north northwest), NNE (north northeast), and SE edges of the halo. This is a very nice globular.

FIGURE 44-1.

NGC 5904 (M 5) (60' field width)

Image reproduced from Digitized Sky Survey courtesy Palomar Observatory and Space Telescope Science Institute

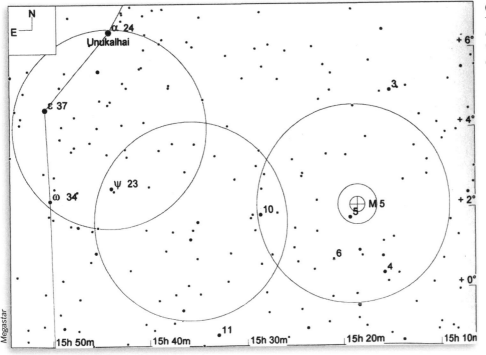

CHART 44-3.

NGC 5904 (M 5) (12° field width; 5° finder circles; 1° eyepiece circle; LM 9.0)

M 16 (NGC 6611)	★★★	❂❂❂	OC	MBUDR
Chart 44-4	Figure 44-2	m6.0, 6.0'	18h 18.7m	-13° 48'

NGC 6611, better known as Messier 16 or M 16, is a bright, medium-size open cluster. The cluster is embedded in and surrounded by the emission nebula IC 4703. Together, these two objects are known as the Eagle Nebula. Philippe Loys De Chéseaux discovered the open cluster in 1745 or 1746, but apparently did not observe the nebula associated with it. Charles Messier independently re-discovered the open cluster on the night of 3 June 1764, and was the first to discover the emission nebula, which he described as a "faint glow." Oddly, although William and Caroline Herschel subsequently cataloged the open cluster, they made no mention of the associated nebulosity, which should have been quite prominent in the 18" instrument they used. Accordingly, the nebulosity now labeled IC 4703 did not appear in John Herschel's General Catalog (GC) or Dreyer's New General Catalog (NGC). It was not until 1908 that IC 4703 was added to the second Index Catalog (IC) supplement to the NGC.

M 16 is a prominent object with any finder or binocular. Although M 16 lies in Serpens, it's easiest to find it by working from Sagittarius, following a chain of easy binocular objects. Start with M 25, and look 1.7° NNE for M 18 and a further 1° N for M 17. With M 17 centered in the binocular or finder, M 16 is prominently visible about 2.4° N.

With our 50mm binoculars, M 16 is a small, bright patch of haze with one m7 star near the center of the nebulosity. At 90X in our

FIGURE 44-2.

NGC 6611 (M 16) (60' field width)

Image reproduced from Digitized Sky Survey courtesy Anglo-Australian Observatory and Space Telescope Science Institute

CHART 44-4.

NGC 6611 (M 16) (10° field width; 5° finder circles; 1° eyepiece circle; LM 9.0)

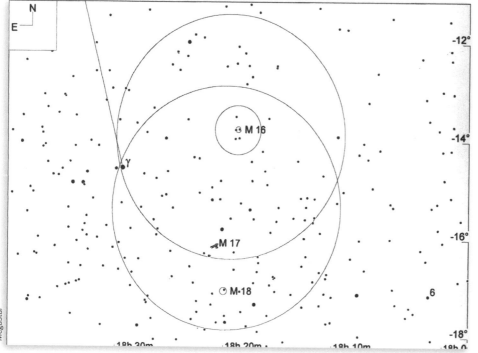

10" reflector, M 16 presents a loose, scattered appearance of about three dozen m8 through m12 stars within a 12' extent. The cluster is embedded in and surrounded by the much fainter nebulosity of IC 4703, the bulk of which lies S of the cluster. The apex of a much darker triangular area extends from the N to nearly touch the cluster. Using our Ultrablock narrowband filter greatly increases the visible extent of the nebulosity and the detail visible in it.

IC 4756	★★	✪✪✪	OC	MBUDR
Chart 44-5	Figure 44-3	m4.6, 52.0'	18h 38.9m	+05° 26'

IC 4756 is a very large, bright, very loose open cluster. The easiest way to find IC 4756 is to start at the bright naked-eye star m3.3 58-η (eta) Serpentis. From there, work your way 11° NE to m4.6 63-θ (theta) Serpentis (Alya), which is dimly visible to the naked eye. IC 4756 lies 4.5° WNW of Alya, where it is visible in a 50mm finder or binocular as a large, rather faint patch of haze with a few faint stars embedded. (As long as you're in the neighborhood, you might as well split Alya, which is one of the double stars featured in this chapter.)

This cluster is nearly a full degree in diameter, four times the area of the full moon. In our 10" Dob even using our low-power eyepiece (42X, 1.7° true field), this object looks more like a random collection of bright field stars than an open cluster. About 75 stars from m8 to m11 are visible, most of which lie within a 45' circle defined by two prominent m6 stars lying on the NW and SE edges of the circle. Although we haven't had an opportunity to try it, we suspect this object would be much more impressive in a giant binocular or our Orion Starblast 4.5" richest-field telescope (3.5° true field).

Image reproduced from Digitized Sky Survey courtesy Palomar Observatory and Space Telescope Science Institute

FIGURE 44-3.

IC 4756 (60' field width)

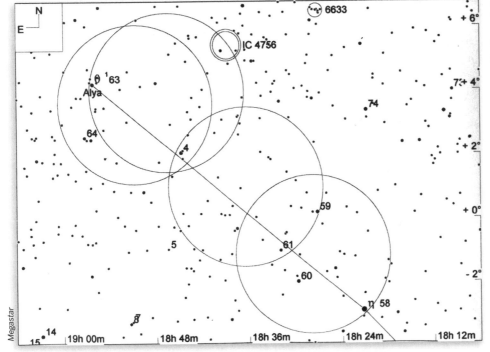

CHART 44-5.

IC 4756 (10° field width; 5° finder circles; 1° eyepiece circle; LM 9.0)

Multiple Stars

13-delta (STF 1954AB)	★★	✦✦✦		MS	UD
Chart 44-1, 44-6		m4.2/5.2, 4.0", PA 173° (2006)		15h 34.8m	+10° 32'

The double star 13-δ (delta) Serpentis is relatively easy to find.
It lies 4.7° NNW of the bright naked-eye star m2.6 24-α (alpha)
Serpentis (Unukalhai). At 90X in our 10" Dob, 13-δ is a close pair of
similar brightness. Both are warm white.

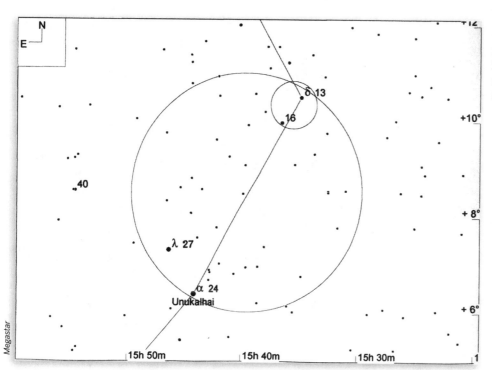

CHART 44-6.

13-delta Serpentis (STF 1954AB) (10° field width; 5° finder circle; 1° eyepiece circle; LM 9.0)

63-theta1 (STF 2417AB)	★★	✧✧✧		MS	UD
Chart 44-2, 44-5		m4.6/4.9, 22.3", PA 104° (2004)		18h 56.2m	+04° 12'

63-θ^1 (theta^1) Serpentis (Alya) is a bright, wide double star that can be split even with a 50mm binocular. Alya is relatively easy to find. Start at the bright naked-eye star m3.3 58-η (eta) Serpentis. From there, work your way 11° NE as shown in Chart 44-5 until you reach m4.6 Alya, which is dimly visible to the naked eye. At 90X in our 10" Dob, Alya is a twin set of blue-white stars.

45

Sextans, The Sextant

NAME: Sextans (SEX-tanz)

SEASON: Winter

CULMINATION: midnight, 21 February

ABBREVIATION: Sex

GENITIVE: Sextantis (sex-TAN-tis)

NEIGHBORS: Crt, Hya, Leo

BINOCULAR OBJECTS: none

URBAN OBJECTS: none

Sextans is an extremely faint, mid-size equatorial constellation that ranks 47th in size among the 88 constellations. Sextans covers 314 square degrees of the celestial sphere, or about 0.8%. Its brightest star is barely fourth magnitude. It has only one of those, and only four of the fifth magnitude. The ancients considered the area of Sextans to be just a dark patch of sky, and included it in none of their constellations. Most modern amateur astronomers have seen Sextans dozens or hundreds of times—it lies just south of Regulus in Leo—but have no idea they've seen it.

Sextans is a modern constellation, created in the late 17th century by Johannes Hevelius, and named in honor of one of his most frequently used instruments. That must have been a real stretch. Most constellations bear little resemblance to what they were named for, but Sextans looks just like a whole lot of nothing. In addition to its utter lack of bright stars, Sextans has few interesting multiple stars, and very few DSOs bright enough to be interesting in typical amateur instruments. The one exception is the rather ordinary spindle galaxy, NGC 3115.

Sextans culminates at midnight on 21 February, and, for observers at mid-northern latitudes, is best placed for evening viewing from mid-winter through late spring.

TABLE 45-1.

Featured star clusters, nebulae, and galaxies in Sextans

Object	Type	Mv	Size	RA	Dec	M	B	U	D	R	Notes
NGC 3115	Gx	9.9	7.2 x 2.4	10 05.2	-07 43					◉	Class S0- sp; SB 10.8

442

CHART 45-1.

The constellation Sextans (field width 40°)

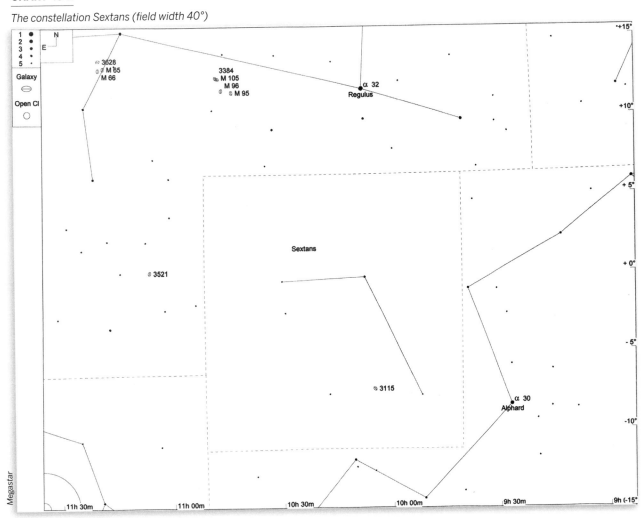

Clusters, Nebulae, and Galaxies

NGC 3115	★★	❂❂❂	GX	MBUDR
Chart 45-2	Figure 45-1	m9.9, 7.2' x 2.4'	10h 05.2m	-07° 43'

NGC 3115 is a large, bright spindle galaxy with high surface brightness. Although it lies in a region that's devoid of bright stars, NGC 3115 is relatively easy to find. Start in Hydra at the bright naked-eye star m2.0 30-α (alpha) and shift your finder 6.2° E until m5.1 8-γ (gamma) Sextantis comes into view. That star and m5.2 22-ε (epsilon) Sextantis 6.2° to its dead E are both dimly visible to the naked eye. NGC 3115 lies halfway on the line from 8-γ to 22-ε, and is visible in a low-power eyepiece.

At 90X in our 10" reflector, NGC 3115 is visible as a bright extended lenticular core with a non-stellar nucleus embedded in a 1' x 4' halo elongated NE-SW (from northeast to southwest).

FIGURE 45-1.

NGC 3115 (60' field width)

Image reproduced from Digitized Sky Survey courtesy Anglo-Australian Observatory and Space Telescope Science Institute

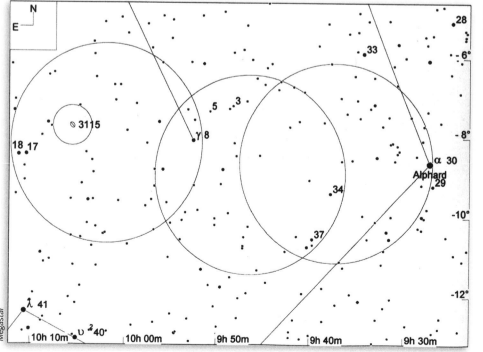

CHART 45-2.

NGC 3115 (12° field width; 5° finder circles; 1° eyepiece circle; LM 9.0)

Multiple Stars

There are no featured multiple stars in Sextans.

46

Taurus, The Bull

NAME: Taurus (TAWR-us)

SEASON: Autumn

CULMINATION: midnight, 30 November

ABBREVIATION: Tau

GENITIVE: Tauri (TAWR-ee)

NEIGHBORS: Ari, Aur, Cet, Eri, Gem, Ori, Per

BINOCULAR OBJECTS: NGC 1432 (M 45), Mel 25, NGC 1647, NGC 1746, NGC 1807, NGC 1817, NGC 1952 (M 1)

URBAN OBJECTS: NGC 1432 (M 45), Hyades, NGC 1647, NGC 1807, NGC 1817

Taurus is a large, prominent north-equatorial constellation that ranks 17th in size among the 88 constellations. Taurus covers 797 square degrees of the celestial sphere, or about 1.9%.

Taurus is an ancient constellation, one of Ptolemy's original 48, and was probably ancient even then. Some scientists believe that Taurus as a constellation is shown in cave paintings at Lascaux, which have been dated to 16,500 years ago. In Greek mythology, Taurus represents the bull form assumed by Zeus to woo the princess Europa, as well as the Cretan Bull that was one of the Twelve Labors of Herakles.

Taurus is rich in DSOs, including two spectacular open clusters, the Pleiades and the Hyades. Taurus is also home to NGC 1952, better known as the Crab Nebula, the remnant of the supernova of 1054, which was reported by Chinese astronomers to be bright enough to be visible during the day. The Crab Nebula is notable

for at least two other reasons as well. It was the first object cataloged by Charles Messier in his list of objects that could easily be mistaken for comets, and it is the only DSO easily visible to amateur astronomers that is close enough to us and expanding fast enough that it actually changes visibly over the short time of a human lifespan.

Taurus culminates at midnight on 30 November, and, for observers at mid-northern latitudes, is best placed for evening viewing from mid-autumn through mid-winter.

TABLE 46-1.

Featured star clusters, nebulae, and galaxies in Taurus

Object	Type	Mv	Size	RA	Dec	M	B	U	D	R	Notes
NGC 1432	OC	1.6	110.0	03 47.0	+24 07	◉	◉	◉			M 45; Cr 42; Mel 22; Class II 3 r (Trumpler) or I 3 r n (modern)
NGC 1514	PN	10.0	1.9	04 09.3	+30 47					◉	Class 3+2
Hyades	OC	0.5	330.0	04 27.0	+16 00			◉	◉		Cr 50; Mel 25; Class II 3 m
NGC 1647	OC	6.4	45.0	04 45.9	+19 08			◉	◉		Cr 54; Mel 26; Class II 2 r
NGC 1746	OC	6.1	41.0	05 03.6	+23 49				◉		Cr 57; Mel 28; Class III 2 p
NGC 1807	OC	7.0	17.0	05 10.8	+16 31			◉	◉		Class II 2 p
NGC 1817	OC	7.7	15.0	05 12.5	+16 41			◉	◉		Cr 60; Class IV 2 r
NGC 1952	SR	8.4	6.0 x 4.0	05 34.5	+22 01	◉	◉				M 1

TABLE 46-2.

Featured multiple stars in Taurus

Object	Pair	M1	M2	Sep	PA	Year	RA	Dec	UO	DS	Notes
59-chi	STF 528	5.4	8.5	19.1	25	2004	04 22.6	+25 37		◉	
118	STF 716AB	5.8	6.7	4.7	208	2002	05 29.3	+25 09		◉	

CHART 46-1.

The constellation Taurus (field width 50°)

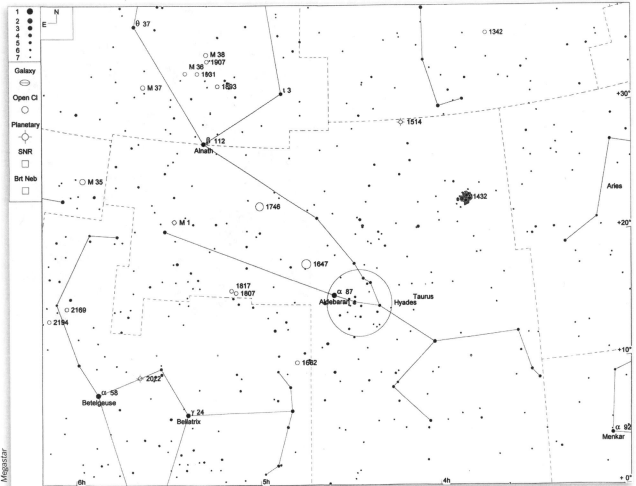

Clusters, Nebulae, and Galaxies

M 45 (NGC 1432)	★★★★	✦✦✦✦	OC	MBUDR
Chart 46-1	Figure 46-1	m1.6, 110.0'	03h 47.0m	+24° 07'

NGC 1432, better known as Messier 45, M 45, the Pleiades, Subaru, or the Seven Sisters, is a huge, extremely bright open cluster, the most prominent open cluster visible in the night sky. The Pleiades has been known since antiquity, and has appeared in cave art and lithographs since before the dawn of recorded history.

Although it is known as the Seven Sisters, only six of the Pleiades stars are easily visible to the naked eye—Alcyone (m2.9), Atlas (m3.6), Electra (m3.7), Maia (m3.9), Merope (m4.2), and Taygeta (m4.3). The seventh star, Pleione (m5.1), is visible to many observers under very dark skies. The eighth and ninth stars, Celaeno (m5.5) and Asterope (m5.8) may be glimpsed under very dark, transparent skies, particularly by young people or others with excellent night vision. Interestingly, the logo of the Japanese automobile company Subaru uses only six stars rather than the traditional seven, and the arrangement of those stars on the logo bears no relation to the actual arrangement of the stars in the Pleiades.

Charles Messier observed and cataloged the Pleiades on the night of 4 March 1769, not because he thought this cluster could be confused with a comet (the original purpose of his list) but because the publication deadline for his catalog was approaching fast and Messier wanted some additional objects to fill out the 41 objects he'd already cataloged to a nice, round number of 45 objects. Accordingly, Messier added four objects that are visible to the naked eye and had been known since antiquity—the Great Nebula in Orion as objects M 42 and M 43, the Beehive Cluster (Praesepe) in Cancer as M 44, and the Pleiades as M 45.

M 45 is a prominent naked-eye object, and is spectacular with even slight optical aid. At nearly 2° in extent, M 45 is far too large to view as a whole in anything other than a binocular or a richest-field telescope at low magnification. However, the nine brightest stars in the Pleiades are grouped closely enough to be viewed at the same time if the eyepiece has a true field of view of just over 1° (62'). Perhaps by coincidence, the 1.25" 25mm or 26mm Plössl eyepiece usually bundled with the ubiquitous 6", 8", and 10" Chinese and Taiwanese Dobsonian scopes has a true field just wide enough to take in the brightest stars of the Pleiades.

With our 50mm binoculars, the Pleiades is a magnificent sight. The dozen or so brightest stars are scattered across a 2° field like diamonds on black velvet. Dozens of dimmer stars fill in the cluster. Although there is nebulosity associated with this cluster, no hint of it is visible in our 50mm binoculars. In our 10" Dob with our 2" 30mm eyepiece (42X with a true field of 1.7°), M 45 is spectacular. The bright blue-white stars of the cluster, particularly Merope, are surrounded by faint wisps of bluish nebulosity. (The nebulosity around Merope is assigned its own catalog number, NGC 1435.) More than 100 dimmer stars are visible, many of them double or triple, arranged in numerous chains and arcs. M 45 is simply magnificent.

FIGURE 46-1.

NGC 1432 (M 45) (60' field width)

Image reproduced from Digitized Sky Survey courtesy Palomar Observatory and Space Telescope Science Institute

NGC 1514	★★	⊙⊙⊙	PN	MBUDR
Chart 46-2	Figure 46-2	m10.0, 1.9'	04h 09.3m	+30° 47'

NGC 1514 is a bright but otherwise rather ordinary planetary nebula, notable only for its unusually bright central star. Although it lies just inside the border of Taurus, it's easiest to find NGC 1514 by starting at the bright naked-eye star m2.9 44-ζ (zeta) Persei (Atik), which lies about 8° N of the Pleiades. Place Atik on the WNW (west northwest) edge of your finder field and look 3.4° ESE for a N-S (from north to south) pair of m8 stars separated by 17'. NGC 1514 lies halfway between those stars.

At 125X and 180X in our 10" Newtonian, NGC 1514 shows a bright 90" disc with a very bright central star. No detail is visible in the disc. Using our Orion Ultrablock narrowband filter does not increase the visible extent of the disc, but does reveal slight darkening of the annulus on the SSW (south southwest) and NNE (north northeast) edges.

FIGURE 46-2.

NGC 1514 (60' field width)

Image reproduced from Digitized Sky Survey courtesy Palomar Observatory and Space Telescope Science Institute

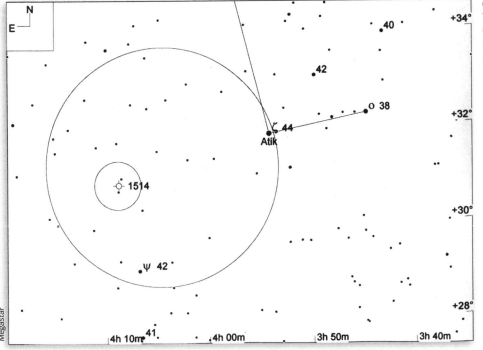

CHART 46-2.

NGC 1514 (10° field width; 5° finder circle; 1° eyepiece circle; LM 9.0)

Hyades	★★★	✷✷✷	OC	MBUDR
Chart 46-1		m0.5, 330.0'	04h 27.0m	+16° 00'

The Hyades, also known as Collinder 50 or Melotte 25, is a gigantic, bright open cluster. At about 150 light years distant, the Hyades are closer to us than any other true open cluster. At about 5.5° extent, the Hyades is much too large to view with anything more than a 7X50 or 10X50 binocular. (So large, in fact, that we couldn't find an image that did the Hyades justice.) Using any instrument with more magnification or a smaller field of view than a binocular simply turns the Hyades into a random collection of bright field stars, losing its cluster nature entirely.

The very bright visual magnitude of the Hyades is due mostly to the presence of the red-orange first-magnitude star Aldebaran in the cluster, although Aldebaran is actually a foreground star rather than a true cluster member. The Hyades centers on the prominent V shape at the head of the bull, formed by Aldebaran and the other bright stars in the cluster. With our 50mm binoculars, the Hyades is a magnificent sight. Most of the bright stars and all of the dimmer ones appear pure white, but five of the brightest stars—Aldebaran, γ (gamma), δ (delta), ε (epsilon), and θ^1 (theta^1)—are distinctly orange.

NGC 1647	★★	✷✷✷✷	OC	MBUDR
Chart 46-3	Figure 46-3	m6.4, 45.0'	04h 45.9m	+19° 08'

NGC 1647 is a large, bright, loose open cluster. To locate NGC 1647, put Aldebaran on the SE edge of your binocular or finder field and look 3.5° NE. With our 50mm binoculars, NGC 1647 is a large, moderately faint patch of haze lying just N of the prominent m6.0 field star visible near the bottom of Figure 46-3. Three or four m9 stars are visible with averted vision, with several others flickering in and out on the edge of visibility. Even at 42X (true field 1.7°) in our 10" reflector, this cluster is not well detached from the background field stars, making it very difficult to determine its boundaries. At 90X, NGC 1647 is a moderately rich, but very loose and scattered open cluster. About 50 stars are visible in a 45' area, the brightest of which are a dozen or so m9/10 stars, with most of the remainder m11/12. Many of the stars are wide doubles or triples, and many are arranged in short arcs, chains, and clumps.

FIGURE 46-3.

NGC 1647 (60' field width)

Image reproduced from Digitized Sky Survey courtesy Palomar Observatory and Space Telescope Science Institute

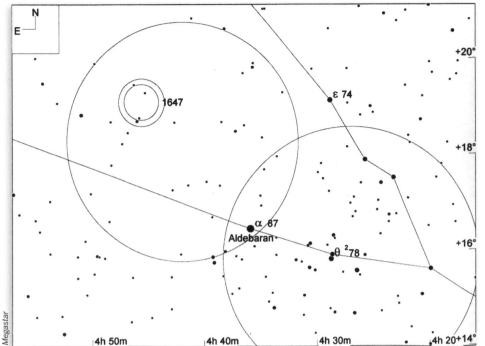

Megastar

NGC 1746	★★	❀❀❀	OC	MBUDR
Chart 46-4	Figure 46-4	m6.1, 41.0'	05h 03.6m	+23° 49'

NGC 1746 is another large, bright, very loose and scattered open cluster. The easiest way to locate NGC 1746 is to start at NGC 1647 and pan your finder or binocular about 6° NE. With our 50mm binoculars, NGC 1746 is a large, moderately faint patch of haze lying just W of the 30' N-S arc of m7/8 stars visible on the left side of Figure 46-4. That arc continues, extending N and then W with the string of three m8 stars visible in the top right quadrant of Figure 46-4.

Even at 42X (true field 1.7°) in our 10" reflector, the borders of this cluster are not well defined. At 90X, NGC 1647 is moderately rich, but extremely loose and scattered. 70+ stars are visible in a 40' area, with two concentrations that were assigned their own NGC numbers, NGC 1750 to the W and NGC 1758 to the E.

FIGURE 46-4.

NGC 1746 (60' field width)

Image reproduced from Digitized Sky Survey courtesy Palomar Observatory and Space Telescope Science Institute

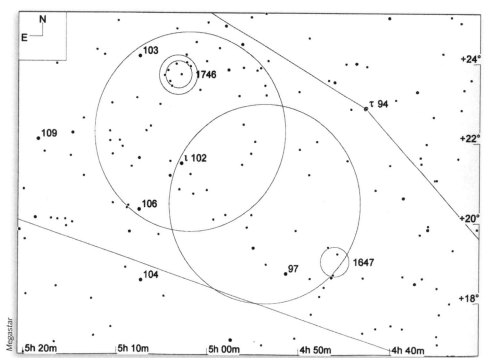

NGC 1807	★★	☾☾☾	OC	MBUDR
Chart 46-5	Figure 46-5	m7.0, 17.0'	05h 10.8m	+16° 31'

NGC 1817	★★	☾☾☾	OC	MBUDR
Chart 46-5	Figure 46-5	m7.7, 15.0'	05h 12.5m	+16° 41'

NGC 1807 and NGC 1817 are a close pair of open clusters that are similar in size and magnitude but much different in appearance. The easiest way to locate NGC 1807 and NGC 1817 is to sweep your binocular or finder field about 8.5° dead E from Aldebaran until the prominent 1.8° E-W arc of three m5 stars (11- and 15-Tauri and SAO 94377) shown in Chart 46-5 comes into view. NGC 1807 and NGC 1817 lie about 30' N of SAO 94377, the E star in that line. The centers of the two clusters are only about 30' apart, so both fit comfortably in a low-power eyepiece field.

The W cluster, NGC 1807, is larger and brighter than the E cluster, NGC 1817, but much sparser. With our 50mm binoculars both clusters are visible as moderately large, moderately bright hazy patches just N of the E-W line of m5 stars. Three m8 stars are visible in NGC 1807, two near the center of the nebulosity and one on the S edge. No stars are visible in NGC 1817.

At 90X in our 10" reflector, both clusters just fit within the field of view. NGC 1807 has 10+ m9/10 stars, most arranged in a N-S line that divides the cluster in half E-W. A prominent triangle of two m9 and one m10 stars lies at the center of the cluster. Another 15+ m11/12 stars are scattered across the 15' extent of the cluster. The most striking feature of NGC 1817 is the prominent 10' NNW-SSE line of three m9 stars that lies along the W edge of the cluster. About three dozen much fainter stars extend E from that line, covering an extent of 15' or so. It's difficult to determine where the N, S, and E boundaries of the cluster lie.

FIGURE 46-5.

NGC 1807 and NGC 1817 (60' field width)

Image reproduced from Digitized Sky Survey courtesy Palomar Observatory and Space Telescope Science Institute

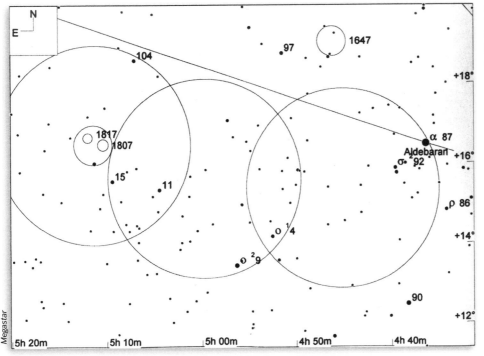

CHART 46-5.

NGC 1807 and NGC 1817 (12° field width; 5° finder circles; 1° eyepiece circles; LM 9.0)

M 1 (NGC 1952)	★★	◊◊◊◊	SR	MBUDR
Chart 46-6	Figure 46-6	m8.4, 6.0' x 4.0'	05h 34.5m	+22° 01'

NGC 1952, better known as Messier 1, M 1, or the Crab Nebula, is the remnant of the of the spectacular supernova of 1054, which was so bright that it was visible during the day for several months and was visible to the naked eye at night for more than a year. M 1 was also the object that got Charles Messier started on his famous list of objects that could be mistaken for comets.

The English astronomer John Bevis discovered and cataloged M 1 in 1731. Charles Messier, unaware of that discovery, independently rediscovered M 1 on the night of 28 August 1758. Messier was scanning the sky that night, looking for the predicted return of Comet Halley, when he happened across a nebulosity that looked very much like a comet. Excited at the possibility of finding a new comet, Messier continued to observe M 1 over the next several nights, hoping to track its proper motion against the background stars. Alas, he eventually determined that M 1 wasn't moving, which meant that it couldn't be a comet. Disgusted at the time he'd wasted on this non-comet, on the night of 12 September 1758 Messier finally recorded and cataloged the position of M 1 as the first object in his new catalog—a list of objects that could easily be mistaken for new comets.

M 1 is easy to find. Identify the bright naked-eye star m1.7 112-β (beta) Tauri (Alnath), which is the S star in the pentagon of Auriga, and look 7.9° SSE for the bright star m3.0 123-ζ (zeta) Tauri. M 1 lies 1.1° NW of 123-ζ.

The Astronomical League Binocular Messier Club lists M 1 as a Challenge object (the most difficult category) for 35mm and 50mm binoculars, and as a Tougher object (the middle category) for an 80mm binocular. We have never seen M 1 in any binocular or finder.

At 90X in our 10" reflector, M 1 is a moderately bright 4'x5' rectangular patch of haze extending NW-SE. With averted vision, some very faint mottling is visible in the bright parts of the nebulosity, but no other detail is visible, at least with a 10" scope.

Although you might expect that the nebulosity of a supernova remnant would be enhanced by a narrowband or line filter just as most planetary nebulae (nova remnants) are, that turns out not to be the case. Unlike planetary nebulae, which are line emitters, supernova remnants are broadband emitters. Using a narrowband or O-III filter on M 1 simply dims it without revealing any additional extent or detail.

FIGURE 46-6.

NGC 1952 (M 1) (60' field width)

Image reproduced from Digitized Sky Survey courtesy Palomar Observatory and Space Telescope Science Institute

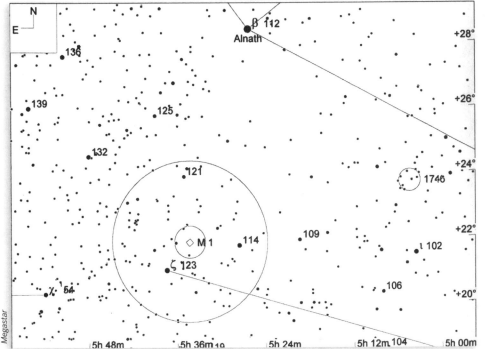

Multiple Stars

59-chi (STF 528)	★★	✹✹✹	**MS**	**UD**
Chart 46-7		m5.4/8.5, 19.1", PA 25° (2004)	04h 22.6m	+25° 37'

59-χ (chi) Tauri (STF 528) is a rather pedestrian double star. It's relatively easy to find, lying 6.6° N of the bright naked-eye star m3.5 74-ε (epsilon) Tauri. To locate 59-χ, put 74-ε near the SE edge of your finder and locate the prominent group of bright stars shown in Chart 46-7. Pan N to put that group near the S edge of your finder and identify 59-χ, which appears prominently near the N edge of the finder field. At 90X in our 10" reflector, the primary appears cool white and the companion white.

118 (STF 716AB)	★★	✹✹✹	**MS**	**UD**
Chart 46-8		m5.8/6.7, 4.7", PA 208° (2002)	05h 29.3m	+25° 09'

118-Tauri (STF 716AB) is another ordinary double star. To locate 118-Tauri, begin at the bright naked-eye star m1.7 112-β (beta) Tauri (Alnath), the S star in the pentagon of Auriga. Look 7.9° SSE for the bright star m3.0 123-ζ (zeta) Tauri. Place 123-ζ on the SSE edge of your finder field, and look 4.5° NNW for m5.8 118-Tauri, which appears very prominent in the finder. At 90X in our 10" reflector, the primary appears warm white and the companion white.

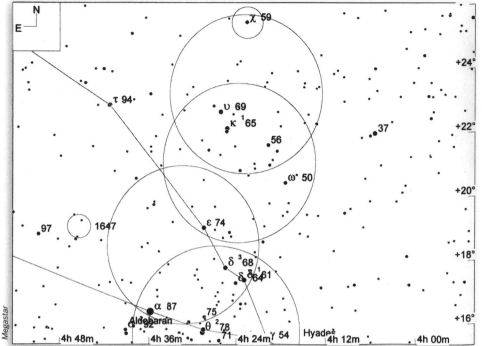

CHART 46-7.

*59-chi Tauri (STF 528) (15°
field width; 5° finder circles; 1°
eyepiece circle; LM 9.0)*

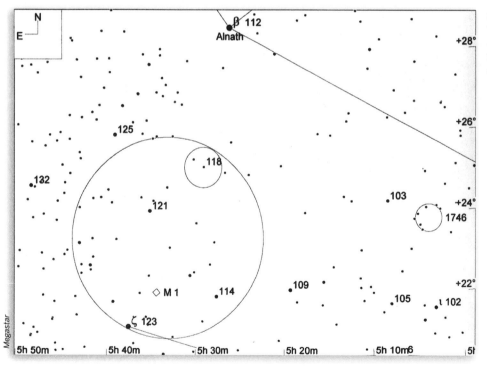

CHART 46-8.

*118-Tauri (STF 716AB) (12°
field width; 5° finder circle; 1°
eyepiece circle; LM 8.0)*

47

Triangulum, The Triangle

NAME: Triangulum (try-ANG-you-lum)

SEASON: Autumn

CULMINATION: midnight, 23 October

ABBREVIATION: Tri

GENITIVE: Trianguli (try-ANG-you-lee)

NEIGHBORS: And, Ari, Per, Psc

BINOCULAR OBJECTS: NGC 0598 (M 33)

URBAN OBJECTS: none

Triangulum is a small, faint mid-northerly constellation that ranks 78th in size among the 88 constellations. Triangulum covers only 132 square degrees of the celestial sphere, or about 0.3%. Triangulum is devoid of bright stars. It has only two third-magnitude stars, one of the fourth magnitude, and seven of the fifth magnitude. Despite its lack of bright stars, however, Triangulum is reasonably prominent because its triangular shape lies far from any bright stars. Triangulum is easy to find, lying nestled SE of the bright stars of Andromeda and SW of the prominent constellation Perseus.

Triangulum is an ancient constellation, one of Ptolemy's original 48 from 1,900 years ago, and old even then. Despite its antiquity, little mythology is associated with this constellation. Although Triangulum is home to many very faint galaxies and other DSOs, with the exception of NGC 598 (M 33, the Pinwheel Galaxy), all are too dim to be satisfying targets for typical amateur instruments.

Triangulum culminates at midnight on 23 October, and, for observers at mid-northern latitudes, is best placed for evening viewing from late summer through mid-winter.

TABLE 47-1.

Featured star clusters, nebulae, and galaxies in Triangulum

Object	Type	Mv	Size	RA	Dec	M	B	U	D	R	Notes
NGC 598	Gx	6.3	65.6 x 38.0	01 33.8	+30 40	◉	◉				M 33; Class SA(s)cd; SB 13.9

TABLE 47-2.

Featured multiple stars in Triangulum

Object	Pair	M1	M2	Sep	PA	Year	RA	Dec	UO	DS	Notes
6-iota	STF 227	5.3	6.7	3.9	69	2002	02 12.4	+30 18		◉	

CHART 47-1.

The constellation Triangulum (field width 30°)

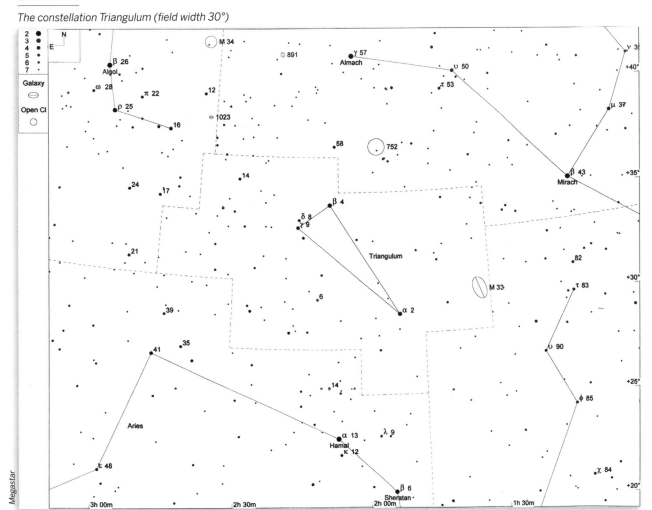

Clusters, Nebulae, and Galaxies

M 33 (NGC 598)	★★	⊙⊙⊙	GX	MBUDR
Chart 47-2	Figure 47-1	m6.3, 65.6' x 38.0'	01h 33.8m	+30° 40'

NGC 598, better known as Messier 33, M 33, or the Pinwheel Galaxy, is a fine spiral galaxy. With the Milky Way itself and the Andromeda Galaxies, M 33 is part of the Local Group of Galaxies, all of which are gravitationally bound. M 33 is one of the more elusive objects on Messier's list. Although its visual magnitude is a relatively bright 6.3, its large size means it has low surface brightness, about 13.9. Under very dark, transparent skies when it is at high elevation, M 33 may be a naked-eye object, but even minor light pollution or haziness may render it completely invisible.

As a naked-eye object, M 33 has probably been known since antiquity, but it was first recorded in the mid-17th century by the Italian astronomer Giovanni Batista Hodierna. Charles Messier, unaware of Hodierna's earlier observation, independently rediscovered and cataloged M 33 on the night of 25 August 1764, describing M 33 as a "whitish light of almost even density" without stars.

M 33 is relatively easy to find, lying about one third of the way along and just SW of the line from m3.4 2-α (alpha) Trianguli to m2.1 43-β (beta) Andromedae (Mirach). The Astronomical League Binocular Messier Club rates M 33 as a Tougher object (the middle category) with a 35/50mm or 80mm binocular, which is probably a reasonable compromise. In our experience, the appearance of M 33 is dramatically affected by observing conditions. On a perfect night—one with very transparent skies and no light pollution—

FIGURE 47-1.

NGC 598 (M 33) (60' field width)

Image reproduced from Digitized Sky Survey courtesy Palomar Observatory and Space Telescope Science Institute

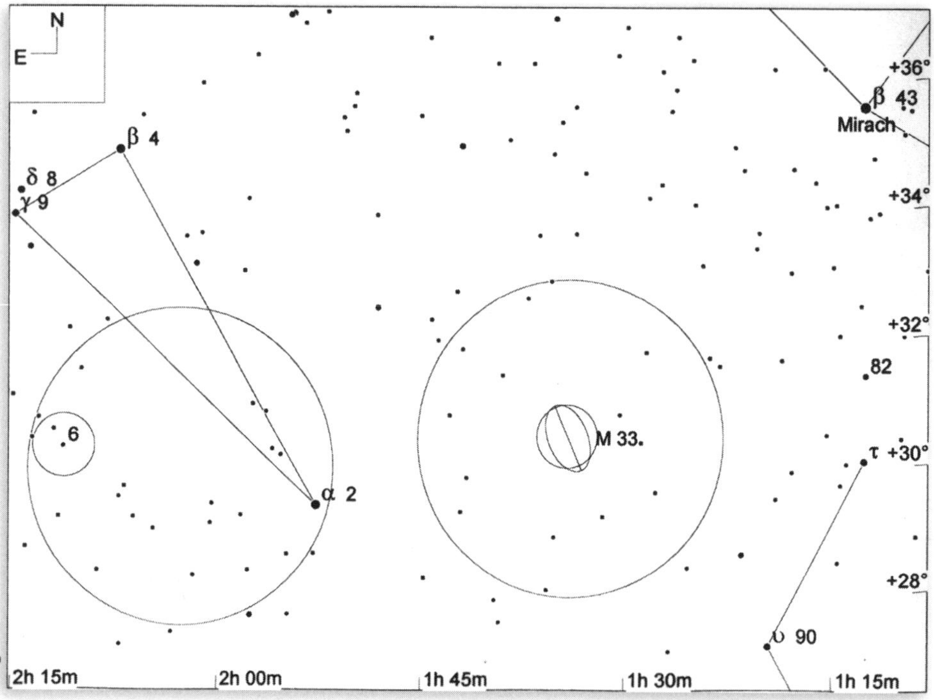

CHART 47-2.

NGC 598 (M 33) and 6-Trianguli (STF 227) (15° field width; 5° finder circles; 1° eyepiece circles; LM 8.0)

M 33 is an easy object with even a 50mm binocular. On even an average night, M 33 may be entirely invisible even with a larger instrument.

We were fortunate enough to bag M 33 on just such a perfect night, when it was at high elevation. With our 50mm binocular, the core of M 33 was visible as a distinctly fuzzy star with tendrils of extremely faint nebulosity surrounding it. At 42X in our 10" reflector, M 33 presented a broad, diffuse core with some mottling and a non-stellar nucleus visible, embedded in a faint, irregular 30' x 50' oval halo extending NNE-SSW (from north northeast to south southwest). Averted vision revealed hints of the spiral structure and dark lanes visible in Figure 47-1.

Multiple Stars

6-iota (STF 227)	★★★	✺✺✺✺		MS	UD
Chart 47-2		m5.3/6.7, 3.9", PA 69° (2002)		02h 12.4m	+30° 18'

6-ι (iota) Trianguli is a very nice double star. It's easy to find m5.2 6-ι by placing m3.4 2-α (alpha) Trianguli on the W edge of your finder field and looking 4.2° E for a prominent star near the W edge of the finder field. At 125X in our 10" Dob, 6-ι presents a nice color-contrast double. The m5.3 primary is distinctly yellow-white, and the close companion a pretty blue-white.

48

Ursa Major, The Larger Bear

NAME: Ursa Major (ER-suh MAY-jur)

SEASON: Spring

CULMINATION: midnight, 11 March

ABBREVIATION: UMa

GENITIVE: Ursae Majoris (ER-sigh muh-JOR-is)

NEIGHBORS: Boo, Cam, Com, CVn, Dra, Leo, LMi, Lyn

BINOCULAR OBJECTS: NGC 3031 (M 81), NGC 3034 (M 82), NGC 3556 (M 108), NGC 3587 (M 97), NGC 3992 (M 109), Win 4 (M 40), NGC 5457 (M 101)

URBAN OBJECTS: NGC 3031 (M 81), NGC 3034 (M 82)

Ursa Major is a large, bright, northerly constellation that ranks 3rd in size among the 88 constellations. Ursa Major covers 1,280 square degrees of the celestial sphere, or about 3.1%. The brightest stars in Ursa Major form the prominent asterism known in the United States as The Big Dipper and in Britain as The Plough. Ursa Major, along with Orion and the Pleiades, is one of the few celestial objects that appears in the Bible.

Ursa Major is an ancient constellation, one of Ptolemy's original 48, and even in Ptolemy's time had been recognized as a constellation for at least three thousand years. Most ancient cultures—including the Chinese, the Egyptians, the Greeks and Romans, the early Britons, and the American Indians—associated Ursa Major with a bear in one form or another. Because bears have short tails, the "handle" stars in the Big Dipper—Alioth, Mizar, and Alkaid—were often, although not aways, associated with cubs following the mother bear.

When we look at Ursa Major, we're looking far above the plane of the Milky Way galaxy, toward intergalactic space. Accordingly, Ursa Major is rich in galaxies, including the bright pair of Messier galaxies, M 81/M 82, and the Pinwheel Galaxy, M 101, as well as dozens of other galaxies that are bright enough to be visible in amateur telescopes. Ursa Major is also home to the bright planetary nebula M 97, otherwise known as the Owl Nebula for its appearance in large amateur telescopes, and to the double star Mizar-Alcor, the best-known double star in the night sky, with a separation of 11.8 arcminutes, nearly half the diameter of the full moon.

All seven of the Messier objects in Ursa Major are listed by the Astronomical League Binocular Messier Club, but most are relatively difficult objects. With a 50mm binocular, M 40, M 81, and M 82 are categorized as Tougher objects (the middle category), and M 97 and M 101 as Challenge objects (the most difficult category). M 108 and M 109 are not listed for 50mm binoculars. With an 80mm binocular, M 40, M 81, and M 82 are categorized as Easy objects, M 97 and M 101 as Tougher objects, and M 108 and M 109 as Challenge objects.

Ursa Major culminates at midnight on 11 March. Because it is so large—the northern boundary lies at 73° N declination and the southern boundary at 28° N declination—the best times of year to observe objects in Ursa Major depend on their positions within the constellation and your own latitude. Far northerly objects are circumpolar for northern observers. For example, for an observer at 50° N latitude, M 81 and M 82, at about 69° N declination, are at 29° elevation or higher at any hour on any night of the year. Conversely, for more southerly observers, southerly objects in Ursa Major such as NGC 3941, at about 37° N declination, are often too low to observe, if not below the horizon. For observers at mid-northern latitudes, most of the objects in Ursa Major are best placed for evening viewing from mid-winter through late summer.

TABLE 48-1.

Featured star clusters, nebulae, and galaxies in Ursa Major

Object	Type	Mv	Size	RA	Dec	M	B	U	D	R	Notes
NGC 2841	Gx	10.1	8.1 x 3.5	09 22.0	+50 59					◉	Class SA(r)b:; SB 12.5
NGC 3079	Gx	11.5	8.0 x 1.4	10 01.9	+55 41					◉	Class SB(s)c sp; SB 13.1
NGC 3031	Gx	7.9	27.1 x 14.2	09 55.6	+69 04	◉	◉	◉			M 81; Class SA(s)ab; SB 12.4
NGC 3034	Gx	9.3	11.3 x 4.2	09 55.9	+69 41	◉	◉	◉			M 82; Class IO sp
NGC 3184	Gx	10.4	7.4 x 6.9	10 18.3	+41 25					◉	Class SAB(rs)cd; SB 13.7
NGC 3556	Gx	10.7	8.7 x 2.2	11 11.5	+55 40	◉	◉				M 108; Class SB(s)cd sp; SB 13.7
NGC 3587	PN	12.0	3.4	11 14.8	+55 01	◉	◉				M 97; Class 3a
NGC 3992	Gx	10.6	7.6 x 4.6	11 57.6	+53 22	◉	◉				M 109; Class SB(rs)bc; SB 11.2
NGC 4026	Gx	11.7	5.2 x 1.4	11 59.4	+50 58					◉	Class S0 sp; SB 10.9
NGC 4088	Gx	11.2	5.3 x 2.1	12 05.6	+50 32					◉	Class SAB(rs)bc
NGC 4157	Gx	12.2	7.7 x 1.3	12 11.1	+50 29					◉	Class SAB(s)b? sp; SB 13.1
NGC 3877	Gx	11.8	5.8 x 1.2	11 46.1	+47 30					◉	Class SA(s)c:
NGC 3941	Gx	11.3	3.7 x 2.3	11 52.9	+36 59					◉	Class SB(s)0ˆ; SB 11.2
Winnecke 4	MS	9.5	---	12 22.2	+58 05	◉	◉				M 40; double star; "Messier Mistake"
NGC 4605	Gx	10.9	5.7 x 2.1	12 40.0	+61 37					◉	Class SB(s)c pec; SB 12.2
NGC 5457	Gx	8.3	28.9 x 26.9	14 03.2	+54 21	◉	◉				M 101; Class SAB(rs)cd; SB 14.0

TABLE 48-2.

Featured multiple stars in Ursa Major

Object	Pair	M1	M2	Sep	PA	Year	RA	Dec	UO	DS	Notes
79-zeta	STF 1744AB	2.2	3.9	14.3	153	2003	13 23.9	+54 55	◉	◉	Mizar
79-zeta	STF 1744AC	2.2	4.0	708.5	71	1991	13 23.9	+54 55	◉	◉	Mizar and Alcor

CHART 48-1.

The constellation Ursa Major (field width 50°)

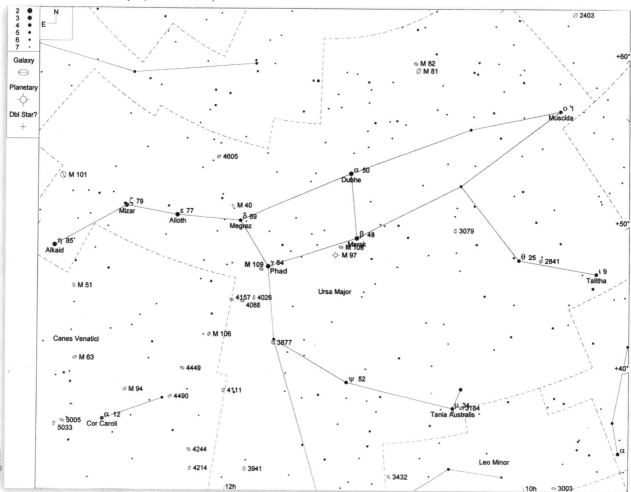

Clusters, Nebulae, and Galaxies

NGC 2841	★★	✪✪✪	GX	MBUDR
Chart 48-2	Figure 48-1	m10.1, 8.1' x 3.5'	09h 22.0m	+50° 59'

NGC 2841 is a pretty spiral galaxy with relatively high surface brightness. NGC 2841 is easy to find. Place the bright naked-eye star m3.2 25-θ (theta) Ursae Majoris on the NE edge of your finder field, as shown in Chart 48-2, and look for an m6.2 star that appears prominently near the center of the field and has a wide m8 companion about 4' NE. NGC 2841 lies 21' SSE of that m6 star, just 4.5' WSW (west southwest) of an m8 field star.

At 90X in our 10" reflector, NGC 2841 shows a bright 2' x 5' NW-SE (from northwest to southeast) halo with a much brighter extended core and non-stellar nucleus.

FIGURE 48-1.

NGC 2841 (60' field width)

Image reproduced from Digitized Sky Survey courtesy Palomar Observatory and Space Telescope Science Institute

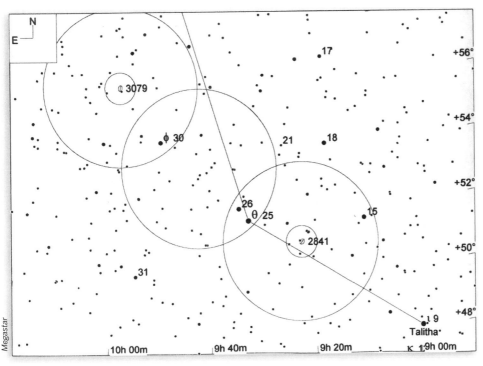

CHART 48-2.

NGC 2841 and NGC 3079 (15° field width; 5° finder circles; 1° eyepiece circles; LM 9.0)

NGC 3079	★★	◔◔◔	GX	MBUDR
Chart 48-2	Figure 48-2	m11.5, 8.0' x 1.4'	10h 01.9m	+55° 41'

NGC 3079 is a large, moderately faint edge-on galaxy. To locate NGC 3079, put the bright naked-eye star m3.2 25-θ (theta) Ursae Majoris on the SW edge of your finder field, as shown in Chart 48-2, and look for m4.6 30-φ (phi) UMa, which appears prominently 3.8° to the NE. Put 30-φ on the SW edge of your finder field. NGC 3079 lies 2.2° NE, approximately centered in the field, where it should be visible in your low-power eyepiece as a moderately bright streak of light.

At 90X in our 10" reflector, NGC 2841 shows a bright 1.5' x 6' halo elongated NNW-SSE (from north northwest to south southeast) with a much brighter extended core and a non-stellar nucleus.

FIGURE 48-2.

NGC 3079 (60' field width)

Image reproduced from Digitized Sky Survey courtesy Palomar Observatory and Space Telescope Science Institute

M 81 (NGC 3031)	★★★	◔◔◔◔	GX	MBUDR
Chart 48-3	Figure 48-3	m7.9, 27.1' x 14.2'	09h 55.6m	+69° 04'

M 82 (NGC 3034)	★★★	◔◔◔◔	GX	MBUDR
Chart 48-3	Figure 48-3	m9.3, 11.3' x 4.2'	09h 55.9m	+69° 41'

NGC 3031—better known as Messier 81, M 81, or Bode's Galaxy—and NGC 3034 (Messier 82, M 82, the Cigar Galaxy), are large, bright galaxies, visible with only a binocular. The German astronomer Johann Elert Bode discovered M 81 and its close companion M 82 on 31 December 1774. Charles Messier observed these two galaxies on the night of 9 February 1781 and added them to his catalog.

M 81 and M 82 are easy to find. They lie 10.3° NW of Dubhe (the top dipper star opposite the handle), and can be found by extending the 10.5° line from Phad (the bottom Dipper star on the handle side) to Dubhe by about the same distance. Both are visible in a 50mm binocular or finder, so it takes only moments to locate them.

The Astronomical League Binocular Messier Club rates M 81 and M 82 as Easy objects with an 80mm binocular and as Tougher objects (the middle category) with a 50mm binocular, although we've never had any difficulty seeing them with our 50mm binoculars or finder. With averted vision we see M 81 as a small, fairly faint patch of nebulosity and M 82 as a small, somewhat brighter, noticeably elongated streak.

At 90X in our 10" reflector, M 81 and M 82 both fit (barely) into the same 45' field of view. M 81 shows a bright 7' x 14' oval halo elongated NW-SE with a very bright 1.5' x 2' oval core and a very small but non-stellar nucleus. Averted vision reveals extremely faint hints of the NW spiral arm and very faint but distinct mottling in the inner halo. M 82 shows a bright, irregular 2' x 10' halo elongated ENE-WSW with very prominent mottling and dark patches. A prominent NNW-SSE dark lane divides the halo into larger ENE and smaller WSW segments. The ENE segment terminates abruptly, while the WSW segment gradually fades to invisibility.

FIGURE 48-3.

NGC 3031 (M 81) at bottom and NGC 3034 (M 82) (60' field width)

Image reproduced from Digitized Sky Survey courtesy Palomar Observatory and Space Telescope Science Institute

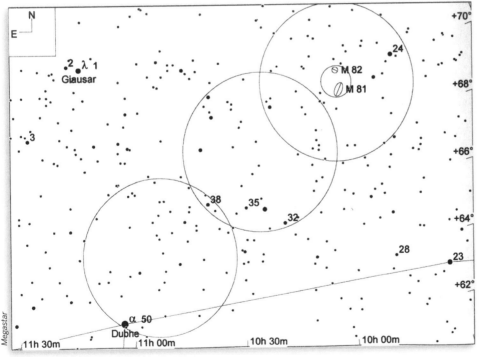

CHART 48-3.

NGC 3031 (M 81) and NGC 3034 (M 82) (15° field width; 5° finder circles; 1° eyepiece circle; LM 9.0)

NGC 3184	★	✿✿✿✿	GX	MBUDR
Chart 48-4	Figure 48-4	m10.4, 7.4' x 6.9'	10h 18.3m	+41° 25'

NGC 3184 is a spiral galaxy that we see nearly face-on. NGC 3184 is easy to find, lying 45' dead W of the bright naked-eye star m3.1 34-μ (mu) Ursae Majoris (Tania Australis), the S member of the bright Tania pair, and just 11' E of an m6.9 star that is quite prominent in the finder. Unfortunately, although it's easy enough to find, NGC 3184 isn't much to look at. At 90X in our 10" reflector, NGC 3184 shows a dim, featureless 5' halo of even brightness that looks more like a very dim comet than a galaxy. We had a chance to look at NGC 3184 in an observing buddy's 17.5" reflector, where it was it was still a dim, featureless blob, albeit a bit larger. We're not sure how this one made the RASC Finest NGCs list.

FIGURE 48-4.

NGC 3184 (60' field width)

Image reproduced from Digitized Sky Survey courtesy Palomar Observatory and Space Telescope Science Institute

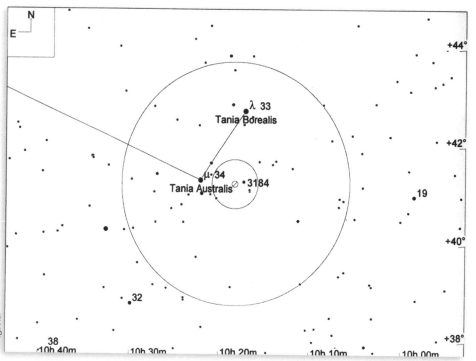

CHART 48-4.

NGC 3184 (10° field width; 5° finder circle; 1° eyepiece circle; LM 9.0)

M 108 (NGC 3556)	★★★	◊◊◊◊	GX	MBUDR
Chart 48-5	Figure 48-5	m10.7, 8.7' x 2.2'	11h 11.5m	+55° 40'

NGC 3556, better known as Messier 108 or M 108, is a fine edge-on galaxy, reminiscent of a smaller, dimmer version of M 82. Pierre Méchain discovered M 108 on the night of 18 or 19 February 1781, two or three nights after he discovered M 97, and reported his discovery to Charles Messier. Messier observed M 108 on the night of 24 March 1781, the same night he determined the position of M 97, and added it to the manuscript version of his catalog as object number 98. For some reason, Messier did not determine the exact position of this object that night, and so omitted it in the published version of his catalog. Apparently, Messier later returned to this object and obtained an accurate position for it, annotating the published version of his catalog to add its location. When others were expanding Messier's original catalog of 103 objects to include seven other objects that Messier had observed but not included, NGC 3556 ended up being assigned M 108.

M 108 is easy to find, lying 1.5° ESE of m2.4 48-β (beta) Ursae Majoris (Merak), the SW star in the bowl of the Big Dipper. We have never been able to bag M 108 with any binocular. At 90X in our 10" reflector, M 108 is a pretty sight. The moderately faint, irregular 2' x 6' halo extends E-W with prominent mottling and dark patches. A bright extended core lies offset toward the W edge of the halo and is broken into two sections, the smaller to the W and the larger toward the center of the halo.

FIGURE 48-5.

NGC 3556 (M 108) at the upper right and NGC 3587 (M 97) (60' field width)

Image reproduced from Digitized Sky Survey courtesy Palomar Observatory and Space Telescope Science Institute

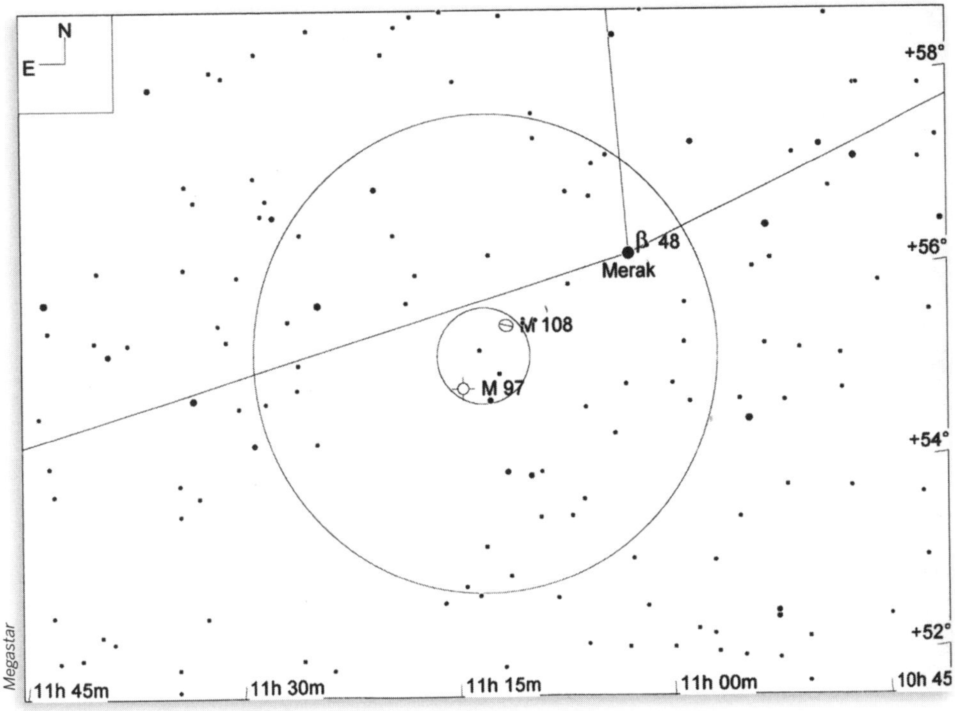

CHART 48-5.

NGC 3556 (M 108) and NGC 3587 (M 97) (10° field width; 5° finder circle; 1° eyepiece circle; LM 9.0)

M 97 (NGC 3587)	★★	✧✧✧✧	PN	MBUDR
Chart 48-5	Figure 48-5	m12.0, 3.4'	11h 14.8m	+55° 01'

NGC 3587, better known as Messier 97, M 97, or the Owl Nebula, is a very large, complex planetary nebula. As is often true of planetary nebula, its visual magnitude is much brighter than its photographic magnitude of 12.0 suggests, perhaps 9.5 to 10. Despite that, its large size and accordingly low surface brightness makes M 97 one of the more elusive Messier objects. Pierre Méchain discovered M 97 on the night of 16 February 1781 and reported this discovery to Charles Messier. Messier observed and cataloged M 97 on the night of 24 March 1781.

M 97 is very easy to find, lying 2.3° ESE of m2.4 48-β (beta) Ursae Majoris (Merak) and 48' SE of M 108. Under very dark, transparent skies with our 50mm finder or binoculars and averted vision, M 97 is a small, extremely faint patch of haze. At 90X in our 10" Dob, M 97 is a large, relatively faint circular patch of nebulosity with no surface detail visible. No hint is visible of the dark areas that form the eyes of the owl and are prominent in photographs. Using our Orion Ultrablock narrowband filter greatly increases the contrast of the nebula with the background sky, but does not increase its visible extent or detail. At 180X and using averted imagination, just a hint of the dark "eye" patches is visible.

M 109 (NGC 3992)	★★	✧✧✧✧	GX	MBUDR
Chart 48-7	Figure 48-6	m10.6, 7.6' x 4.6'	11h 57.6m	+53° 22'

NGC 3992, better known as Messier 109 or M 109, is a relatively faint spiral galaxy that we see from an oblique angle. Pierre Méchain discovered M 109 on the night of 12 March 1781 and reported his discovery to Charles Messier. Messier observed and cataloged M 109 on the night of 24 March 1781. Unfortunately, we don't know for sure which object Méchain and Messier were referring to.

The object currently accepted as M 109, NGC 3992, lies 38' ESE of m2.4 64-γ (gamma) Ursae Majoris (Phad), the SE star in the bowl of the Big Dipper. Méchain did not give an exact position for M 109, noting only that it lay "close to Gamma in the Great Bear" and adding that he "could not yet determine" its position, which seems odd if he had been referring to NGC 3992. Referring to M 109, Messier noted that it lay at the same right ascension as Gamma and one degree "more south." We can't even be certain that Méchain and Messier were talking about the same object. From their descriptions, it's possible although unlikely that Méchain was observing NGC 3992, but given his comment on right ascension it seems much more likely that Messier was actually observing NGC 3953, which lies 1.4° dead S of Phad.

NGC 3992 would have been a very difficult object for the telescopes used by Méchain and particularly by Messier, whose favorite instrument was an f/11 90mm refractor, much like our grab-'n-go scope but without the advantages of coated lenses or modern eyepieces. At 42X in our 10" reflector, we have observed both NGC 3992 and NGC 3953 in the same field. NGC 3953 has noticeably higher surface brightness—we estimate nearly half a magnitude—and we think it very likely that Méchain and Messier in fact observed NGC 3953 rather than NGC 3992.

Still, NGC 3992 is now universally recognized as M 109, so that's what we'll call it. The Astronomical League lists M 109 as a Challenge object (the most difficult category) with an 80mm binocular. We've certainly never seen even a hint of it with our 50mm binoculars or finder. At 90X in our 10" reflector, M 109 shows a faint, diffuse 4' x 7' halo elongated ENE-WSW that contains a nearly stellar core. Averted vision shows extremely faint mottling or granularity in the parts of the halo nearest the core.

FIGURE 48-6.

NGC 3992 (M 109) (60' field width)

Image reproduced from Digitized Sky Survey courtesy Palomar Observatory and Space Telescope Science Institute

CHART 48-6.

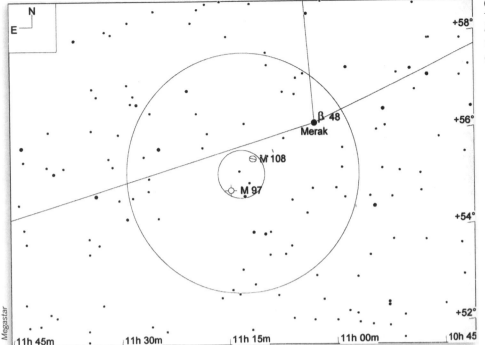

NGC 4026	★★	๑๑๑	GX	MBUDR
Chart 48-7	Figure 48-7	m11.7, 5.2' x 1.4'	11h 59.4m	+50° 58'

NGC 4026 is an edge-on galaxy with very high surface brightness. NGC 4026 is relatively easy to find, lying 2.9° SSE of m2.4 64-γ (gamma) Ursae Majoris (Phad), the SE star in the bowl of the Big Dipper. NGC 4026 is invisible in a 50mm finder, but quite prominent in a low-power eyepiece, where it appears as a bright N-S streak of light. At 90X in our 10" reflector, NGC 4026 shows a bright 0.75' x 3' N-S halo, extremely elongated, with a small, very bright oval core.

FIGURE 48-7.

NGC 4026 (60' field width)

Image reproduced from Digitized Sky Survey courtesy Palomar Observatory and Space Telescope Science Institute

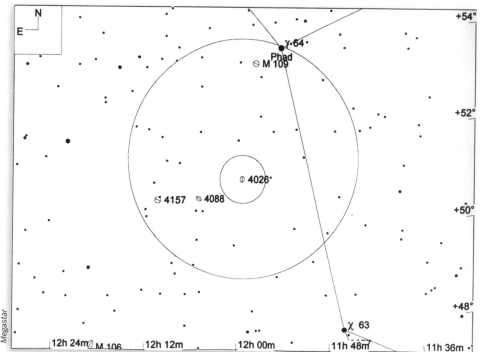

CHART 48-7.

*NGC 4026 (10° field width; 5°
finder circle; 1° eyepiece circle;
LM 9.0)*

NGC 4088	★★	✵✵✵	GX	MBUDR
Chart 48-7, 48-8	Figure 48-8	m11.2, 5.3' x 2.1'	12h 05.6m	+50° 32'

NGC 4088 is an irregular galaxy with fairly low surface brightness.
By starting at NGC 4026, it's relatively easy to locate NGC 4088,
which lies 1.1° ESE. (We use our low-power eyepiece, which in our
10" reflector provides a true field of 1.7°; both of these galaxies fit
easily into one field.) At 90X in our 10" Dob, NGC 4088 shows a
moderately bright 2' x 5' NE-SW halo, with slight brightening to an
extended core. No mottling or other detail is visible.

FIGURE 48-8.

NGC 4088 (right) and NGC 4157 (60' field width)

Image reproduced from Digitized Sky Survey courtesy Palomar Observatory and Space
Telescope Science Institute

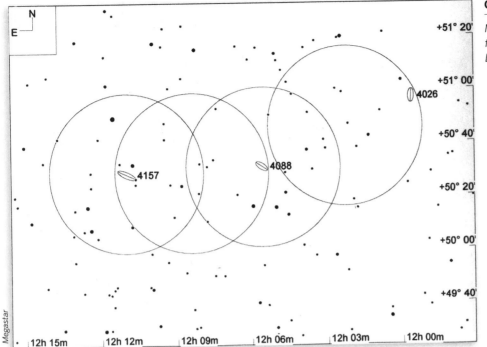

NGC 4157	★★	໑໑໑	GX	MBUDR
Chart 48-8	Figure 48-8	m12.2, 7.7' x 1.3'	12h 11.1m	+50° 29'

NGC 4157 is an edge-on galaxy with fairly low surface brightness.
It's easiest to find NGC 4157 by starting from NGC 4088. NGC 4157
lies 53' ESE. (Once again, we use our low-power eyepiece, which in
our 10" reflector provides a true field of 1.7°; NGC 4088 and NGC
4157 both fit easily into one field.) At 90X in our 10" Dob, NGC 4157
shows a moderately bright 0.75' x 4' halo, highly elongated ENE-
WSE, with significant brightening to an extended core. No mottling
or other detail is visible.

NGC 3877	★★	໑໑໑໑	GX	MBUDR
Chart 48-9	Figure 48-9	m11.8, 5.8' x 1.2'	11h 46.1m	+47° 30'

NGC 3877 is an edge-on galaxy with fairly low surface brightness.
NGC 3877 is very easy to find, lying 17' S of the naked-eye star
m3.7 63-χ (chi) UMa. At 90X in our 10" Dob, NGC 3877 shows a
faint 1' x 5' NE-SW halo that brightens gradually to a concentrated
central core. No mottling or other detail is visible.

FIGURE 48-9.

NGC 3877 (60' field width)

Image reproduced from Digitized Sky Survey courtesy Palomar Observatory and Space
Telescope Science Institute

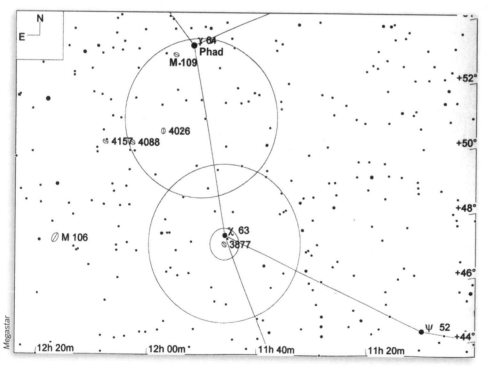

CHART 48-9.

*NGC 3877 (15° field width; 5°
finder circles; 1° eyepiece circle;
LM 9.0)*

NGC 3941	★★	◐◐	GX	MBUDR
Chart 48-10	Figure 48-10	m11.3, 3.7' x 2.3'	11h 52.9m	+36° 59'

NGC 3941 is a small but bright oval galaxy. Finding it requires a bit of work. Begin by identifying the naked-eye pair m3.5 54-ν (nu) Ursae Majoris (Alula Borealis) and m3.8 53-ξ (xi) Ursae Majoris (Alula Australis). Both are located just off the S edge of Chart 48-1. Place those two stars on the W edge of your finder field and pan slightly N of E until the bright pair m5.3 61-UMa and m5.8 62-UMa come into view on the E edge of your finder field. Place 61-UMa on the SW edge of your finder field and look near the NE edge of the field for two m6.5 stars, which appear very prominently. NGC 3941 forms the SW apex of a scalene triangle with those two stars, 45' dead S of the N star and 31' WNW of the S star. At 90X in our 10" Dob, NGC 3941 shows a bright 1' x 2' N-S halo that brightens quickly to a very bright concentrated central core. No mottling or other detail is visible.

FIGURE 48-10.

NGC 3941 (60' field width)

Image reproduced from Digitized Sky Survey courtesy Palomar Observatory and Space Telescope Science Institute

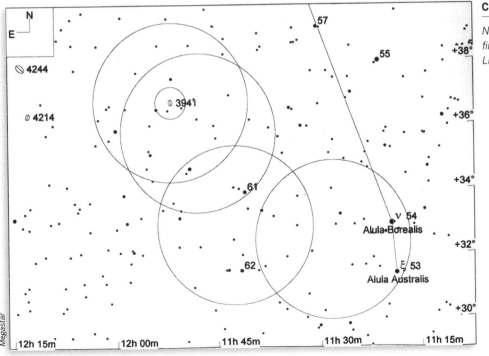

CHART 48-10.

NGC 3941 (15° field width; 5° finder circles; 1° eyepiece circle; LM 9.0)

M 40 (Winnecke 4)	★	✦✦✦✦	(double star)	MBUDR
Chart 48-11	Figure 48-11	m9.5, ---'	12h 22.2m	+58° 05'

Winnecke 4, better known as Messier 40 or M 40, is one of Messier's "mistakes." Charles Messier observed and cataloged M 40 on the night of 24 October 1764, as he was searching for a nebula that Hevelius had previously reported in the vicinity. As it turned out, there was no nebula at the reported location, or indeed anywhere nearby. (We think Hevelius was probably misled by flare from m5.5 70-UMa, which lies only 17' SW of M 40; the primitive instruments of the time lacked modern anti-reflection coatings, and flare from nearby bright objects was a major problem.) Messier could see no nebulosity, but logged the position of a close pair of m9 stars as M 40.

M 40 is very easy to find, lying 1.4° NE of m3.3 69-δ (delta) Ursae Majoris (Megrez), the NE star in the bowl of the Big Dipper, and 17' NE of the aforementioned m5.5 70-UMa, which appears very prominently in the finder. At 90X in our 10" reflector, M 40 is a closely matched pair of m9 stars.

FIGURE 48-11.

Winnecke 4 (M 40) (60' field width)

Image reproduced from Digitized Sky Survey courtesy Palomar Observatory and Space Telescope Science Institute

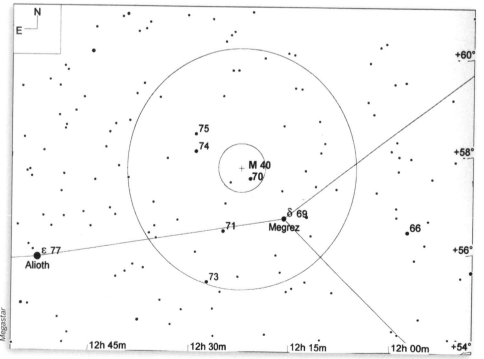

CHART 48-11.

Winnecke 4 (M 40) (10° field width; 5° finder circle; 1° eyepiece circle; LM 9.0)

NGC 4605	★★	⊘⊘⊘	GX	MBUDR
Chart 48-12	Figure 48-12	m10.9, 5.7' x 2.1'	12h 40.0m	+61° 37'

NGC 4605 is an edge-on galaxy with relatively high surface brightness. It's relatively easy to find, lying 5.5° ENE of m3.3 69-δ (delta) Ursae Majoris (Megrez), the NE star in the bowl of the Big Dipper. To locate NGC 4605, place Megrez on the SW edge of your finder field and look for a prominent pair of bright stars near the center of the field, m5.4 74-UMa and m6.1 75-UMa. Place that pair on the SW edge of the finder field and look for four m6 stars that appear prominently in the finder. The S two of these stars form a 1.1° SE-NW line. Extend that line 35' further to the NW to center NGC 4605 in your eyepiece. At 90X in our 10" reflector, NGC 4605 shows a bright 1.5' x 5' WNW-ESE halo with a very bright concentrated thin central core without a visible nucleus. No mottling or other detail is visible.

FIGURE 48-12.

NGC 4605 (60' field width)

Image reproduced from Digitized Sky Survey courtesy Palomar Observatory and Space Telescope Science Institute

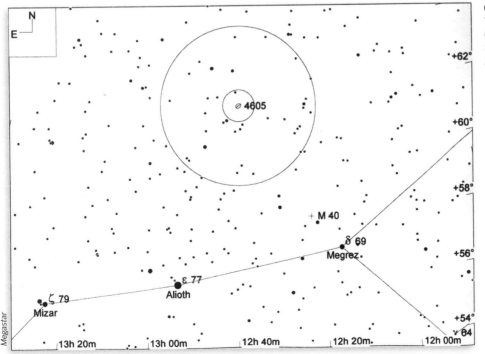

CHART 48-12.

NGC 4605 (15° field width; 5° finder circle; 1° eyepiece circle; LM 9.0)

M 101 (NGC 5457)	★★★	✸✸✸	GX	MBUDR
Chart 48-13	Figure 48-13	m8.3, 28.9' x 26.9'	14h 03.2m	+54° 21'

NGC 5457, better known as Messier 101, M 101, or the Pinwheel Galaxy (not to be confused with M 33, the Pinwheel Galaxy in Triangulum) is a large, moderately bright face-on spiral galaxy. Pierre Méchain discovered M 101 sometime in March 1781 and communicated that discovery to Charles Messier, who observed and cataloged M 101 on the night of 27 March 1781.

Despite its relatively bright magnitude of 8.3, the large extent of M 101—about the size of the full moon—means that it has very low surface brightness, usually stated as 14.0 or higher. Accordingly, M 101 can be one of the more elusive Messier objects. On a night of less than perfect transparency or with even a sliver of moon in the sky, M 101 may be entirely invisible. Conversely, on a dark, transparent night, M 101 can reveal a wealth of detail even if your scope is of moderate aperture.

Although you can locate M 101 by a star hop along the arc of m5/6 stars that extends to the E and ESE of Mizar, it's easiest to find M 101 by using the geometric relationship of its position relative to Mizar and Alkaid, with which it forms the NE apex of a nearly equilateral 6° triangle. Simply point your finder about where M 101 should be, and look for it in your low-power eyepiece. If you don't see M 101, the night is probably too hazy.

On a dark, transparent night at 90X in our 10" reflector, M 101 is a beautiful galaxy, full of detail. A bright 2' circular core is surrounded by much fainter nebulosity. Averted vision reveals very

FIGURE 48-13.

NGC 5457 (M 101) (60' field width)

Image reproduced from Digitized Sky Survey courtesy Palomar Observatory and Space Telescope Science Institute

CHART 48-13.

NGC 5457 (M 101) (12° field width; 5° finder circle; 1° eyepiece circle; LM 9.0)

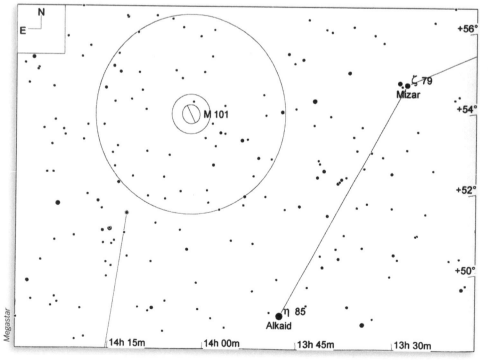

faint, mottled spiral arms curving clockwise around the core. The brightest spiral arm originates on the SE edge of the core and curls E and then N until it dims to invisibility. A second, considerably fainter, spiral arm originates on the W side of the core, which curves S and then E.

Multiple Stars

79-zeta (STF 1744AC)	★★★	✧✧✧✧	MS	UD
Chart 48-1, 48-14		m2.2/4.0, 708.5", PA 71° (1991)	13h 23.9m	+54° 55'

79-zeta (STF 1744AB)	★★★	✧✧✧✧	MS	UD
Chart 48-1, 48-14		m2.2/3.9, 14.3", PA 153° (2003)	13h 23.9m	+54° 55'

The double star 79-ζ (zeta) Ursae Majoris and 80-UMa, better known as Mizar and Alcor, is the most famous of double stars, known even to non-astronomers. Actually, it's a triple star. The widely-separated primary and secondary, m2.2 Mizar and m4.0 Alcor, are separated by nearly 12 arcminutes, and can be split with the naked eye. Mizar is itself double, with an m3.9 companion lying 14.3" to the SSE.

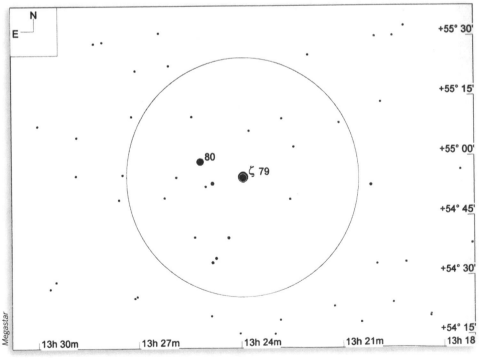

CHART 48-14.

Mizar (79-UMa) and Alcor (80-UMa) (STF 1744) (2° field width; 1° eyepiece circle; LM 12.0)

49

Virgo, The Virgin .

NAME:	Virgo (VUR-goh)
SEASON:	Spring
CULMINATION:	midnight, 12 April
ABBREVIATION:	Vir
GENITIVE:	Virginis (vur-GIN-is)
NEIGHBORS:	Boo, Com, Crt, Crv, Hya, Leo, Lib, Ser

BINOCULAR OBJECTS: NGC 4303 (M 61), NGC 4374 (M 84), NGC 4406 (M 86), NGC 4472 (M 49), NGC 4486 (M 87), NGC 4552 (M 89), NGC 4569 (M 90), NGC 4579 (M 58), NGC 4594 (M 104), NGC 4621 (M 59), NGC 4649 (M 60)

URBAN OBJECTS: NGC 4374 (M 84), NGC 4406 (M 86), NGC 4486 (M 87), NGC 4594 (M 104)

Virgo is a large, bright equatorial constellation that ranks 2nd in size among the 88 constellations. Virgo covers 1,294 square degrees of the celestial sphere, or about 3.1%.

Virgo is an ancient constellation, one of Ptolemy's original 48. Virgo is probably the subject of more myths and legends than any other constellations. Across time and cultures, Virgo has usually been associated in one form or another with virgins, maidens, purity, and fertility. In Greek mythology, Virgo is most often associated with Persephone, the daughter of Zeus and Demeter. When Persephone was kidnapped and taken to the underworld by Hades, Demeter threw a fit, ruining crops and causing widespread famine. To placate Demeter, Zeus recalled Persephone from the underworld by promising Hades that Persephone would return for a short time each year. Demeter accepted that compromise and again allowed the crops to grow, averting a continued famine. Other Greek myths associate Virgo with Astraea, the virgin daughter of Zeus and Themis. Astraea was the goddess of justice, and in that guise is closely associated with the bordering constellation Libra, the Scales of Justice.

Virgo is the realm of galaxies, home to the Virgo Cluster of galaxies. In most constellations, the hard part is tracking down an object and getting it into the eyepiece. In Virgo, there are so many galaxies so close together that the problem is often identifying which particular galaxy you are looking at among the many others visible in the eyepiece. In the most densely populated parts of the Virgo Cluster—or the "Virgo Clutter" as many experienced observers call it—literally a dozen or more galaxies may be visible simultaneously in the field of a low-power eyepiece. In most constellations, you star-hop with your finder to work your way to the object you want to view. In Virgo, you galaxy-hop with your eyepiece, first identifying one galaxy for sure and then working from it to identify other nearby galaxies.

We recommend that you observe the objects in this chapter in the order we present them, because it's relatively easy to hop from one object to the next but it can be very difficult to locate and identify a particular object unless you first locate and identify the intermediate objects. It's very easy to get lost in the forest of galaxies in Virgo, making it difficult or impossible to identify the individual trees. If you do get lost, just take a deep breath, return to the last galaxy you've identified with certainty, and proceed from there.

Although all 11 of the Messier objects in Virgo appear on the Astronomical League Binocular Messier Club lists, none of them are easy objects. For 50mm binoculars, AL considers only two of those objects possible. M 49 is rated a Tougher object (the middle category), and M 104 a Challenge object (the most difficult category.) All 11 objects are considered possible with an 80mm binocular, but none of them are easy even with that size instrument. M 49, M 60, M 61, M 87, and M 104 are Tougher objects, and M 58, M 59, M 84, M 86, M 89, and M 90 are Challenge objects.

Virgo culminates at midnight on 12 April, and, for observers at mid-northern latitudes, is best placed for evening viewing from early spring through mid-summer.

TABLE 49-1.

Featured star clusters, nebulae, and galaxies in Virgo

Object	Type	Mv	Size	RA	Dec	M	B	U	D	R	Notes
NGC 4762	Gx	10.2	8.8 x 1.7	12 52.9	+11 14					◉	Class SB(r)0^? sp; SB 12.4
NGC 4621	Gx	10.6	5.3 x 3.2	12 42.0	+11 39	◉	◉				M 59; Class E5; SB 11.9
NGC 4649	Gx	9.8	7.4 x 6.0	12 43.7	+11 33	◉	◉				M 60; Class E2; SB 12.2
NGC 4579	Gx	9.6	5.9 x 4.7	12 37.7	+11 49	◉	◉				M 58; Class SAB(rs)b; SB 12.4
NGC 4567/8	Gx	12.1 / 11.7	3.3 x 2.0 / 4.8 x 2.0	12 36.5	+11 16					◉	Class SA(rs)bc (both)
NGC 4552	Gx	10.7	3.5 x 3.5	12 35.7	+12 33	◉	◉				M 89; Class E0-1
NGC 4569	Gx	10.3	9.6 x 4.3	12 36.8	+13 10	◉	◉				M 90; Class SAB(rs)ab
NGC 4486	Gx	9.6	7.4 x 6.0	12 30.8	+12 23	◉	◉	◉			M 87; Class E+0-1 pec; SB 12.6
NGC 4438	Gx	11.0	8.6 x 3.1	12 27.8	+13 01					◉	Class SA(s)0/a pec: SB 13.6
NGC 4374	Gx	10.1	6.4 x 5.5	12 25.0	+12 53	◉	◉	◉			M 84; class E1; SB 11.9
NGC 4406	Gx	9.9	8.9 x 5.7	12 26.2	+12 57	◉	◉	◉			M 86; Class E3; SB 13.1
NGC 4388	Gx	11.8	7.6 x 1.4	12 25.8	+12 40					◉	Class SA(s)b: sp; SB 13.3
NGC 4216	Gx	11.0	8.7 x 1.7	12 15.9	+13 09					◉	Class SAB(s)b:; SB 12.1
NGC 4472	Gx	9.4	9.3 x 7.0	12 29.8	+08 01	◉	◉				M 49; Class E2; SB 12.5
NGC 4526	Gx	10.7	7.2 x 2.3	12 34.0	+07 42					◉	Class SAB(s)0^
NGC 4535	Gx	9.9	7.1 x 5.0	12 34.3	+08 12					◉	Class SAB(s)c; SB 13.7
NGC 4303	Gx	10.2	6.5 x 5.7	12 21.9	+04 28	◉	◉				M 61; Class SAB(rs)bc; SB 12.8
NGC 4517	Gx	11.1	11.2 x 1.5	12 32.7	+00 07					◉	Class SA(s)cd: sp; SB 14.0
NGC 4699	Gx	10.4	4.0 x 2.8	12 49.0	-08 40					◉	Class SAB(rs)b; SB 10.9
NGC 4594	Gx	9.0	8.8 x 3.5	12 40.0	-11 38	◉	◉	◉			M 104; Class SA(s)a sp; SB 11.6
NGC 5746	Gx	11.3	7.5 x 1.3	14 44.9	+01 57					◉	Class SAB(rs)b? sp; SB 13.8

TABLE 49-2.

Featured multiple stars in Virgo

Object	Pair	M1	M2	Sep	PA	Year	RA	Dec	UO	DS	Notes
gamma*	STF 1670AB	3.5	3.5	0.4	86	2006	12 41.7	-01 27		◉	Porrima

CHART 49-1.

The constellation Virgo (field width 50°)

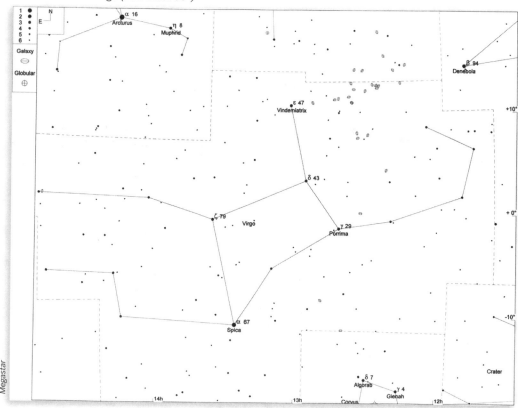

CHART 49-2.

The Virgo Galaxy Cluster (15° field width; 5° finder circle (for scale); LM 6.0)

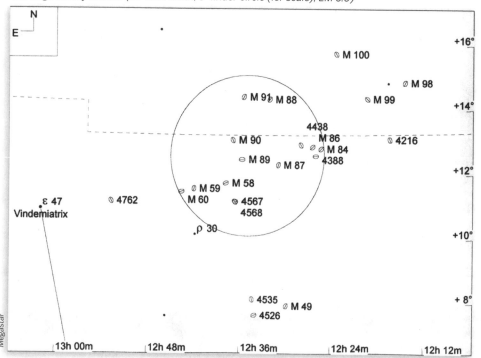

Clusters, Nebulae, and Galaxies

NGC 4762	★★	✺✺✺	GX	MBUDR
Chart 49-3	Figure 49-1	m10.2, 8.8' x 1.7'	12h 52.9m	+11° 14'

NGC 4762 is a mid-size spindle galaxy with relatively high surface brightness that we see nearly edge-on. NGC 4762 is relatively easy to find, lying 2.3° W of the bright naked-eye star m2.8 47-ε (epsilon) Virginis (Vindemiatrix). Place Vindemiatrix near the E edge of your finder field, and look for two bright stars in the N half of the field, m6.1 34-Virginis and m6.2 41-Vir, which appear very prominently in the finder. NGC 4762 forms the S apex of a nearly equilateral triangle with those two stars, and will be located near the center of your finder field.

At 90X in our 10" reflector, NGC 4762 shows a very bright 0.5' x 5' halo, extremely elongated NNE-SSW (from north northeast to south southwest) with a slightly brighter circular core.

FIGURE 49-1.

NGC 4762, with NGC 4754 and NGC 4733 (lower right) (60' field width)

Image reproduced from Digitized Sky Survey courtesy Palomar Observatory and Space Telescope Science Institute

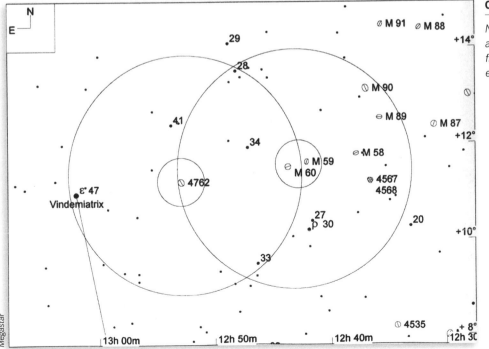

CHART 49-3.

NGC 4762, NGC 4621 (M 59), and NGC 4649 (M 60) (10° field width; 5° finder circles; 1° eyepiece circles; LM 9.0)

M 59 (NGC 4621)	★★★	✪✪✪	GX	MBUDR
Chart 49-3	Figure 49-2	m10.6, 5.3' x 3.2'	12h 42.0m	+11° 39'

M 60 (NGC 4649)	★★★	✪✪✪	GX	MBUDR
Chart 49-3	Figure 49-2	m9.8, 7.4' x 6.0'	12h 43.7m	+11° 33'

NGC 4621, better known as Messier 59 or M 59, and NGC 4649 (Messier 60, M 60) are bright elliptical galaxies. The German astronomer Johann Gottfried Köhler discovered M 59 and M 60 on the night of 11 April 1779, while he was observing a comet. Charles Messier, observing the same comet and unaware of Köhler's discovery, independently rediscovered M 59 and M 60 four nights later, on 15 April 1779, just after he discovered M 58 just a degree to the W.

M 59 and M 60 are relatively easy to find. Begin with m2.8 47-ε (epsilon) Virginis (Vindemiatrix) on the W edge of your finder field and NGC 4762 centered in your eyepiece. NGC 4762 lies exactly halfway along the line from Vindemiatrix to M 60, which will lie near the E edge of the finder field, although it will be invisible in the finder. Look for two bright stars in the N half of your finder field, m6.2 41-Vir and m6.1 34-Vir. These two stars form a 1.7° E-W line that points directly to M 59 and M 60. The center of that galaxy pair lies 1.2° W of 34-Virginis. Another way to locate the galaxy pair geometrically is to note that M 59 forms the NW apex of an isosceles triangle with m6.1 34-Virginis 1.3° to its ENE (east northeast) and m4.9 30-ϱ (rho) Virginis about the same distance to its S. Both galaxies are bright patches of nebulosity in a low power eyepiece.

At 90X in our 10" reflector (44' true field), M 59 and M 60 fit comfortably into the same field of view. Despite its dimmer cataloged surface brightness (12.2, versus 11.9 for M 59) and its larger extent, M 60 is somewhat brighter in the eyepiece than M 59, which lies just 25' to the W. M 59 shows a bright 1.5' x 2.5' oval halo elongated NNW-SSE (from north northwest to south southeast) that brightens gradually to a concentrated circular core with a non-stellar nucleus. M 60 shows a bright 3' circular halo that brightens quickly to a large, circular central core without a visible nucleus. NGC 4647, just to the upper right of M 60 in Figure 49-2, with averted vision shows a much fainter, diffuse 1.5' circular halo with slight central brightening.

FIGURE 49-2.

*NGC 4649 (M 60) at left and
NGC 4621 (M 59) (60' field width)*

Image reproduced from Digitized Sky Survey courtesy Palomar Observatory and Space Telescope Science Institute

M 58 (NGC 4579)	★★★	✧✧✧	GX	MBUDR
Chart 49-4	Figure 49-3	m9.6, 5.9' x 4.7'	12h 37.7m	+11° 49'

NGC 4579, better known as Messier 58 or M 58, is a bright spiral galaxy. Charles Messier discovered and cataloged M 58 on the night of 15 April 1779, while observing the comet of that year.

Once you have located M 59 and M 60, M 58 is relatively easy to find, lying 1.1° W of M 59 and just 7' dead E of a lonely m8 star. Use an eyepiece hop, simply by drifting your eyepiece field on the line from M 60 to M 59 about 1.1° to the W until M 58 comes into view. (Actually, in our 10" Dob, our 2" GSO SuperView 30mm finder eyepiece gives us a 1.7° field of view, which is wide enough to encompass M 59, M 60, and M 58 in the same field of view.)

At 90X in our 10" reflector, M 58 shows a moderately faint, diffuse 3' x 4' oval halo elongated ENE-WSW with a much brighter 2' x 3' condensed inner halo that brightens dramatically to a 0.5' x 2' core with a prominent stellar nucleus. Averted vision shows very faint mottling within the inner halo.

FIGURE 49-3.

NGC 4579 (M 58) at upper left, and the galaxy pair NGC 4567/NGC 4568 at lower right (60' field width)

Image reproduced from Digitized Sky Survey courtesy Palomar Observatory and Space Telescope Science Institute

CHART 49-4.

NGC 4579 (M 58) (10° field width; 5° finder circle; 1° eyepiece circle; LM 9.0)

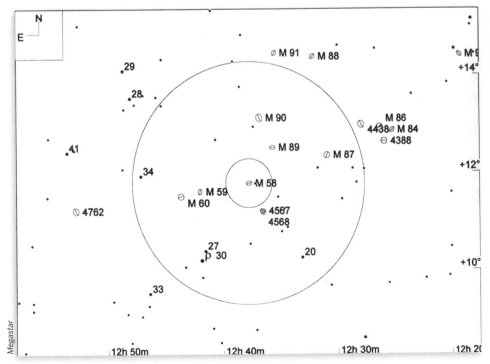

NGC 4567/4568	★	⦿⦿⦿		GX	MBUDR
Chart 49-5	Figure 49-3	m12.1/11.7, 3.3' x 2.0'/4.8 x 2.0		12h 36.5m	+11° 16'

NGC 4567/4568, sometimes called the Siamese Twin Galaxies, are a pair of galaxies that appear to be touching at the E tips of their halos. This galaxy pair is relatively easy to find, working from M 58. Put M 58 on the NE edge of your low-power eyepiece field and look 37' SSW for this galaxy pair, which lies about 13' SE of a prominent m8 star and may be invisible with direct vision in smaller scopes. At 90X in our 10" reflector with averted vision, these are unimpressive objects. NGC 4567 (the N galaxy) shows a fairly faint 0.75' x 1.5' E-W oval halo with slight central brightening, but no other detail visible. NGC 4568 shows a uniform, faint 1' x 2' NE-SW halo without any visible variation in brightness.

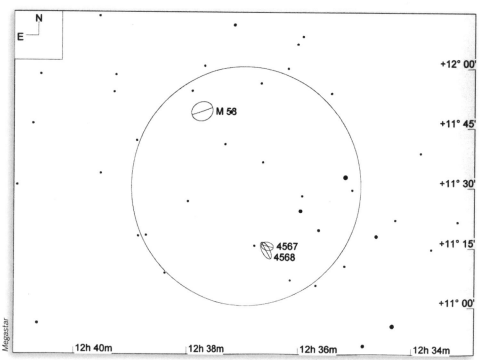

CHART 49-5.

NGC 4567/4568 (2° field width; 1° eyepiece circle; LM 12.0)

M 89 (NGC 4552)	★★	⦿⦿⦿		GX	MBUDR
Chart 49-6	Figure 49-4	m10.7, 3.5' x 3.5'		12h 35.7m	+12° 33'

M 90 (NGC 4569)	★★★	⦿⦿⦿		GX	MBUDR
Chart 49-6	Figure 49-4	m10.3, 9.6' x 4.3'		12h 36.8m	+13° 10'

NGC 4552, better known as Messier 89 or M 89, and NGC 4569 (Messier 90, M 90) are a close pair of bright galaxies. Charles Messier discovered and cataloged M 89 on his "big night" of 18 March 1781, the same night he discovered and cataloged the Virgo galaxies M 84, M 85, M 86, M 87, M 88, and M 90, not to mention the galaxy M 91 in Coma Berenices, and the globular cluster M 92 in Hercules.

Working from M 58, M 89 is relatively easy to find, lying just 53' NW of M 58 and about 13' ENE of an m9 star that appears prominent in the eyepiece. M 90 is equally easy to find in the eyepiece by looking 40' NNE of M 89.

At 90X in our 10" reflector (44' true field), M 89 and M 90 both fit within the same field of view, although just barely. The elliptical galaxy M 89 has a relatively bright 1.5' diffuse circular halo with significant brightening to a well-concentrated core without a visible nucleus. The spiral galaxy M 90 has a relatively faint 2' x 6' oval halo elongated NNE-SSW with a large, bright, concentrated 1' x 3' core. Averted vision shows very faint mottling in the core, particularly toward the W edge.

FIGURE 49-4.

NGC 4552 (M 89) and NGC 4569 (M 90) (60' field width)

Image reproduced from Digitized Sky Survey courtesy Palomar Observatory and Space Telescope Science Institute

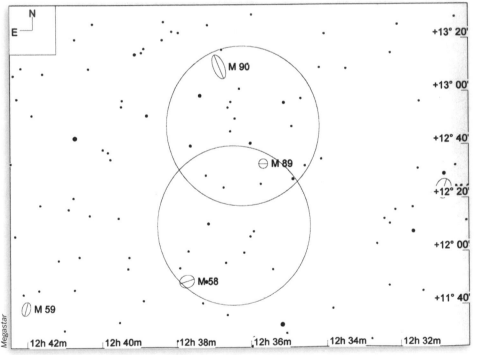

CHART 49-6.

NGC 4552 (M 89) and NGC 4569 (M 90) (3° field width; 1° eyepiece circles; LM 12.0)

M 87 (NGC 4486)	★★★	✷✷✷	GX	MBUDR
Chart 49-7	Figure 49-5	m9.6, 7.4' x 6.0'	12h 30.8m	+12° 23'

NGC 4486, better known as Messier 87 or M 87, is a bright elliptical galaxy. Charles Messier discovered and cataloged M 87 on his "big night" of 18 March 1781.

Working from M 89, M 87 is relatively easy to find, lying just 1.2° W of M 89 and about 5' S of an m8 star that appears prominent in the eyepiece. At 90X in our 10" reflector, M 87 has a very bright 30" core without a nucleus, surrounded by a bright 2' circular inner halo that gradually fades to invisibility.

FIGURE 49-5.

NGC 4486 (M 87) (60' field width)

Image reproduced from Digitized Sky Survey courtesy Palomar Observatory and Space Telescope Science Institute

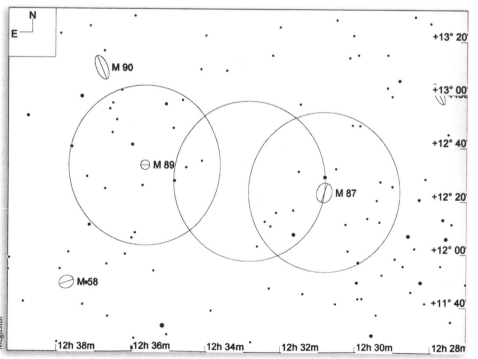

CHART 49-7.

NGC 4486 (M 87) (3° field width; 1° eyepiece circles; LM 12.0)

NGC 4438	★★	❖❖❖	GX	MBUDR
Chart 49-8	Figure 49-6	m11.0, 8.6' x 3.1'	12h 27.8m	+13° 01'

NGC 4438 is a moderately bright galaxy. Working from M 87, NGC 4438 is easy to find, lying 1.3° NW of M 87 and about 15' W of an m9 star that appears prominent in the eyepiece. At 90X in our 10" reflector, NGC 4438 shows a relatively bright 1.5' x 2.5' diffuse oval halo elongated NNE-SSW with a very bright, concentrated 0.5' x 1.5' core centered on a stellar nucleus. NGC 4435, about 4' NNW, shows a bright stellar nucleus in a moderately faint, uniform 0.5' x 1.5' oval halo elongated N-S.

FIGURE 49-6.

*NGC 4438, NGC 4406 (M 86),
NGC 4374 (M 84), and NGC 4388 (60' field width)*

Image reproduced from Digitized Sky Survey courtesy Palomar Observatory and Space Telescope Science Institute

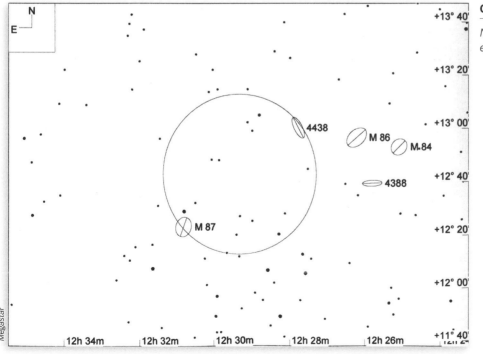

CHART 49-8.

NGC 4438 (3° field width; 1° eyepiece circle; LM 12.0)

M 84 (NGC 4374)	★★	❂❂❂	GX	MBUDR
Chart 49-9	Figure 49-6	m10.1, 6.4' x 5.5'	12h 25.0m	+12° 53'

M 86 (NGC 4406)	★★	❂❂❂	GX	MBUDR
Chart 49-9	Figure 49-6	m9.9, 8.9' x 5.7'	12h 26.2m	+12° 57'

NGC 4374, better known as Messier 84 or M 84, and NGC 4406 (Messier 86, M 86) are a close pair of bright elliptical galaxies. Charles Messier discovered and cataloged M 84 and M 86 on the night of 18 March 1781, the same night he discovered and cataloged numerous other galaxies in Virgo and Coma Berenices.

M 84 and M 86 are relatively easy to find. Their centers lie on a 17' E-W line. The center of that line lies 1.4° WNW of the very bright galaxy M 87, and just half a degree W of the previous object, NGC 4438. Although both of these galaxies are very bright and very prominent in the eyepiece, unfortunately they're just not very

interesting to look at. At 90X in our 10" reflector (44' true field), M 84 and M 86 both fit within the same field of view, along with the preceding object, NGC 4438, the next object, NGC 4388, and several other galaxies that are not covered in this book. M 84 and M 86 are by far the brightest objects in the field, bright enough that it's easy to overlook the fainter galaxies that are also present. Like most ellipticals, M 84 and M 86 present a rather uniform appearance. M 84 shows a 2' x 2.5' NW-SE oval halo with gradual brightening to a large, concentrated circular core. M 86 appears very similar to M 84, but about 50% larger.

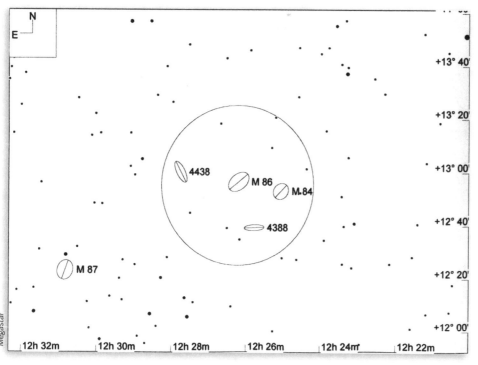

CHART 49-9.

NGC 4406 (M 86), NGC 4374 (M 84), and NGC 4388 (3° field width; 1° eyepiece circle; LM 12.0)

NGC 4388	★★★	❂❂❂	GX	MBUDR
Chart 49-9	Figure 49-6	m11.8, 7.6' x 1.4'	12h 25.8m	+12° 40'

NGC 4388 is a very nice edge-on spiral galaxy. It's considerably fainter than M 84 or M 86, but shows more detail. Once you have M 84 and M 86 in the eyepiece, NGC 4388 is easy to find. It forms the S apex of a 17' equilateral triangle with those other two galaxies,

and is visible in the same eyepiece field. At 90X in our 10" reflector, NGC 4388 shows a moderately faint 0.5' x 3.5' E-W halo that brightens noticeably to an extended core with a stellar nucleus.

NGC 4216	★★★	✷✷✷	GX	MBUDR
Chart 49-10	Figure 49-7	m11.0, 8.7' x 1.7'	12h 15.9m	+13° 09'

NGC 4216 is a nice spindle galaxy with moderately high surface brightness. It's relatively easy to locate NGC 4216 by working W from the M 86 group. With M 86 centered in your eyepiece, drift the finder about half a field W, looking for the bright stars m5.1 6-Com and m5.9 12-Vir, which are very prominent in the finder and form a 4.7° N-S line. NGC 4216 lies just over a third of the way along the line from 6-Com to 12-Vir, and is visible in a low-power finder eyepiece. At 90X in our 10" reflector, NGC 4216 shows a bright 0.5' x 5.5' NNE-SSW halo with a large, bright extended core and no visible nucleus.

FIGURE 49-7.

NGC 4216 (60' field width)

Image reproduced from Digitized Sky Survey courtesy Palomar Observatory and Space Telescope Science Institute

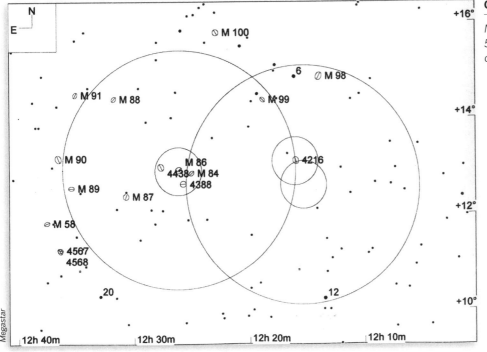

CHART 49-10.

NGC 4216 (10° field width; 5° finder circles; 1° eyepiece circles; LM 9.0)

NGC 4472 (M 49)	★★	❂❂❂	GX	MBUDR
Chart 49-11	Figure 49-8	m9.4, 9.3' x 7.0'	12h 29.8m	+08° 01'

NGC 4472, better known as Messier 49 or M 49, is another very bright elliptical galaxy. Charles Messier discovered and cataloged M 49 on the night of 19 February 1771, making it the first of the Virgo Cluster to be discovered.

M 49 is easy to locate by a short star hop. Begin by placing the bright naked-eye star m2.8 47-ε (epsilon) Virginis (Vindemiatrix) on the ENE edge of your finder field, as shown in Chart 49-11. Look 4.1° WSW of Vindemiatrix for the m5.7 star 33-Virginis, which appears very prominent in the finder. Drift the finder field on that line WSW until 33-Vir is on the NE edge of the finder field. Look for two m6 stars that appear very prominently toward the W and WSW edge of the finder field and form a 1.3° NW-SE line. M 49 lies directly on that line just over halfway from the N star to the S star. On an extremely dark, transparent night, M 49 may be very faintly visible in a 50mm finder.

Like M 84 and M 86, M 49 is very bright but otherwise disappointing. At 90X in our 10" reflector, M 49 shows a bright, diffuse 4' circular halo with a large, bright extended core and no visible nucleus.

FIGURE 49-8.

NGC 4472 (M 49) (60' field width)

Image reproduced from Digitized Sky Survey courtesy Palomar Observatory and Space Telescope Science Institute

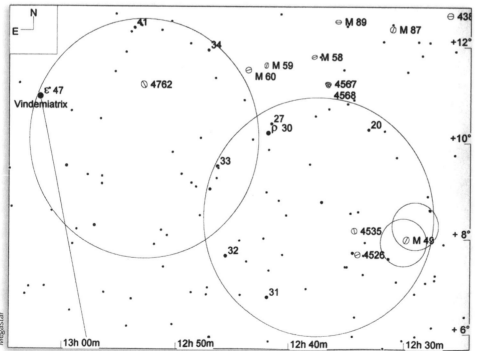

CHART 49-11.

NGC 4472 (M 49) (10° field width; 5° finder circles; 1° eyepiece circles; LM 9.0)

NGC 4526	★★	◌◌◌	GX	MBUDR
Chart 49-12	Figure 49-9	m10.7, 7.2' x 2.3'	12h 34.0m	+07° 42'

NGC 4535	★★	◌◌◌	GX	MBUDR
Chart 49-12	Figure 49-9	m9.9, 7.1' x 5.0'	12h 34.3m	+08° 12'

NGC 4526 and NGC 4535 are an interesting close galaxy pair. NGC 4526 is a relatively prominent edge-on galaxy, and NGC 4536 a much dimmer spiral galaxy that we see nearly face-on. Both of these galaxies are easy to find by working from M 49. With M 49 centered in the eyepiece, look in your finder 1.1° ESE of M 49 for a pair of m7 stars that are very prominent in the finder and form a 15' E-W line. NGC 4526 lies halfway between those two stars. NGC 4535 lies 30' N of NGC 4526, with both visible in the same field of a medium- to high-power eyepiece.

At 90X in our 10" reflector, NGC 4526 shows a moderately faint 1' x 3' lenticular halo elongated WNW-ESE that brightens rapidly to a large, bright inner halo with a small, nearly stellar core. NGC 4535 is larger but much fainter, showing a very faint, diffuse 3' x 5' oval NNE-SSW halo with no visible detail.

FIGURE 49-9.

NGC 4526 at lower center and NGC 4535 (60' field width)

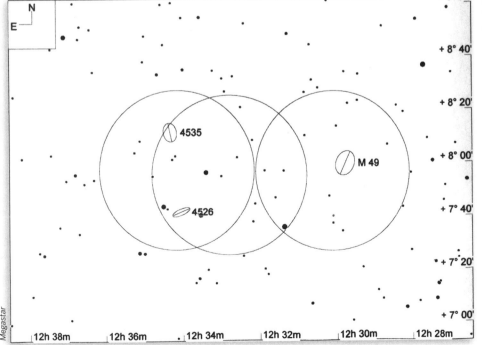

CHART 49-12.

NGC 4526 and NGC 4535 (3° field width; 1° eyepiece circles; LM 12.0)

M 61 (NGC 4303)	★★★	✸✸✸	GX	MBUDR
Chart 49-13	Figure 49-10	m10.2, 6.5' x 5.7'	12h 21.9m	+04° 28'

NGC 4303, better known as Messier 61 or M 61, is a bright spiral galaxy that we see nearly face-on. The Italian astronomer Barnabus Oriani discovered M 61 on 5 May 1779 as he was searching that area of the sky for the comet of 1779. Charles Messier also observed M 61 on that same night, but mistook it for the comet. Messier continued to observe the object on the following nights, and finally decided on the night of 11 May that the object was fixed against the background stars and therefore could not be a comet. Messier's observing notes described M 61 as a "Nebula, very faint & difficult to perceive. M. Messier mistook this nebula for the Comet of 1779, on the 5th, 6th & 11th of May; on the 11th he recognized that this was not the Comet, but a nebula which was located on its path & in the same point of the sky."

M 61 is relatively easy to find. Start by identifying the naked-eye star m3.9 15-η (eta) Virginis (Zaniah), which lies 5.5° W of the bright naked-eye star m2.7 29-γ (gamma) Virginis (Porrima). Place Zaniah at the S edge of your finder field and look 4.0° dead N for an m5.0 star, which appears very prominent in the finder. M 61 lies 1.2° NNE of that star.

At 90X in our 10" reflector, M 61 shows a bright, nearly circular 4' halo with significant mottling that brightens dramatically to a concentrated 30" core. Averted vision shows hints of the spiral arms, particularly to the SE, and a prominent dark lane on the E border of the core.

FIGURE 49-10.

NGC 4303 (M 61) (60' field width)

Image reproduced from Digitized Sky Survey courtesy Palomar Observatory and Space Telescope Science Institute

CHART 49-13.

NGC 4303 (M 61) (15° field width; 5° finder circles; 1° eyepiece circle; LM 9.0)

NGC 4517	★	◊◊◊	GX	MBUDR
Chart 49-14	Figure 49-11	m11.1, 11.2' x 1.5'	12h 32.7m	+00° 07'

NGC 4517 is a very faint edge-on galaxy. NGC 4517 is relatively easy to find, lying 2.7° NW of the bright naked-eye star m2.7 29-γ (gamma) Virginis (Porrima). Place Porrima on the SE edge of your finder field to put NGC 4517 near the center of the field and then use averted vision with your low-power finder eyepiece to locate the galaxy. At 90X in our 10" reflector with averted vision, NGC 4517 shows a faint, diffuse 1' x 5' halo elongated E-W, without a visible core.

FIGURE 49-11.

NGC 4517 (60' field width)

Image reproduced from Digitized Sky Survey courtesy Palomar Observatory and Space Telescope Science Institute

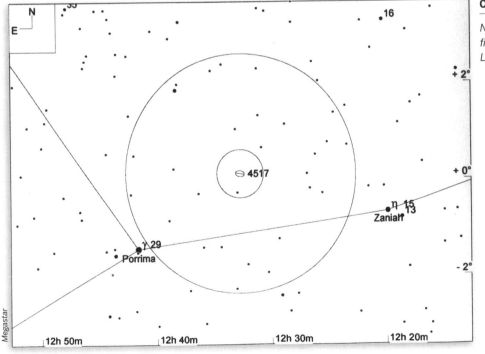

CHART 49-14.

NGC 4517 (10° field width; 5° finder circle; 1° eyepiece circle; LM 9.0)

NGC 4699	★★	◌◌◌	GX	MBUDR
Chart 49-15	Figure 49-12	m10.4, 4.0' x 2.8'	12h 49.0m	-08° 40'

NGC 4699 is a bright elliptical galaxy. NGC 4699 lies 9.2° WNW of the bright naked-eye star m1.1 67-α (alpha) Virginis (Spica), and is relatively easy to find by doing a short star hop from Spica. Place Spica on the E edge of your finder field and look about 4° W for a pair of bright stars, m5.2 49-Virginis and m6.0 50-Virginis, which form a 37' NE-SW line. Place the 49/50-Virginis pair on the E edge of your finder field, and look in the NW quadrant of the field for m4.8 40-ψ (psi) Virginis, which stands out brilliantly among the much dimmer stars in the field. NGC 4699 lies just 1.6° NW of 40-ψ, and is visible with direct vision in your low-power eyepiece. At 90X in our 10" reflector, NGC 4699 shows a bright 2' x 3' oval halo, elongated NE-SW, surrounding a very bright concentrated oval core that brightens sharply to a stellar nucleus. With averted vision, the core appears mottled or granular.

FIGURE 49-12.

NGC 4699 (60' field width)

Image reproduced from Digitized Sky Survey courtesy Anglo-Australian Observatory and Space Telescope Science Institute

CHART 49-15.

NGC 4699 (15° field width; 5° finder circles; 1° eyepiece circle; LM 9.0)

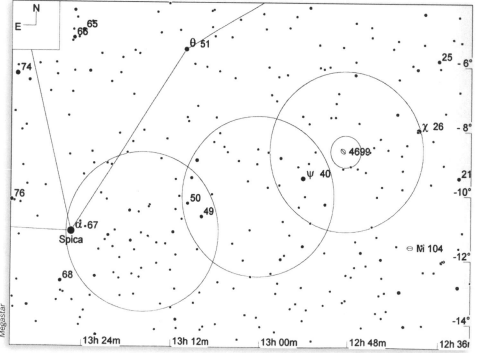

M 104 (NGC 4594)	★★★	☉☉☉	GX	MBUDR
Chart 49-16	Figure 49-13	m9.0, 8.8' x 3.5'	12h 40.0m	-11° 38'

NGC 4594, better known as Messier 104, M 104, or the Sombrero Galaxy, is a very bright edge-on galaxy with a very prominent dark lane that divides it longitudinally. M 104 is the first object that did not appear in the final published version of Charles Messier's catalog. Pierre Méchain discovered M 104 on 11 May 1781 and communicated news of the discovery to Messier, who added M 104 as a handwritten annotation to his published catalog.

M 104 is relatively easy to find. It lies 11.1° W of Spica and 5.5° NNE of the bright naked-eye star m3.0 7-δ (delta) Corvi (Algorab), the NE star in the pattern of Corvus, in an area of the sky that includes many m5 and m6 stars that serve as prominent sign-posts for locating M 104. If you've just observed NGC 4699, the easiest way to get to M 104 is to make the star hop shown in Chart 49-16. Place m4.7 26-χ (chi) Virginis on the NNW edge of your finder field and m5.5 21-Virginis on the W edge of your finder field, and then move the finder field slightly SW until 21-Vir is on the NW edge of the field. In the S half of the field, look for a very bright pair of m6 stars that form a 1.9° E-W line. M 104 lies 1.4° NNW of the E star and 2.0° NE of the W star.

M 104 may be very faintly visible with averted vision in a 50mm finder, and is extremely bright in a low-power eyepiece. At 90X in our 10" reflector, M 104 shows a very bright lens-shaped 2' x 6' E-W halo with a very prominent dark lane separating the halo into large N and much smaller S sections. The dark lane forms the brim of the sombrero, with the N section as the crown.

FIGURE 49-13.

NGC 4594 (M 104) (60' field width)

Image reproduced from Digitized Sky Survey courtesy Anglo-Australian Observatory and Space Telescope Science Institute

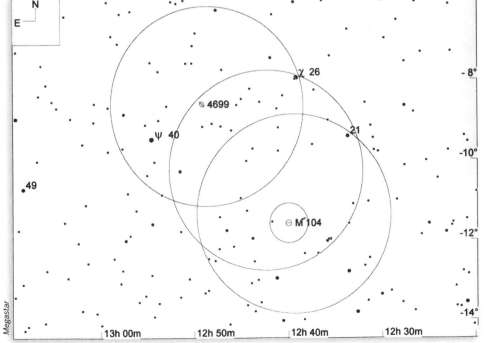

CHART 49-16.

NGC 4594 (M 104) (12° field width; 5° finder circles; 1° eyepiece circle; LM 9.0)

NGC 5746	★★★	✦✦✦	GX	MBUDR
Chart 49-17, 49-18	Figure 49-14	m11.3, 7.5' x 1.3'	14h 44.9m	+01° 57'

NGC 5746 is a bright edge-on galaxy. If you can identify m3.7 109-Virginis, you can find NGC 5746 in about 10 seconds. Unfortunately, although 109-Vir is easily visible to the naked eye, there are enough other nearby stars of similar brightness that it can be hard to be sure that you're actually looking at 109-Vir rather than some other star. If you're uncertain, it's easy enough to make a star hop from the bright naked-eye star m2.6 27-β (beta) Librae (Zubeneschamali). Put Zubeneschamali on the E edge of your finder field and look 4.0° W for the m5.0 star 19-δ (delta) Lib. Move the finder to put 19-δ on the S edge of the field, and look 4.3° NNW for m4.5 16-Librae. Put 16-Librae on the S edge of the finder and look 2.6° NW for m4.9 11-Librae. Put 11-Librae on the S edge of the finder field, and look 4.3° NNW for m3.7 109-Virginis. NGC 5746 lies just 20' W of 109-Virginis. At 90X in our 10" Dob, NGC 5746 shows an extremely elongated 0.5' x 6' N-S halo with a dense, bright 2' core with a central bulge, offset toward the W edge of the halo.

FIGURE 49-14.

NGC 5746 (60' field width)

Image reproduced from Digitized Sky Survey courtesy Palomar Observatory and Space Telescope Science Institute

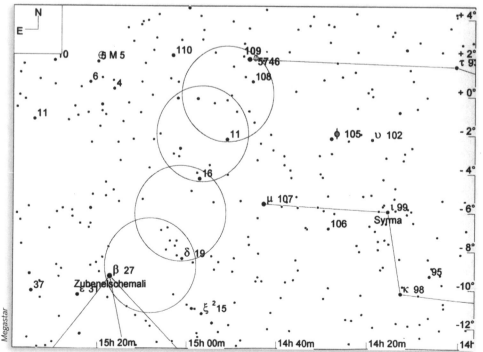

CHART 49-17.

NGC 5746 overview (25° field width; 5° finder circles; LM 7.0)

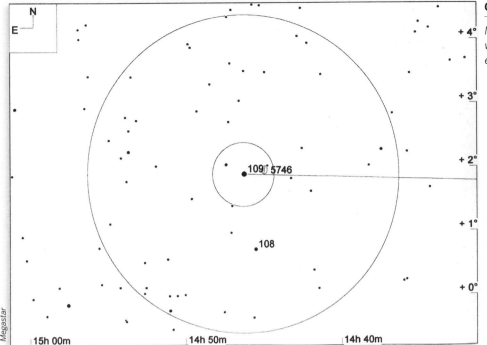

CHART 49-18.

NGC 5746 detail (7.5° field width; 5° finder circle; 1° eyepiece circle; LM 9.0)

Multiple Stars

gamma (STF 1670AB)	★	✹✹✹✹	MS	UD
Chart 49-1		m3.5/3.5, 0.4", PA 86° (2006)	12h 41.7m	-01° 27'

29-γ (gamma) Viriginis (Porrima) is a double star, but cannot currently be split in amateur telescopes. When the Astronomical League Double Star Club added this star to its list, the separation was much larger, and the double could be split in amateur instruments of modest size. As of 2007, the separation is much too small to allow this star to be split even in large instruments because even on the steadiest night of the year atmospheric turbulence swamps the small separation. When we viewed Porrima with our 10" reflector in summer 2006 at 125X, 250X and 500X (!), it showed not even the slightest hint of being double.

The Astronomical League Double Star Club recommends that you do the best you can, which in this case means that you'll record your observation as a single star that doesn't show even a bulge or slight elongation.

50

Vulpecula, The Little Fox

NAME: Vulpecula (vul-PECK-you-luh)

SEASON: Summer

CULMINATION: midnight, 26 July

ABBREVIATION: Vul

GENITIVE: Vulpeculae (vul-PECK-you-lye)

NEIGHBORS: Cyg, Del, Her, Lyr, Peg, Sge

BINOCULAR OBJECTS: Cr 399, NGC 6823, NGC 6853 (M 27), NGC 6940

URBAN OBJECTS: Cr 399, NGC 6853 (M 27), NGC 6940

Vulpecula is an extremely faint, mid-size mid-northerly constellation that ranks 55th in size among the 88 constellations. Vulpecula covers 268 square degrees of the celestial sphere, or about 0.6%. Its brightest star is barely fourth magnitude, and all of its other stars are fifth magnitude or dimmer. The ancients considered the area of Vulpecula to be just a dark patch of sky, and included it in none of their constellations. Most modern amateur astronomers have seen Vulpecula dozens or hundreds of times—it lies just south of the bright constellations Cygnus and Lyra—but have no idea they've seen it.

Vulpecula is a modern constellation, created in the late 17th century by Johannes Hevelius, and therefore has no mythology associated with it. Despite its lack of bright stars, Vulpecula is home to several interesting DSOs, including the large, bright planetary nebula NGC 6853 (M 27) and the famous Coathanger Cluster, which indeed looks exactly like a celestial coathanger.

Vulpecula culminates at midnight on 26 July, and, for observers at mid-northern latitudes, is best placed for evening viewing from early summer through mid-autumn.

TABLE 50-1.

Featured star clusters, nebulae, and galaxies in Vulpecula

Object	Type	Mv	Size	RA	Dec	M	B	U	D	R	Notes
Cr 399	OC	3.6	60.0	19 26.2	+20 06		◉	◉			Coathanger Cluster; Bracchi's Cluster; Class III 3 m
NGC 6802	OC	8.8	5.0	19 30.6	+20 16				◉		Cr 400; Class I 1 m
NGC 6823	OC	7.1	12.0	19 43.2	+23 18			◉			Cr 405; Class I 3 m n
NGC 6853	PN	7.6	6.7	19 59.6	+22 43	◉	◉	◉			M 27; Class 3+2
NGC 6940	OC	6.3	31.0	20 34.6	+28 19		◉	◉	◉		Cr 424; Mel 232; Class III 2 r

CHART 50-1.

The constellation Vulpecula (field width 30°)

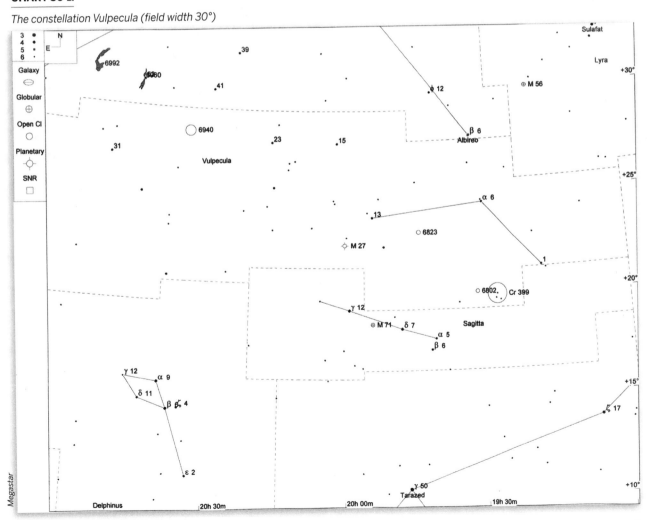

Clusters, Nebulae, and Galaxies

Collinder 399	★★★	✧✧✧✧	OC	MBUDR
Chart 50-2, 50-3	Figure 50-1	m3.6, 60.0'	19h 26.2m	+20° 06'

Collinder 399, better known as Brocchi's Cluster or the Coathanger Cluster, was long thought to be an asterism rather than a true open cluster, but it has been known since the 1970s that several of the brightest stars in this cluster share a common proper motion, and are therefore gravitationally bound as a true open cluster. Cr 399 was overlooked by Messier, Herschel, and other early catalogers of DSOs, probably because its large size made it difficult to observe as one object in the telescopes of the time, and because its appearance suggests that it is an asterism rather than a discrete object.

The Coathanger Cluster is easy to find. It lies just over a third of the way along a direct line from Altair in Aquila to Vega in Lyra. The easiest way to locate the Coathanger is just to point your finder or binocular about where it should be and look for a prominent coathanger shape.

With our 50mm binocular, the "hanger" part of the cluster is a prominent E-W (from east to west) 58' line of four to six m6/7 stars that are aligned so exactly they might have been drawn in with a straightedge. (The m6.3 star 7-Vulpeculae lies 20' E on the same line and another m7.1 star lies 8.3' W. These two stars are sometimes considered part of the Coathanger.) The "hook" of the Coathanger is made up of three m5/6 stars that lie directly S of the center of the hanger section, and are "open" to the W. Figure 50-1 shows the hook section below the three E stars of the hanger section. The fourth, W star of the hanger is just out of view to the right.

At 42X in our 10" reflector, the cluster includes about 3 dozen stars m9 and dimmer, most of which are S of the hanger section. The area to the immediate N of the hanger section is nearly devoid of stars. The S star in the hook, 4-Vulpeculae is a pretty triple star.

FIGURE 50-1.

Collinder 399 (The Coathanger Cluster) (60' field width)

Image reproduced from Digitized Sky Survey courtesy Palomar Observatory and Space Telescope Science Institute

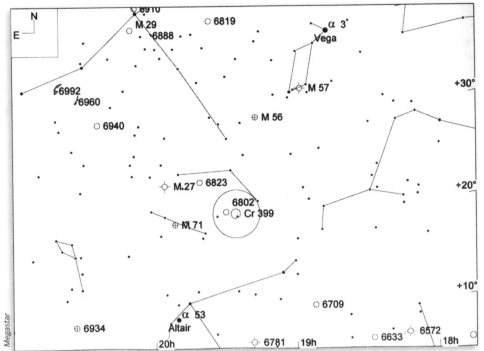

CHART 50-2.

Collinder 399 overview (50° field width; 5° finder circle; LM 5.0)

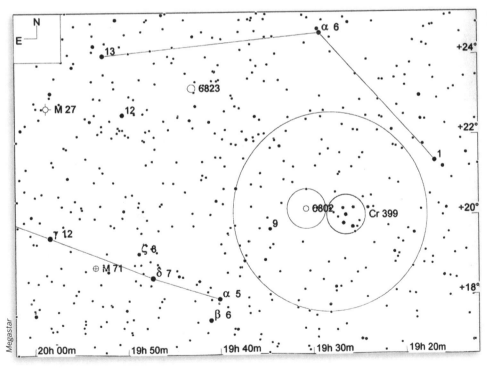

CHART 50-3.

Collinder 399 and NGC 6802 (12° field width; 5° finder circle; 1° eyepiece circle; LM 9.0)

NGC 6802	★★	✦✦✦✦	OC	MBUDR
Chart 50-3	Figure 50-2	m8.8, 5.0'	19h 30.6m	+20° 16'

NGC 6802 is a small, dim open cluster that lies 18' dead E of 7-Vul, the easternmost star in the E-W line of six stars that form the hanger portion of the Coathanger Cluster. At 90X in our 10" reflector, NGC 6802 is a small, faint hazy patch with four or five embedded m11 through m13 stars visible.

FIGURE 50-2.

NGC 6802 (60' field width)

NGC 6823	★★	✦✦✦	OC	MBUDR
Chart 50-4	Figure 50-3	m7.1, 12.0'	19h 43.2m	+23° 18'

NGC 6823 is small, dim open cluster, despite its relatively large, bright catalog data. The easiest way to locate NGC 6823 is to place the Coathanger Cluster on the S edge of your binocular or finder field and look for m4.4 6-α (alpha) Vulpeculae near the N edge of the field. Move the field about a full field E until the prominent stars m4.9 12-Vul and m4.6 13-Vul come into view, as shown in Chart 50-4. NGC 6823 forms the W apex of a scalene triangle with 12- and 13-Vulpeculae, making it easy to judge its position in a finder or binocular.

With our 50mm binocular, NGC 6823 is a small, moderately faint hazy patch with two embedded m9 stars faintly visible with averted vision. At 90X in our 10" Dob, direct vision shows a tight group of six m9/10 stars lies near the center of the cluster. Averted vision reveals a dozen or so more very faint stars scattered about the central group.

FIGURE 50-3.

NGC 6823 (60' field width)

CHART 50-4.

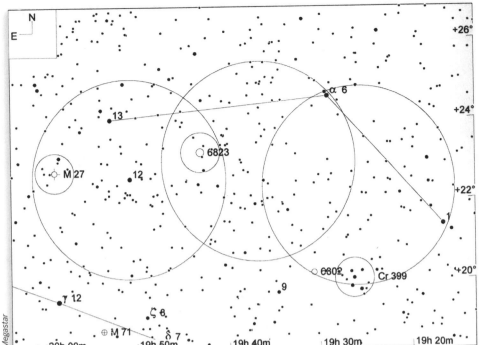

M 27 (NGC 6853)	★★★★	❂❂❂	PN	MBUDR
Chart 50-4	Figure 50-4	m7.6, 6.7'	19h 59.6m	+22° 43'

NGC 6853, better known as Messier 27, M 27, or the Dumbbell Nebula, is a large, bright planetary nebula, by far the best planetary nebula visible to us. Charles Messier discovered and cataloged M 27 on the night of 12 July 1764, and in doing so discovered a new class of object.

Although it is located far from any bright naked-eye stars, M 27 is relatively easy to find. We find it by putting the Coathanger Cluster at the S edge of our binocular or finder field and looking about 4.5° for m4.4 6-α (alpha) Vulpeculae near the N edge of the field. Move the field about a full field E until the prominent stars m4.9 12-Vul and m4.6 13-Vul come into view, as shown in Chart 50-4, and then continue past them about half a field until M 27 pops into view.

M 27 is an easy object with any binocular. With our 50mm binoculars, M 27 is a large, bright nebulous patch lying 24' SSE of the m5.7 star 14-Vulpeculae, which appears very prominent in the finder. At 90X in our 10" Dob, M 27 is a spectacular object. It looks to us more like a peanut or a bowtie than a dumbbell. The bright core of the nebula is made up of two triangular lobes extending 6' N-S and joined at their points at a small darker area within which the central star is faintly visible with averted vision. Even without filtration, averted vision also reveals the faint halo that lies E and W of the brighter core, increasing the visible extent of the nebula to about a 6' circle. A narrowband filter greatly increases the contrast of the nebulosity, particularly in the dimmer halo sections, and reveals subtle wisps and whorls within the nebulosity, but does not greatly increase the visible extent of the object.

FIGURE 50-4.

NGC 6853 (M 27) (60' field width)

Image reproduced from Digitized Sky Survey courtesy Palomar Observatory and Space Telescope Science Institute

NGC 6940	★★★	✺✺✺	OC	MBUDR
Chart 50-5	Figure 50-5	m6.3, 31.0'	20h 34.6m	+28° 19'

NGC 6940 is a large, bright open cluster. Although you can locate it by a long star-hop, it's much easier to locate it using its geometric relationship to nearby bright stars. NGC 6940 actually lies in Vulpecula, but it's easier to find by working from Cygnus. To do so, locate the bright naked-eye stars m2.5 53-ε (epsilon) Cygni and m3.2 64-ζ (zeta) Cygni, which together make up the end of the S wing of the Swan. NGC 6940 forms the SW apex of a right triangle with these two stars, and can easily be found simply by pointing your binocular or finder about where it should be.

With our 50mm binoculars, NGC 6940 is a large, relatively bright nebulous patch lying about 2.4° SSE (south southeast) of the very prominent star m4.0 41-Cygni. Two m8 stars lie on the S edge of the nebulosity and a third m8 star lies near the NE edge. Averted vision shows two or three very faint stars embedded in the nebulosity. At 90X in our 10" reflector, NGC 6940 is a beautiful open cluster. Although it is not well-detached from the surrounding star field, the center of the cluster is well concentrated. More than 100 stars down to m12 are visible, many of which are grouped in arcs, chains, and knots. Three m9 stars form a prominent 6' flattened triangle near the S and SW edge of the cluster. Most of the cluster stars are grouped in a wide E-W band that crosses the center of the cluster. The S half of the cluster is relatively densely populated, with the N half more sparsely populated.

FIGURE 50-5.

NGC 6940 (60' field width)

Image reproduced from Digitized Sky Survey courtesy Palomar Observatory and Space Telescope Science Institute

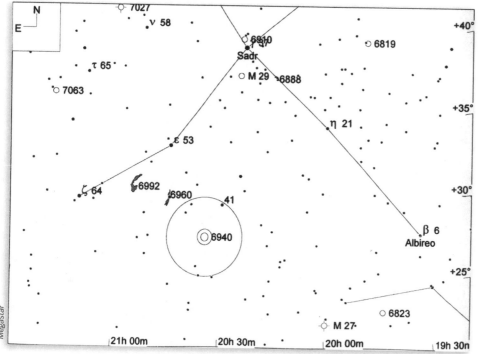

CHART 50-5.

NGC 6940 (30° field width; 5° finder circle; LM 6.0)

Multiple Stars

There are no featured multiple stars in Vulpecula.

INDEX

79-zeta Ursae Majoris (STF 1744AB), 463, 479
79-zeta Ursae Majoris (STF 1744AC), 463, 479
86-gamma Ceti (STF 299AB), 178, 183
86-zeta Piscium (STF 100AB), 376, 380–381
94-Aquarii (STF 2998), 76, 83
95-Herculis (STF 2264), 257, 262–263
113-alpha Piscium (STF 202AB), 376, 380
118-Tauri (STF 716AB), 447, 455–456

A

aberration, chromatic, 41
absolute magnitude, 17
accessories, miscellaneous, 62
Adhara, 140, 145
ADS 6255, 383, 391
ADS 9913 (8-beta Scorpii), 417, 421–422
ADS 9951 (14-nu Scorpii), 417, 422
Aitken Double Star Catalog (ADS), 18
Albireo (STF 43Aa-B), 215, 227
Alcor (STF 1744AC), 463, 479
Alcyone, 448
Alderblick, 40
Alfirk (STF 2806Aa-B), 167, 175
Algedi (STFA 51AE), 146, 149
Algieba (STF 1424AB), 281, 289
Algol (26-beta Perseii), 363, 374
Algorab (7-delta Corvi), 208, 212
Alkalurops (STF 28), 102, 106
Almach (STF 205A-BC), 66, 73
Alnitak (50-zeta Orionis), 342, 355
Alpha Perseii Association (Collinder 39), 362, 370
alphabet, Greek, 19
Alrakis (STF 2130AB), 235, 240–241
Alrescha (STF 202AB), 376, 380
Alya (STF 2417AB), 434, 441
amplification factor, Barlow lens, 46
Andromeda constellation, 66–73
Andromeda Galaxy, Great, 29
angle, position, 18
anti-reflection coatings, binoculars, 39–40
aperture, binoculars, 38
apochromatic, Barlow lens, 46

apparent field of view (AFoV), eyepiece, 49
apparent visual magnitude, 17
Aquarius constellation, 74–83
Aquila constellation, 84–89
arcminutes, 14, 16
arcseconds, 14, 16
Argo Navis, 382
Aries constellation, 90–93
Arschkopf, 252
ascension, right, 17
Asterope, 448
Astrolite, 45
Astronomical League, 30
Astronomy Hacks, 42, 64
Astronomy magazine, 64
Atlas, 448
atlases, 57–62
Auriga constellation, 94–101

B

Baraff, Gene, 64
Barlow lens, 46–47
Barnard 33 (Horsehead Nebula), 347
Barnard 86 (Ink Spot), 410
Bartsch, Jakob
    Camelopardalis, 108
    Monoceros constellation, 314
Bausch and Lomb, 40
Bayer, Johann, Draco, 234
Bayer designation, 19, 20
Beehive Cluster (NGC 2632), 118, 120
Belt of Orion, 343–344
Bessel, Friedrich Wilhelm, STF 2758AB, 228
Bevis, John
    M 1, 454
    M 35, 252
Big Dipper (STF 1821), 105
binary star, 14
Binocular Messier Club, 31
binoculars, 38–41
blinking DSOs, 114
Blinking Planetary Nebula (NGC 6826), 214, 226–227
blue-green light, 26
Blue Snowball Nebula (NGC 7662), 66, 72
blue stragglers, 24

Bode, Johann Elert
    M 53, 193
    M 64, 192
    M 81, 466
    M 82, 466
    M 92, 259
Bode's Galaxy (NGC 3031), 463, 466–467
Bonanno, Emil, 61
Boötes constellation, 102–107
Bopp, Thomas, M 70, 400
Bridal Veil Nebula (NGC 6992/ 6995/6960), 214, 222
bright nebulae (BN), 26
Bright Star Atlas, 58–59
brightness, surface, 21
Brocchi's Cluster (Collinder 399), 500, 502–503
Bubble Nebula (NGC 7635), 151, 153–154
bulls-eye finders, 54
Burgess/TMB Planetary Series, 52
Butterfly Cluster (NGC 6405), 416, 420

C

Camelopardalis constellation, 108–117
Camel's Eye Nebula (NGC 1501), 108, 112–113
Cancer constellation, 118–123
Canes Venatici constellation, 124–139
Canis Major constellation, 140–145
Canon, 41
Capricornus constellation, 146–149
Carina constellation, 382
Cartes du Ciel software, 18, 55–57
cases, gear, 62–63
Cassini, Giovanni, M 50, 321
Cassiopeia constellation, 150–165
Castor (66-alpha Geminorum), 250, 255
catalog numbers, 20
catalogs, star, 18, 20
Cat's Eye Nebula (NGC 6543), 234, 239–240
Celaeno, 448
Celestron, 40, 41, 44, 45, 47
Cepheus constellation, 166–177
Cetus constellation, 178–183
chairs, 54–55
chapter conventions, 34–37

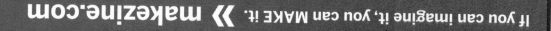
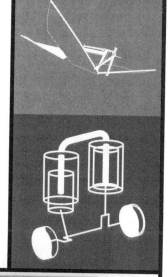